The Engineering Communication Manual

The Engineering Communication Manual

Richard House

Richard Layton

Jessica Livingston

Sean Moseley

New York Oxford
OXFORD UNIVERSITY PRESS

Oxford University Press is a department of the University of Oxford.
It furthers the University's objective of excellence in research,
scholarship, and education by publishing worldwide.

Oxford New York
Auckland Cape Town Dar es Salaam Hong Kong Karachi
Kuala Lumpur Madrid Melbourne Mexico City Nairobi
New Delhi Shanghai Taipei Toronto

With offices in
Argentina Austria Brazil Chile Czech Republic France Greece
Guatemala Hungary Italy Japan Poland Portugal Singapore
South Korea Switzerland Thailand Turkey Ukraine Vietnam

For titles covered by Section 112 of the US Higher Education
Opportunity Act, please visit www.oup.com/us/he for the
latest information about pricing and alternate formats.

Published by Oxford University Press
198 Madison Avenue, New York, New York 10016
http://www.oup.com

Library of Congress Cataloging-in-Publication Data
House, Richard (English professor)
The Engineering Communication Manual / Richard House, Richard Layton,
Jessica Livingston, Sean Moseley.
pages cm
Includes bibliographical references and index.
ISBN 978-0-19-933910-5 (alk. paper)
1. Communication of technical information--Handbooks, manuals, etc. I. Layton,
Richard A. (Mechanical engineer) II. Livingston, Jessica, 1975- III. Moseley,
Sean. IV. Title.
T10.5.H68 2017
808.06'662--dc23

2015013932

Printing number: 9 8 7 6

Printed in the United States of America
on acid-free paper

Brief contents

Table of contents

GENRES

PROCESSES

COMPONENTS

VISUALS

MEDIA

Index of documents

1 Planning your communication

2 Understanding your audience

3 Meeting your ethical obligations

4 Accommodating global and cultural differences

5 Designing documents for users

6 Engineers

7 Technicians and technical staff

14 Proposing

15 Instructing

16 Applying for a job

17 Researching

18 Drafting 19 Revising

20 Collaborating

21 Meeting

22 Headings

23 Paragraphs

24 Sentences

26 Summaries

27 Front and back matter

28 Graphs

29 Illustrations

30 Tables, equations, and code

31 Print pages

32 Talks

33 Presentation slides

34 Posters

Preface

This book begins with the premise that communication, whether written, spoken, or visual, is a primary, essential, and routine engineering activity and therefore a crucial skill of successful engineers.

As a description of engineers' work, this is uncontroversial. In the world of industry, our former students tell us that they spend many of their working days writing and speaking to their colleagues, clients, and other stakeholders. Even in the engineering departments and colleges where student engineers are educated, engineering writing and speaking drive scholarly research, from grant proposal to conference talk to journal article.

Listening to many engineering students talk about writing and speaking, one would never know that this is the case. Recently, one of us overheard a student engineer proclaim to a friend, "I just don't do writing." Such students no longer labor under the delusion that they won't need to write in their careers (and know better than some engineers in past generations, who hoped that they'd employ secretaries to handle the writing and speaking on their behalf). Our students have become highly aware of the importance of their communication skills, and yet this awareness rarely changes the way they see the process of learning those skills. They still come to engineering writing expecting it to be bothersome, difficult, nerve-wracking, and above all frustrating, because they're sure that they aren't any good at it. Students enroll in an engineering major today with fantasies of becoming Tony Stark, the inventor and raconteur in Marvel Comics' Iron Man armor, or perhaps an equivalent real-world tech celebrity like Elon Musk of Tesla Motors and SpaceX. When it comes to writing and speaking, though, many still seem to identify with Dilbert, the profession's awkward, inarticulate mascot.

The Engineering Communication Manual (ECM) was undertaken by a team of two mechanical engineers (Layton and Moseley) and two English professors (House and Livingston) who believe that clear engineering thinking is inseparable from the practice of communicating that thinking to others. We wanted a book that would:

- Supply sound writing advice in technical courses requiring reports, proposals, talks, or other communication tasks, and
- Represent engineering accurately and positively in dedicated technical communication courses taught by writing instructors.

To meet these needs, we have endeavored to show engineering communication in its most positive and distinctive light, informed by the profession's guiding values and aesthetic principles. The *ECM* will help student engineers to create rule-governed, data-driven, efficient, and economical prose, data displays, and documents. From the selection of the content to the binding, all of our own writing and design decisions have been conducted with these goals in mind.

Authentic engineering writing . . .

Many students have intentionally specialized in math and science since high school, sometimes deliberately avoiding writing-intensive courses. They may also attribute the same mindset to their engineering professors, having seen communication lumped into a category of "soft skills" that are secondary to technical content.

To earn credibility with this audience, the book is designed around sample documents drawn from authentic engineering practice—most contributed by our former students from their current positions as engineers in industry. Their contributions show tomorrow's engineers how compelling technical work is advanced by effective writing, speaking, and visual design.

Individual authors' names have been changed, and proprietary information has been altered or concealed to protect the interests of the companies and government agencies that have generously supplied us with documents. Wherever possible, though, we've shown where those documents originated, so that student engineers can see direct connections between the work they're now doing and what they'll do one day in the workplace at Cummins or Freescale.

. . . in rhetorical, ethical, and global perspectives

Our first priority is to showcase engineers being effective writers, speakers and designers. Nevertheless, we do think that there's something to be said for rhetoricians, philosophers, and other practitioners of the humanities and social sciences. The *ECM* is informed by the most important research in fields from technical and professional communication to engineering and business ethics. With the encouragement of our reviewers, though, we have worked to reduce the jargon from these fields to the bare minimum, balancing research-based best practices with accessibility.

Some scholarly fields and findings make brief cameo appearances—the psychology of visual perception, for instance, in the module on graphs. Others, however, are more fundamental to our entire approach. The *ECM*'s first section views the art of engineering communication through the disciplines of rhetoric and ethics. These are the fields that are necessary for a full picture of engineering in its full interpersonal complexity, with the engineer consistently facing competing demands:

- How can I write a technical proposal that simultaneously meets the needs of managers, engineers, and technicians?
- Why is an experimental report organized so differently for a technical journal than it is for an "in-house" audience within my company?
- When reporting information about my own company to external audiences, how much information can and should I disclose ethically?

How to use this book

The *ECM* is designed in a modular, non-sequential format. Instructors can combine units of content to meet the learning objectives of any assignment. Students can locate on their own relevant sections within the appropriate modules.

For example, consider an assignment to write an executive summary. An instructor might choose to focus on: the audience, by assigning reading from Module 8: *Executives*; the revision process, with reading from Module: 19 *Revising*; and the makeup of an executive summary, with reading from Module 26*: Summaries*. Instructors can expand or contract this list to meet their own objectives.

EXECUTIVE SUMMARY	POTENTIAL READING IN MODULES:	TO FOCUS ON:
	8 *Executives*	Audiences
	19 *Revising*	Processes
	26 *Summaries*	Components

For an experimental report assignment, an instructor might focus on: the report genre, with reading from Module 11: *Reporting in a research community*; on document components, with reading from Module 22: *Headings*; on visuals, with reading from Module 28: *Graphs*; and on the print medium, with reading from Module 31:

Print pages. In an assignment for first-year students, an instructor might replace Module 22: *Headings* with Module 24: *Sentences* and omit the visuals altogether.

EXPERIMENT REPORT	POTENTIAL READING IN MODULES:	TO FOCUS ON:
	11 *Reporting in a research community*	Genres
	22 *Headings*	Components
	28 *Graphs*	Visuals
	31 *Print pages*	Media

For an assigned talk given by upper-level students expecting a project client to be in the audience, an instructor might focus on: the audience, with reading from Module 9: *Clients*; on team process, with reading from Module 21: *Meeting*; on delivering an effective talk, with reading from Module 32: *Talks*; and on creating effective slides, with reading from Module 33: *Presentation slides*. In contrast, an assignment for a first-year student team talk might focus only on the modules for talks and presentation slides. Of course, students seeking greater depth can easily locate other relevant modules.

GIVING A TALK	POTENTIAL READING IN MODULES:	TO FOCUS ON:
	9 *Clients*	Audiences
	21 *Meeting*	Processes
	32 *Talks*	Media
	33 *Presentation slides*	Media

For a résumé and cover letter assignment—a mainstay of technical communication courses—the focus would include: the genre, with reading from Module 16: *Applying for a job*; and a cover letter's basic component, the paragraph, with reading from Module 23: *Paragraphs*.

RÉSUMÉ AND COVER LETTER	POTENTIAL READING IN MODULES:	TO FOCUS ON:
	16 *Applying for a job*	Genres
	23 *Paragraphs*	Components

The next example is a memo assignment responding to an ethics case study. The subject requires a focus on: the case context, with reading from Module 3: *Meeting your ethical obligations*; and the typically relevant audiences, with readings from Module 9: *Clients* and Module 10: *The public and the public sector*. The memo genre can be explored with reading from Module 13: *Corresponding* and the writing process with reading from Module 18: *Drafting*.

ETHICS CASE STUDY MEMO	POTENTIAL READING IN MODULES:	TO FOCUS ON:
	3 *Meeting your ethical obligations*	Contexts
	9 *Clients*	Audiences
	10 *The public and the public sector*	Audiences
	13 *Corresponding*	Genres
	18 *Drafting*	Processes

To illustrate how an assignment might differ for students at different levels, a design report assigned to first-year students might focus on the drafting process (18) and the sentence component (24). In contrast, a capstone design proposal might include a broader selection of material, including the proposing genre (14), how to collaborate (20), and both headings (22) and summaries (26) as components of document design.

1ST-YEAR DESIGN REPORT	POTENTIAL READING IN MODULES:	TO FOCUS ON:
	18 *Drafting*	Processes
	24 *Sentences*	Components

CAPSTONE DESIGN REPORT	POTENTIAL READING IN MODULES:	TO FOCUS ON:
	14 *Proposing*	Genres
	20 *Collaborating*	Processes
	22 *Headings*	Components
	26 *Summaries*	Components

To instructors, we emphasize the importance of articulating learning objectives and using them to design assignments. To students, we emphasize the importance of finding relevant sections and reading for understanding.

Examples for critical reflection

In 2013, *The Journal of Engineering Education* published a study of how engineering students use textbooks. Christine Lee and her coauthors conclude that engineering students spend little time reading the text passages; instead, they page through the book in search of an example of the kind of problem that they are trying to solve, applying "the solution pathway in the example problem to their problem at hand without critical reflection" (279).

We have designed the *ECM* in anticipation of engineering students bringing their habit of looking for "plug-and-chug" solutions to technical problems—locating an equation or model that approximately matches the problem that they are trying to solve—but we have also attempted to redirect this search into more reflective rhetorical analysis and problem solving. Student engineers will quickly find sample documents in each module, but their features are not simply presented as a template to be imitated. Rather, systematic annotations of each sample document analyze and interpret their components as purposeful, rhetorical responses to the demands of the content and the audience. The most effective technical communication makes complex information usable without oversimplifying it; this has been the goal of our design and commentary throughout.

FAULT CODE 1682 (Air-assisted)
Aftertreatment Diesel Exhaust Fluid Dosing Unit Input Lines - Condition Exists

The title states the fault code number that is the subject of the document. The subtitle describes the relevant hardware (the "dosing unit").

Overview

CODE	REASON	EFFECT
Fault Code: 1682 PID: SPN: 3362 FMI: 31 LAMP: Amber SRT:	Aftertreatment Diesel Exhaust Fluid Dosing Unit Input Lines - Condition Exists. An error has been detected by the aftertreatment diesel exhaust fluid dosing unit.	Diesel exhaust fluid injection into the SCR aftertreatment system is disabled.

A table immediately summarizes the main points of the document: the fault code number, what it means, and its effect on engine performance.

Circuit Description

For an air assisted SCR aftertreatment system, the aftertreatment diesel exhaust fluid dosing unit requires air pressure from the OEM air tanks. The diesel exhaust fluid dosing unit precisely measures the amount of diesel exhaust fluid (DEF or Urea) to be injected into the aftertreatment system. The diesel exhaust fluid dosing unit has three primary cycles. A priming cycle at initial engine start makes sure that diesel exhaust fluid is available at the diesel exhaust fluid dosing unit. During the dosing cycle, the diesel exhaust fluid is being delivered to the aftertreatment nozzle. A purge cycle occurs when the engine is turned off. The purge cycle makes sure that all the diesel exhaust fluid is removed from the diesel exhaust fluid line and aftertreatment nozzle.

Liberal use of acronyms, abbreviations, and jargon indicates an intended audience already well acquainted with the system being described.

Component Location

The aftertreatment diesel exhaust fluid dosing unit location is OEM dependent. Refer to the OEM service manual for more information.

Conditions for Running the Diagnostics

This diagnostic consists of multiple parts, which run when the engine is first started, and make take up to 12 minutes to complete.

The specific condition that causes the fault code to appear is described in one sentence.

Conditions for Setting the Fault Codes

The aftertreatment diesel exhaust fluid dosing unit is **not** able to provide the correct dosing rate to the aftertreatment nozzle.

An example of the Answers First format: a report describing a "fault code" that appears for a type of exhaust aftertreatment failure in a diesel engine. Printed with licensed permission from Cummins, Inc. © 2015 Cummins, Inc., all rights reserved.

Instructor materials and assessment for ABET accreditation

The *ECM* broadly supports the student outcomes defined in ABET's General Criterion 3 (the "a-k" outcomes), with outcome (g), "an ability to communicate effectively," situated within engineering design and problem-solving contexts relevant to the other a-k outcomes. Most notably, individual modules directly address outcome (d)—"an ability to function on multidisciplinary teams"—and outcome (f)—"professional and ethical responsibility."

Assessment of these outcomes is a central goal of our web resources, and a major reason that we are supplying our own syllabi, assignment sheets, in-class activities, and grading rubrics. Like other engineering educators, we try to address ABET (a-k) outcomes in an organic way, so that they support and fit naturally within the more specific technical content and professional skills that define our courses. In assessing student communication outcomes, adopters of the book may find it most convenient to use one of our supplied assignments and the corresponding grading rubric. These rubrics enable the easy scoring of student documents on several different areas of communication competency (audience accommodation, organization, soundness of evidence and argument, correct sentence mechanics, usage, and grammar, etc.). Alternately, when instructors are embedding communication tasks within assignments of their own design, it may be more convenient to borrow individual items from one of these rubrics. For each skill, rubrics differentiate levels of student performance from workplace-ready to non-passing, making the collection of assessment data simple and efficient, whether the instructor's primary expertise is in communication or in engineering.

Acknowledgments

The four of us have spent several years on this project, and our work has benefited immeasurably from the help of many other professionals whose insights and contributions have, in one way or another, made this a better book. (Its remaining shortcomings, of course, belong solely to us.) Our deepest appreciation goes to:

- Nancy Blaine, Christine Mahon, and John Appeldorn, Oxford University Press's all-star engineering editorial team, and Patrick Lynch, Editorial Director of the OUP Higher Education Group
- The professional and scholarly communities—and the many friends and colleagues therein—who supplied us with the ideas

and many of the sample documents that sustained this entire project. We are especially grateful to our friends and colleagues in the IEEE Professional Communication Society and the American Society for Engineering Education (Liberal Education/Engineering & Society and Educational Research & Methods divisions).

- The many reviewers who delivered such insightful feedback on many drafts of the manuscript, including Adam Carberry (Arizona State University), Maria Christian (Oklahoma State University Institute of Technology), Susan Codone (Mercer University), Edward M. Cottrill (University of Massachusetts), Christine Cranford (North Carolina State University), Kim Davis (Carleton University), Jeffrey A. Donnell (Georgia Tech), Henry A. Etlinger (Rochester Institute of Technology), Julie Dyke Ford (New Mexico Tech), Jeffrey M. Foresta (University of California, Irvine), Debbie Hall (Valencia College), Brad Henderson (University of California, Davis), Joanne Lax (Purdue University), Jon A. Leydens (Colorado School of Mines), Christina Matta (University of Wisconsin-Madison), Theresa Merrick Cassidy (Kansas State University), Kenneth W. Miller (St. Cloud State University), Christina J. Moore (St. Edward's University), Cecelia A. Musselman (Northeastern University), Jon Negrelli (Cleveland State University), Kathryn Northcut (Missouri S&T), Karim G. Oweiss (Michigan State University), Mark C. Petzold (St. Cloud State University), Donald R. Riccomini (Santa Clara University), Kenneth C. Ronkowitz (New Jersey Institute of Technology), John A. Roubidoux (Fort Lewis College), Cynthia Ryan (University of Alabama at Birmingham), Gefen Bar-On Santor (University of Ottawa), Brogan Sullivan (University of South Florida), Kirk St. Amant (East Carolina University), and Mary Westervelt (University of Pennsylvania)
- The baristas and kitchen staff of Java Haute, who fed and caffeinated us through the entire writing process
- The many contributors who supplied the sample documents that form the foundation of the book. When we asked for help, so much excellent work poured in that we couldn't use all of it. We are grateful to the organizations that allowed us permission to use their work:
    - From industry: Cummins Inc., Duarte Press, Freescale Semiconductor Inc., GSI (Grain Systems, Inc.), Numerical Concepts, Inc., Sourceable.net, and Strand Associates, Inc.

- From government: the City of Carlsbad, CA, the City of Des Moines, IA, the Congressional Research Service, the Illinois Environmental Protection Agency, the National Highway Traffic Safety Administration, the National Oceanic and Atmospheric Administration, the National Science Foundation, and the US Patent and Trademark Office
- From the academic and nonprofit sector: Engineers for a Sustainable World, the Joint Replacement Surgeons of Indiana Foundation, Inc., the Keck Institute for Space Studies, MIT Lincoln Laboratory, the National Electrical Manufacturers Association, the National Society of Professional Engineers, the Trustees of Princeton University, the Union of Concerned Scientists, and the University of California, Berkeley, Center for Catastrophic Risk Management

We are also especially grateful to the many individuals who contributed documents and assisted in copyright permissions:

- Those who assisted in obtaining permission to reprint wonderful writing in this book include Nancy Duarte; Nazih Khaddaj Mallat (IEEE Region 8); Karen Kaminsky, Patrick Eells, and Alexander Dale (Engineers for a Sustainable World—University of Pittsburgh); and Kevin Otto (Cummins Inc.).
- Rose-Hulman faculty and staff friends who contributed documents include Thomas Adams, Patsy Brackin, Rob Coons, Phil Cornwell, Paul Leisher, Jerry Leturgez, Jay McCormack, Sriram Mohan, Jenny Mueller Price, Renee Rogge, Chandan Rupakheti, Scott Small, Rick Stamper, and Sarah Summers.
- Rose-Hulman student and alumni contributors include Matt Billingsley, Emma Dosmar, Jeff Elliott, Zac Erba, Connor Freeman (Edgile), Becca Hecht, Andrew Hopkins, Dakota Huckaby, Alex Jacoby, Andrew Jordan (GSI Inc.), Luke Kennedy, Alex Mullans, David Pick, David Radue (MIT Lincoln Laboratory), Brandi Sturgill Rodriguez (Strand Associates), Ryan Seale, Noura Sleiman, Katelyn Stenger, Louis Vaught, Giuliana Watson, Lindsey Watterson, and Caroline Winters. Contributions also came from our

school's chapter of Engineers Without Borders and our Human Powered Vehicle Team.

- Everyone else in our close community at Rose-Hulman Institute of Technology—our alumni, students, staff, and faculty colleagues
- Special thanks are due to the colleagues with whom we've taught (and learned) engineering communication—Corey Taylor, Mark Minster, Caroline Carvill, Rebecca Dyer, and Sarah Summers. Meredith Johnson moved on to a new school but left a big influence on what we know and what we teach. Most importantly, we wouldn't know much at all without Julia Williams and Anneliese Watt, Rose-Hulman's original champions for excellence in engineering communication. Most of these friends will recognize their influence on these pages; whether or not they do, we acknowledge it here.

Finally, we'd like to express appreciation and love to our families for their patience and understanding during this project. You made this book possible, too!

About the authors

Richard House is Professor of English at Rose-Hulman Institute of Technology. He received a B.A. from Illinois Wesleyan University and M.A. and Ph.D. from the University of California, Irvine. In addition to engineering communication and pedagogy, he has scholarly interests in sustainability and Shakespeare.

Richard Layton is Professor of Mechanical Engineering and past Director of the Center for the Practice and Scholarship of Education at Rose-Hulman Institute of Technology. He received a B.S. from California State University, Northridge, and an M.S. and Ph.D. from the University of Washington. His areas of scholarship include data visualization, student teaming, and undergraduate retention.

Jessica Livingston is Associate Professor of English at Rose-Hulman Institute of Technology. She received a B.A. from The University of Georgia, an M.A. from the University of Kentucky, and a Ph.D. from the University of Florida. Her areas of interest include humanitarian engineering, the intersections of gender and work in a global economy, and documentary film.

Sean Moseley is Associate Professor of Mechanical Engineering at Rose-Hulman Institute of Technology. He received a B.S. from The Georgia Institute of Technology and an M.S. and Ph.D. from the University of California, Berkeley. His areas of interest include effective engineering education techniques, solid mechanics, and humanitarian engineering.

Contexts

...define your communication objectives and the expectations of those who will read your documents or listen to your talk. Some principles of good writing apply in just about all contexts—you'll always want to be precise and accurate—but most will require adjustment based on the circumstances in which you are working. (For example, audiences usually want the central message stated up front—unless a message involves bad news, when tact requires that groundwork be laid first to ease disappointment.) As in engineering, carefully developed models can help you to understand what a problem requires. Such models will include the information needs of your audience, the types of evidence that support your ideas, and even your professional ethical obligations. Factoring such matters into your writing, you will be prepared to address content successfully, whether it includes difficult technical material or sensitive professional relationships.

Contexts

Planning your communication

1

Engineers are problem-solvers, who achieve professional satisfaction by designing solutions that are effective, appropriate, and even elegant. As problem-solvers, we know that reaching such solutions requires an orderly, systematic process. The engineer and anthropologist Gary Downey has divided the general approach of engineering problem solving into six steps: Given, Find, Diagram, Make Assumptions, Equations, Solve. (As rigorous as this system may seem, Downey's point is that it actually *oversimplifies* authentic engineering practice, which requires a patient, thoughtful effort to define and frame problems in the first place, well before the engineer can even formulate the sort of mathematical problems that might show up on an exam or in a homework assignment.)

Engineering communication involves different kinds of assumptions, models, and analysis. Fundamentally, though, effective writing, listening, and speaking require analytical tools and critical thinking, just like your other engineering work.

A vexing homework problem may sometimes lead to a search for a simple "plug-and-chug" solution—finding the appropriate equation in a textbook, inserting the values specified in the problem, and solving. In practice, of course, real engineering work seldom allows such shortcuts. Even arriving at the right equation can require significant intellectual work.

The "plug-and-chug" approach can be just as tempting when writing or preparing a presentation: find a good document of the type that you're trying to write, duplicate its sections, insert your own content, and submit. "Plug-and-chug" carries the same dangers, though: unless your assumptions and model of the situation are very highly accurate, you're unlikely to end up with the result you want.

Objectives

- To use the rhetorical triangle to describe the defining components of a given communication task or situation—speaker or writer, audience, content, genre, medium, and context
- To describe what the audience will consider sufficient evidence and appropriate proof
- To establish credibility to audience members by demonstrating technical competence as well as sincerity and good will
- To identify the exigence of your content for a given audience—the reason a reader or listener ought to care about what you're presenting
- To determine the appropriate level of technicality for targeted audiences, including both the types of information included and the vocabulary used to convey that information
- To categorize your document or communication act within the overall genres of professional communication—such as reports, proposals, or correspondence—as well as specialized subgenres (progress report, experimental report, design report)

1.1 Assessing the rhetorical situation

A *rhetorical* approach to communication shapes what we write and say by developing our ideas within a holistic, contextual model of the situations in which we're working. Aristotle defined rhetoric as "the art of discerning, in any given situation, the available means of persuasion." We traditionally define such a "given situation" by specifying the *audience* of readers or listeners, the *speaker* or *writer* addressing that audience, and the *content* with which they are concerned. These elements form the corners of the *rhetorical triangle.*

From an engineering viewpoint, we can see this as an optimization problem with several related criteria:

Content — Should be analyzed and reported accurately and completely, with appropriate evidence based on the standards of your technical discipline and specialty area

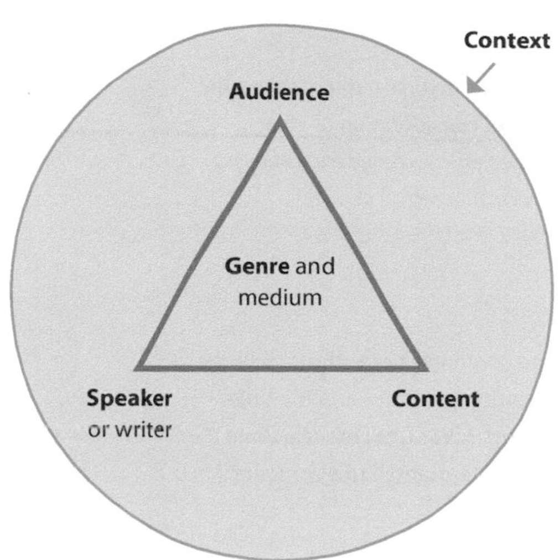

Audience	Should be informed, engaged, and persuaded to adopt your conclusions or your analytical approach
Speaker/writer	Should appear credible, competent, and trustworthy, in command of both the technical content and the audience's needs

As is usually the case when optimizing for multiple criteria, compromises are required. For instance, you may be able to establish your own technical expertise by using advanced jargon known only to specialists in your technical area—but doing so comes at a heavy cost in clarity for your audience. These competing demands, and the possible tradeoffs between them, should guide you through numerous decisions that shape the eventual document:

- Will you translate the concepts into non-expert language or use specialized terms from your technical field or industry? If such terms can be used, do readers need definitions or analogies to help them to understand?
- Will you show mathematical reasoning, refer to a specific result of that reasoning, or keep all of the math "behind the scenes"?
- Are graphs, illustrations, or tables needed to ensure that the most important concepts are also the most prominent ones?

Within the triangle are the categories of documents or communication acts that serve the needs of the writer, audience, and content.

Genres are categories of communication (reports, proposals, memos, letters, and so forth) that have evolved over time to accommodate the various elements of the rhetorical situation. Broad genres—such as experimental reports—are governed by general rules that may be followed with little variation across multiple technical fields or industries.

A *medium* is a material channel through which communication can be delivered—most likely print, electronic, or oral communication. Many tasks require us to manage multiple media simultaneously—presentation slides that accompany a spoken talk, or a written document that may be read onscreen or in hard copy.

Multiple, diverse audiences require careful consideration of what makes effective evidence and the appropriate level of detail.

The committee of authors includes fourteen members and multiple editors: ideas from many people were integrated into a coherent whole.

The authors of the report make recommendations for systems-level changes improving the quality of the U.S. healthcare system.

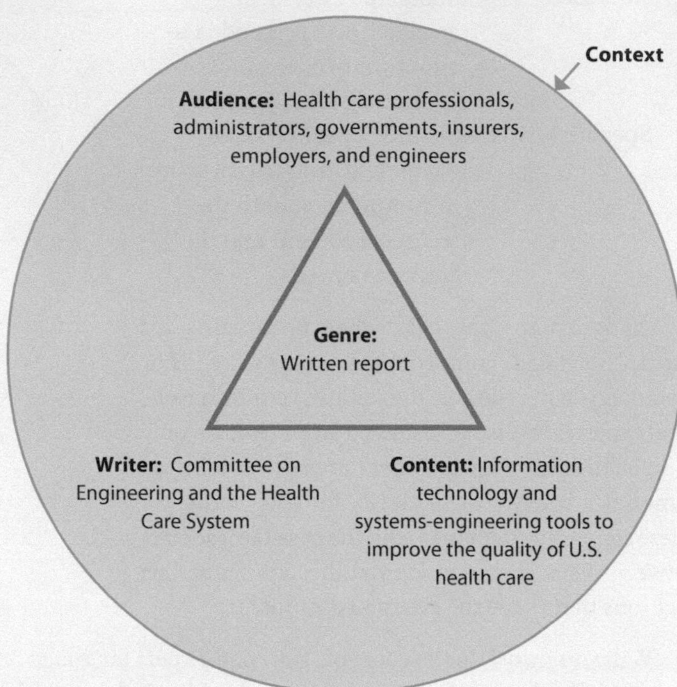

This rhetorical triangle illuminates the complex rhetorical situation being addressed by the report "Building a Better Delivery System: A New Engineering/Health Care Partnership," issued jointly by the US National Academy of Engineering (NAE) and Institute of Medicine (IOM). © 2005 National Academy of Sciences. Used with permission.

1.2 Displaying evidence and reasoning (*logos*)

Technical experts are used to thinking of technical content or subject matter first: after all, technical concerns and details define our profession and the nature of the problems that we're attempting to solve. For this reason, we usually associate technical writing with persuasion on the basis of evidence and logical proof—the reasoning strategies that rhetoricians categorize as *logos*, from the ancient Greek word for "truth."

Varieties of reasoning

Logos may mean "truth"—the root of the English word *logic* and the suffix *-ology* for various forms of scientific study—but this aspect of persuasion involves much more than making true or provable statements. Rather, if you are an engineer planning an argument from this perspective, you should be thinking about what will *count* as sufficient evidence or appropriate proof within your profession or organization.

- Are you starting from a theoretical model or from an experimental finding?
- What degree of confidence do you have in your data?
- Is your goal to propose a solution, or to define a problem?
- Does your evidence need to show that an idea is technically possible? Profitable? Appealing to potential clients or customers?

Such questions show that evidence and reasoning raise complex questions about whether any given inference is warranted. Perhaps most importantly, the discussion of technical and professional issues cannot be reduced to a simple distinction between certain "facts" and mere "opinions"—in any significant professional discussion, expect "the facts" to be embedded in complex questions and disputes about how they should be interpreted and addressed. Engineers will most often work with quantitative evidence, but numbers may not be the best approach for all analysis in all circumstances. Quantities that can easily be measured may not be the ones that matter most, and prominent numbers (revenue projections) can unduly overshadow factors that are harder to quantify (human health and safety).

One need only note that: (1) more than 98,000 Americans die and more than one million patients suffer injuries each year as a result of broken health care processes and system failures (IOM, 2000; Starfield, 2000); (2) little more than half of U.S. patients receive known "best practice" treatments for their illnesses and less than half of physician practices use recommended processes for care (Casalino et al., 2003; McGlynn et al., 2003); and (3) an estimated thirty to forty cents of every dollar spent on health care, or more than a half-trillion dollars per year, is spent on costs associated with "overuse, underuse, misuse, duplication, system failures, unnecessary repetition, poor communication, and inefficiency" (Lawrence, in this volume). Health

Specific figures—98,000 Americans die, 30–40% of healthcare spending is wasted, 43 million Americans are without insurance—are more compelling than generalizations ("millions" of Americans are without health insurance).

Citing sources also demonstrates the credibility of the authors' research.

care costs have been rising at double-digit rates since the late 1990s—roughly three times the rate of inflation—claiming a growing share of every American's income, inflicting economic hardships on many, and decreasing access to care. At the same time, the number of uninsured has risen to more than 43 million, more than one-sixth of the U.S. population under the age of 65 (IOM, 2004a).

The authors provide statistics as evidence to support their argument that the U.S. health care sector warrants engineers' attention to improve its quality and delivery.

1.3 Conveying credibility (*ethos*)

In theory, evidence and reasoning are to be judged on their own merits: a mathematical derivation or the result of an experiment is supposed to be equally valid, regardless of who's making the argument. *Logos* is governed by rules for procedures and inferences that operate the same way for anyone making an argument.

In practice, of course, the messenger matters. You're likely to follow the advice of a trusted confidante or mentor who argues for a particular way of analyzing a data set, based in part on past experience: you've come to know this person's extensive technical knowledge and sound judgment. The same argument will probably not carry the same weight, though, if it comes from a new teammate that you don't know well. Even if the substance is the same, you'll likely perceive it differently: your teammate's levels of knowledge and discretion are uncertain. Is this someone who knows the technical area well, and speaks out only in areas of his or her expertise? Or is it someone who thinks aloud and offers general ideas all the time?

Ethos—the root of the English word "ethics"—accounts for these inevitable judgments about the *character*, *credibility*, and *competence* of the person delivering the argument. Establishing your *ethos* successfully means convincing others that you're trustworthy—that you know what you're talking about, and that your motives are genuine.

Establishing technical credibility

Every engineer wants to be accepted by others as technically proficient. When you submit work to an engineering professor, the keys to your credibility are transparency and completeness—"showing your work." It's obviously desirable to solve a homework problem correctly, arriving at the right answer, but the answer itself isn't all that matters: listing your assumptions and drawing a clear diagram shows your methodical work and your application of the principles that you've been studying. Someone who shows a generally correct analytical approach containing a small error may be less *correct* but more *credible* than someone who submits a correct answer without showing how it was reached.

This strategy of complete, explicit reasoning works to impress engineering professors, who are explicitly evaluating your learning from a position of advanced expertise. For a manager or a client, though, it's often better not to lead with complicated math or computer code, large data sets, or intricate applications of engineering principles. You may need to explain some of these in answering questions, or refer broadly to them in making a point. In the workplace, though, dwelling constantly on advanced technical content may diminish your credibility rather than enhancing it: these are audiences that are most likely to judge your competence by determining whether you can correctly identify and clearly explain big ideas. In earning the respect of a client or a senior executive, technical details are instead more important as a stockpile of evidence to be summoned when needed. (Of course, exceptions exist: some senior executives make a point of mastering the technical minutiae.)

Establishing trustworthiness and sincerity

Your peers and clients need to rely not only on your technical capabilities, but also on your motivations. Asking audiences to believe what you're telling them means asking them to trust you:

- That your proposed solution is based on an honest assessment of the problem, not on a desire to sell expensive technical systems or services
- That your data is complete and accurate
- That your team can deliver what you claim, on schedule and within budget.

As you communicate with others, therefore, you're making implicit arguments not only about your technical expertise, but also about the way that you're *using* that expertise—that your professional work is founded on honest concern for the best interests of your clients and colleagues.

Crediting the funding foundations is a form of transparency while also vouching for the study's merit. Would these prestigious institutions and foundations support the project if it weren't sound?

Citing prior work by IOM makes the present report credible by suggesting that its aims are already well-established and regarded by experts as sound.

With support from the National Science Foundation, National Institutes of Health (NIH), and Robert Wood Johnson Foundation, the National Academy of Engineering (NAE) and Institute of Medicine (IOM) of the National Academies convened a committee of 14 engineers and health care professionals to identify engineering tools and technologies that could help the health system overcome these crises and deliver care that is safe, effective, timely, patient-centered, efficient, and equitable—the six quality aims envisioned in the landmark IOM report, Crossing the Quality Chasm (Box ES-1).

The ethos of the report is initially built on the reputations not only of the National Academy of Engineering and the Institute of Medicine, but also of the funding agencies.

1.4 Accommodating audience needs, values, and priorities (*pathos*)

Pathos traditionally refers to emotions (and especially negative ones such as sadness or pity: one literal meaning of the Greek word is "suffering"). Engineering communication will rarely involve *pathos* in the narrowest form: you will seldom advance professional success by inspiring pity or anger in audiences. Such appeals make *pathos* prominent in marketing and electoral politics, where we frequently mistrust speakers and writers as manipulative and even deceptive.

Nevertheless, you can use *pathos* effectively in an ethical way, without exploiting others' feelings. *Pathos* can also mean "experience"—a meaning present in the English word *empathy*. When you empathize with people, you understand their experience in a way that enables you to share their concerns, feelings, and values. Working with others' values in mind creates a crucial perspective for both your engineering practice and communication, aligning your work and your message with the priorities, concerns, and beliefs of the people to whom you're speaking or writing.

Some values and priorities, such as efficiency, productivity, safety, and the elimination of waste, are sufficiently widespread in engineering that appealing to them is normally a sound strategy. (There are exceptions even to these: for instance, workers at a plant might be nervous that management could attempt to increase efficiency or productivity by instituting harsher working conditions.) Other values are much more dependent on the culture of an organization: one firm might place the highest priority on keeping costs as low as possible, while another might value the highest possible product performance.

Establishing the relevance of the content (exigence)

One of the most important uses of *pathos* is to address the reasons that different audience members and stakeholder groups are most likely to care about the issues you're discussing. *Exigence* is the rhetorical term for a community's motivation to care about a situation, to see it as important and worthy of attention. Think of exigence as the answer to the question *So what?*

Winning over your audience members often means leading them to see a problem with the status quo (or with an anticipated future state of affairs). For instance, if you are arguing for the need to retool a manufacturing plant, you might begin with a few talking points for particular audiences:

Investors	At present, the plant's assembly lines process material less efficiently than competitors' newer plants, leading to more waste, lower hourly productivity, and, ultimately, reduced profits.
Workers	The current equipment is less safe to operate than what is now available.
Local communities	If the assembly lines are allowed to become obsolete, the plant's future—and citizens' jobs—might be jeopardized.

Establishing exigence should never mean that you're making up problems or possible benefits in which you don't really believe. Presumably, your ideas have emerged in response to real needs as you perceive them. Exigence doesn't require that your audience agree with you on all of the major topics—just that they see the content as sufficiently important to merit careful discussion.

The "Six Quality Aims" establish exigence by addressing known concerns of its community of readers.

Safety is a defining concern for engineers and physicians alike.

Healthcare providers worry about underserved patients who aren't getting the care they need.

Timeliness and efficiency are defining values for process engineers.

BOX ES-1
Six Quality Aims for the 21st Century Health Care System

The committee proposes six aims for improvement to address key dimensions in which today's health care system functions at far lower levels than it can and should. Health care should be:

- Safe—avoiding injuries to patients from the care that is intended to help them.
- Effective—providing services based on scientific knowledge to all who could benefit and refraining from providing services to those not likely to benefit (avoiding underuse and overuse, respectively).
- Patient-centered—providing care that is respectful of and responsive to individual patient preferences, needs, and values and ensuring that patient values guide all clinical decisions.
- Timely—reducing waits and sometimes harmful delays for both those who receive and those who give care.
- Efficient—avoiding waste, including waste of equipment, supplies, ideas, and energy.
- Equitable—providing care that does not vary in quality because of personal characteristics such as gender, ethnicity, geographic location, and socioeconomic status.

This report not only identifies its criteria for quality care but also provides definitions for those criteria. © 2005 National Academy of Sciences. Used with permission.

Adjusting the level of technicality

The other initial step in analyzing your audience is to estimate the level and type of technical knowledge that your anticipated reader or listener may possess:

Expert	Deep familiarity with the topic down to its nuances and complexities, leading to recognition by others as an authority
Professional	Practical and/or theoretical knowledge of the topic or technology, sufficient to work with it in some capacity
Novice	Some elementary working knowledge of basic terms, concepts, and principles, or a framework in which they can be made meaningful
General	Lack of previous exposure to the topic beyond that of the average layperson, consumer, or citizen

At higher levels of expertise, audiences will not only be able to understand more detailed content; the information they need to make judgments correctly will likely *require* that they have such content. Using more specialized technical jargon is usually part of meeting those needs efficiently. (Translating high-level concepts into general language may be necessary when you have multiple audiences, but it reduces the conciseness of your work.)

These levels of familiarity with a particular topic involve education, work experience, and specialization. They are, however, often misunderstood: don't conflate expertise with intelligence, or even with university education. You can expect that most experts will be highly intelligent, and that many will have advanced degrees in relevant fields of study. At the same time, being highly educated (or highly paid) doesn't mean that one doesn't have plenty to learn from those with other types of experience. Moreover, all of us have high-level professional knowledge in some areas while remaining novices in others: you might be able to write to your engineering manager at a nearly expert level about control systems, but have to reduce the level when it comes to digital circuit design.

1.5 Writing within genres

Genres—categories of documents that arrange their content in well-known ways—are most important to writers and speakers as a guide to what audiences will expect. Genres can be very broad, or very specific: the main menu of Netflix allows users to browse within general categories of films (such as comedy, drama, romance, science fiction, and documentary), but also leads users to much more specific subgenres within each category. Professional documents and talks can likewise be categorized either broadly or narrowly: one can think about "reports" in general or any one of the many more specific types—progress reports, experimental reports, or corporate annual reports.

Genre, rules, and audience expectations

Genre is important because it governs the expectations of audiences. When you go to an action movie, you bring certain expectations about characters, plot elements, settings, and pacing, and you may be dissatisfied if the film violates too many of these. The audience perspective is actually much the same for readers and listeners in a document. Your undergraduate research advisor will most likely read an experimental report carefully and patiently, considering questions of method and evidence, anticipating that your report won't discuss the practical meaning of major findings until the last section. The same experiment would be discussed very differently in a memo in the industry: there, a supervisor or manager would expect to find clear recommendations stated up front. Those readers would probably also exhibit less patience for detailed data and calculations, and might want to see them attached rather than included in the body of the memo.

The NAE/IOM authors call their document a "report" or "study": it shares the results of a significant and deep exploration of the situation being addressed.

The NAE/NIM report also makes many recommendations to support the stated goals, so it also includes many important features of the "proposal" genre.

Chapter 4 Recommendations

Recommendation 4-1. The committee endorses the recommendations made by the Institute of Medicine Committee on Data Standards for Patient Safety, which called for continued development of health care data standards and a significant increase in the technical and material support provided by the federal government for public-private partnerships in this area.

Recommendation 4-2. The committee endorses the recommendations of the President's Information Technology Advisory Council that call for: (1) application of lessons learned from advances in other fields (e.g., computer infrastructure, privacy issues, and security issues); and (2) increased coordination of federally supported research and development in these areas through the Networking and Information Technology Research and Development Program.

Recommendation 4-3. Research and development in the following areas should be supported:

- human-information/communications technology system interfaces
- voice-recognition systems
- software that improves interoperability and connectivity among systems from different vendors
- systems that spread costs among multiple users
- software dependability in systems critical to health care delivery
- secure, dispersed, multi-agent databases that meet the needs of both providers and patients
- measurement of the impact of information/communications systems on the quality and productivity of health care

This NAE/NIM study makes eighteen specific recommendations in the executive summary, a few of which are shown here.

Many engineers are attracted to the idea that genres are governed by definite rules that can be learned, but this attraction often leads to a misguided desire to "follow the rules" by using some other document as a template and duplicating its features exactly. This is the same "plug-and-chug" approach that can seem attractive in solving technical problems mathematically: find an equation capable of fitting the data that we have, and solve.

Engineering students, of course, know that this approach to problem solving goes wrong more often than we'd like: even if our chosen equation is right in its general form, it may rely on assumptions that don't correspond to the particular situation we're analyzing. The same is often true for documents.

In practice, it can be useful to start writing by studying a document that someone else has produced—as is often done in technical workplaces. This approach will only generate worthwhile results, however, insofar as two criteria are met:

- The original document's situation matches the one you're facing—performing a comparable analytical task or reporting comparable information to the same audience in the same circumstances.
- You are identifying, assessing, and imitating purposeful strategies used by the original document's author—not simply copying surface features.

Adapting and evolving genres: When to break rules

Genres evolve over time, reflecting changing professional practices, new communication technologies, and the demands of new types of content. You don't need to be a historical scholar of technical and professional communication—but you should be aware that standard practices will evolve during your career, and be prepared to change your own approach when necessary.

For such evolution to occur, individual writers obviously have to break from existing conventions. In creative work, the standard advice on rule breaking is that one needs to understand the rules before breaking them. (Shakespeare, Beethoven, Picasso, and the Beatles all began with relatively formulaic work before departing radically from the generic forms that many of their peers continued to follow.) In technical writing, it is especially important to innovate only in careful and purposeful ways: some audiences may even mistrust your content if its presentation differs from what they've come to expect.

Still, a place for calculated risk taking does exist in professional communication, and becoming an excellent technical communicator demands the strategic design and development of your message, well beyond merely following rules correctly.

- If you're preparing a résumé to apply for your first engineering internship, and don't have paid work experience in your engineering discipline, a standard list of employers and job titles won't make for an effective "Experience" section. Instead, you'll want to change that section's scope and focus so that it can include your best, most relevant work.
- A pie chart might be a standard feature of a particular type of report produced in your company. If your data is divided into too many categories, though, you might conclude that a dot plot is more effective at making it clear to readers.
- The discussions and the sample documents in this book reflect both "standard practice" and purposeful, successful work that adapts the rules of the genre to best deliver the content of the communication.

Summary

Assessing the rhetorical situation

- Any communication situation includes the *audience* of readers or listeners, the *speaker or writer* addressing that audience, and the *content* with which they are concerned. Each entails its own major considerations for communication strategy.

Displaying evidence and reasoning (*logos*)

- Providing evidence and logical proof includes thinking about what will count as sufficient evidence or appropriate proof within your profession or organization.

Conveying credibility (*ethos*)

- Establishing your ethos successfully means convincing others that you're technically credible—that you've sought out appropriate evidence and reasoned carefully about it, using accepted analytical techniques.
- Building trust with your audience also means convincing them that your motivations are genuine. As you communicate with others, therefore, you're making implicit arguments not only about your technical expertise, but also about the way that you're using that expertise—that your professional work is founded on honest concern for the best interests of your clients and colleagues.

Accommodating audience needs, values, and priorities (*pathos*)

- Working with others' values in mind creates a crucial perspective for both your engineering practice and communication, aligning your work and your message with the priorities, concerns, and beliefs of the people to whom you're speaking or writing.
- Some values and priorities, such as efficiency, productivity, safety, and the elimination of waste, are sufficiently widespread in engineering that appealing to them is normally a sound strategy. Other values are much more dependent on the culture of an organization.
- Audience members and stakeholder groups will likely have different reasons for caring about the issues you're discussing. *Exigence* is the rhetorical term for a community's motivation to care about a situation, to see it as important and worthy of attention.
- Technical workplace audiences range, on any particular topic, from general knowledge to high levels of expertise that require more detailed content and usually support more specialized language.

Writing within genres

- *Genres* are categories of communication that have evolved over time to accommodate the various elements of the rhetorical situation. Genres are most important to writers and speakers as a guide to what audiences will expect, and as a way of identifying recognized, appropriate ways of organizing and presenting content.
- Genres evolve over time, reflecting changing professional practices, new communication technologies, and the demands of new types of content. You should be aware that standard practices will evolve during your career, and be prepared to change your own approach when necessary.
- In technical writing, it is perhaps even more important to innovate only in careful and purposeful ways: your readers expect you to deliver information clearly, not to challenge their understanding with stylistic experiments.

Understanding your audience

2

As you model the rhetorical situation in order to optimize your communication—working, as with any design, to reach specific goals within a set of constraints—your intended audience is by far the most important variable to consider. When writing for one of your engineering professors, you're able to depend on a reader who normally possesses expert technical knowledge and deep familiarity with the content you're engaging; you seldom have to think about ways of explaining a concept more clearly, raising a problem more tactfully, or soliciting the professor's thoughts in a way that encourages forthright dialogue. Once you enter professional practice as an engineer, however, all of these become routine considerations in your speaking and writing. To prepare effectively for this part of your career, you need to start as a student to develop awareness and sensitivity to the needs of your audiences.

Objectives

- To identify all relevant stakeholders—those who bear risk or stand to benefit from the situation or issues at hand—and the interest or stake that defines each
- To make informed decisions about which stakeholders to engage as audiences in your communication
- To model the likely attitudes, concerns, and responses of audiences based on available information about them
- To distinguish among levels of listening defined by various degrees of awareness and receptiveness to others' experience
- To apply these levels to specific tasks and stages in engineering design experiences
- To choose deliberate techniques to enhance one's own listening
- To recognize the most common forms of bias that can prevent engineers and other decision-makers from hearing all voices relevant to a problem

2.1 Analyzing stakeholder audiences

To think strategically about how to serve audience needs, first consider the broad constituencies within an organization—groups of *stakeholders*.

A stakeholder is anyone who's affected by your work or its results. The term comes from the field of business ethics (where it was constructed as a more inclusive alternative to *shareholders*, those who own shares of the company's stock). Many approaches to business management dictate that a business should manage its affairs so that its investors or owners enjoy maximum returns on investment (ROI). Stakeholder theory, on the other hand, acknowledges that every organization also depends on contributors beyond its investors—employees, clients or customers, suppliers, and so on—and owes obligations to them as well. It can be useful to think of such groups as "investors" in their own right, contributing non-financial resources.

While each stakeholder group may contribute value to the organization in its own way, all have an interest in honest, open dialogue about the organization's affairs, and it's fairly easy to identify the priorities that different stakeholders will likely bring to a discussion:

Customers	The quality and value of products and services; the fairness of prices
Investors	Return on investment; market share; favorable publicity; organizational stability
Workers	Workplace health and safety; compensation; job security; fair and equitable management; autonomy and respect
Suppliers and creditors	Revenue; future business opportunities
Local communities	Public health and safety; jobs and economic growth; environmental impacts
Government	Compliance with laws and regulatory policies; tax revenue

Interviewed

Access to Independence of North County
Acushnet Company
Agua Hedionda Foundation
Army and Navy Academy
Batiquitos Lagoon Foundation
Boy Scouts of America, Santa Margarita District
Boys and Girls Club
Buena Vista Audubon Society
Buena Vista Lagoon Foundation
Building Industry Association of San Diego
Carlsbad Convention and Visitors Bureau
Carlsbad Educational Foundation
Carlsbad High School
Carlsbad Library and Arts Foundation
Carlsbad Lightning Soccer Club
Carlsbad Village Business Association
Carlsbad Village Improvement Partnership
Carlsbad Youth Baseball
Carlsbad-Oceanside Art League (COAL)
Carltas Company (Flowerfields)
Four Seasons Resort Aviara
Friends of Carrillo Ranch
Friends of the Carlsbad Library
Fun 05 Friends Play Group
Gemological Institute of America
Grand Pacific Palisades Resort & Hotel
Green Encinitas
Hoehn Honda
Imagine Carlsbad
Interfaith Community Services
Invitrogen (formerly Life Technologies)
Jim Boylan

This list of interviewees includes advocates of stakeholder interests within the Carlsbad community that could be missed if organizers had not sought them out.

Municipal planners and architects often conduct design charrettes—extended design activities that involve formal gathering of ideas from stakeholders. This report states that its stakeholder interviewees "included local businesses in fields such as biotechnology, automobile sales, real estate, retail, entertainment, hospitality, and manufacturing. Non-profit or volunteer organizations interviewed represented interests such as environmental preservation, the youth, seniors, lagoons, arts and culture, the libraries, bicycling, sports leagues, specific neighborhoods, affordable housing, homelessness, and people with disabilities [. . .] local schools and parent-teacher associations, religious communities, and county or regional bodies."

Constructing an audience profile

Information about stakeholder interests and priorities usually provides the best way to begin analyzing an audience. Before you think about the particular individuals who will first respond to your ideas, you can begin your plan by addressing the broad organizational context, letting you anticipate:

- The broad range of potential audiences
- Their top concerns and the topics of likely discussion
- Which ideas or arguments may be most controversial

You may start to think about audience reactions by focusing on a particular audience member: What will the client (or vice-president, or professor) think? It's useful to remember that such individuals will themselves be thinking about others' responses: executives and administrators are always considering the perspectives of the stakeholders, which tend to outweigh their own subjective personal preferences.

When you do want to focus on specific readers or listeners, construct a profile that includes the following dimensions:

Professional role	What decisions or actions will the reader or listener be contemplating? What are his or her primary job responsibilities?
Organizational role	What is the person's job title? To which colleagues, superiors, and subordinates will he or she be summarizing your ideas?
Education and credentials	What level and type of education does the reader or listener possess? • Academic degrees and majors • Professional credentials, licenses, and certifications • On-the-job training and professional development
Professional experience	What positions has this person previously held? What priorities, concerns, and beliefs might remain from these past roles?

Types of technical expertise	What perspectives and interests inform this person's view of relevant technologies? • Design and underlying theory (engineer) • Detailed maintenance, repair, and modification (technician) • Everyday operation (from novice to "power user")
Cultural and personal background	What do you know about the person's origins and formative experiences? • Geographic (nation, region, local community) • Linguistic (languages other than U.S. English) • Experiences of other kinds of cultures and groups (gender, socioeconomic status, disability)
Attitudes toward topic	Are you aware of past statements, experiences, personal values, or beliefs that suggest a particular attitude toward your content?

Understanding people that you don't know well is seldom easy. In examining these topics and questions, you may notice the potential for inaccurate conclusions based on stereotypes. Be especially hesitant to ascribe beliefs or positions to others based on personal characteristics: Don't make assumptions about your CEO's beliefs, for instance, because she's a woman or because she speaks Spanish.

Conjectures based on professional roles and experiences are usually less perilous. You might err by including too much advanced math in your presentation based on the technical backgrounds of your engineering team, and doing so might force you to come up with some simpler explanations during the course of your talk. Such a misstep, however, won't likely cause your audience to take offense or to perceive you as insensitive.

2.2 Listening to stakeholders

An audience profile resembles other kinds of engineering models. If careful reasoning and evidence inform the development of your model, you can expect reasonable accuracy (at least relative to simple guesswork). Ultimately, though, any model may need to be revised or even abandoned if real-world evidence demonstrates that its predictions aren't sound. In communication, that evidence is obtained by *listening* carefully and thoughtfully to those affected by your work.

Fortunately, listening is a well-defined skill that can be learned, and making it a priority in your studies can make you a more capable professional. Many engineers in leadership positions have been influenced by Stephen Covey's book *The Seven Habits of Highly Effective People*, which suggests that we should monitor our listening, classifying it on the following scale:

> When another person speaks, we're usually "listening" at one of four levels. We may be **ignoring** another person, not really listening at all. We may practice **pretending**, "Yeah. Uh-huh. Right." We may practice **selective listening**, hearing only certain parts of the conversation. We often do this when we're listening to the constant chatter of a preschool child. Or we may even practice **attentive listening**, paying attention and focusing energy on the words that are being said. But very few of us ever practice the fifth level, the highest form of listening, **empathic listening**. [. . .] Empathic listening is so powerful because it gives you accurate data to work with. [. . .] You're listening to understand. (© 2013 Simon & Schuster, Inc. Reprinted with permission.)

Covey's scale is only one of the many frameworks that we can use to rate our level of attention and receptiveness to the ideas and concerns of others. To assess the level of listening that an engineering project requires, and whether the key personnel are successfully engaging with everyone affected by the project, three general levels may suffice.

Low levels: Listening only to technology

At the lowest level of listening, engineers' attention is dominated by specifications, computer models, and quantitative analysis—potentially excluding people's perspectives and social contexts.

Of course, not every engineering project requires extended communication with a wide range of stakeholders: Sometimes an assignment really is dictated purely by specifications that a design must satisfy. This isn't always a failure of your professional skills; you might be working in a large firm, with responsibilities that concern only a part, assembly, or subsystem. In such a case, another team altogether might be devoting all of their effort to analyzing user experience for the system as a whole.

However, it may be just as frequent for engineers to ignore or dismiss important perspectives—choosing, whether consciously or unconsciously, to ignore potential insight into their technical problems. According to many scholars, this habit may be a natural result of the problem-solving processes that dominate our education as engineers. Such processes involve assumptions, diagrams, and physical principles, but not the human agents and contexts in which problems are defined and decisions are made about how to approach them.

As a result, engineers' attention to those people and contexts may be limited. We may ignore information that isn't codified in a quantitative requirement issued by a supervisor, or just listen selectively for the kinds of information that usually occupy our attention: technical feasibility and functionality, potential profit and costs, budgets and deadlines, and so on.

Failing to listen beyond this level can cause serious errors in your professional practice. The worst result, of course, occurs when an engineer ignores a client or stakeholder's input and, in doing so, loses crucial information, leading to the failure of a technical system. More commonly, an engineer might come across as rude or arrogant because he or she is reasoning soundly about valid data but has ignored important perspectives and is therefore solving the wrong problem.

Middle levels: Interpreting user experiences

Adding two basic elements to their engagement with engineering problems and projects, more advanced listeners are:

- *Observant*, perceiving and considering the experiences, needs, and preferences reported by others
- *Open-minded*, letting those factors shape the understanding of a problem and the design of a solution

In recent years, engineers and designers have sought out user perspectives in more systematic ways than was once the case. The design process can be opened to user input, though, at many different levels. Engineering educator Carla Zoltowski has defined a number of these levels: for instance, she points out that collecting information from prospective users at the beginning of the design process is important but doesn't guarantee that those users' needs will be carefully considered in all of the subsequent stages.

The middle levels can be considered, then, as a progression toward more open and more intensive partnerships with users and others who have some stake in your work. The most elementary levels of such involvement might involve a simple questionnaire where stakeholders answer simple questions (responding "yes" or "no," or rating something on a scale—"very simple" to "very difficult," or "very important" to "not at all important").

High levels: Connecting with others' perspectives

At higher levels still, you're no longer *merely* receiving or interpreting information from others: rather, that exchange of information has become part of a more substantial connection, marked by mutual trust and respect.

Experts have defined *empathic design* as the highest level of human-centered design that can emerge from such relationships. Cutting-edge design firms in industry, like IDEO, base their business strategies on creating the conditions for empathic design—"intentionally seek[ing] opportunities to connect with people in meaningful ways and to set aside reactions and behaviors that will interfere with it." Such empathic abilities grow when your professional practice includes attitudes and behaviors like these:

- Thinking of clients and stakeholders as equal partners in your endeavors
- Interacting with those partners socially as well as professionally
- Tempering your own expertise with humility, thinking carefully about what you don't know
- Developing enthusiasm for perspectives and forms of expertise outside of engineering
- Recognizing and accommodating differences in culture and personality

Even when we are trying to listen empathically, we still might make the mistake of listening empathically only to some people. This oversight is often unintentional and can be avoided with conscious reflection. Consider who has the most day-to-day lived experience or who will experience the potential consequences—positive or negative—of your project. These are the stakeholders with whom you need to speak.

Because communication across cultures is particularly challenging, empathic listening strategies have special relevance for engineers and project managers working on international and humanitarian projects; students often gain experience with such projects through groups like Engineers Without Borders and Engineers for a Sustainable World.

(Similar needs arise when designing for poor communities or for people with disabilities.) It is a mistake, though, to consider empathic design methodologies and listening strategies *only* in those specialized contexts: Advancing one's listening abilities to the highest levels can likewise enable success in "standard" business and industry settings.

2.3 Techniques for listening

Like other dimensions of communication, listening is a skill that can be developed. Listening techniques can help to assure others that you're taking their ideas and perspectives seriously, but they are probably even

2.3 Interview Materials and Process

Stakeholder interviews were typically conducted with one interviewer and one note-taker. Training was provided by the City Communication Manager on techniques to use to assure a successful interview. A guide was prepared for use by interviewers to help ensure consistent interview styles and scope, even when different people conducted interviews. The guide provided space to record the date, time, interviewer(s), and interviewee(s); introductory information about the Envision Carlsbad process; and five basic guiding questions as a starting point for dialogue, and moving the interview along. The guiding questions were:

- As a representative of [organization name], what do you value most about the Carlsbad community?
- As you look ahead to the next 10 to 20 years, what are three important challenges and three opportunities you anticipate for the Carlsbad community?
- What would be your organization's advice to the City for addressing these challenges and opportunities?
- What would make Carlsbad a better place to live, work, and play today and in the future?
- What other thoughts can you share about Carlsbad's future?

A note-taker recorded all comments. The body of the report analyzes themes that emerged in these interviews and the appendix of the report includes comments verbatim.

Using the same questions for each interview provides some consistency. At the same time there is the expectation that these interviews will be a dialogue. The interviews were scheduled one hour apart and took approximately 45 minutes.

Stakeholder interviews are not unique to the design charrettes that involve civil engineers: They are a crucial element of requirements gathering and usability testing, used widely by engineers in many disciplines.

The interview process for this urban planning charrette invites contributions from many community groups, using open-ended questions that allow interviewees to steer the direction of the conversation. © 2009 City of Carlsbad, CA. Reprinted with permission.

more valuable in helping you to process information and to consider it carefully.

Showing attentiveness

Take notes so that you can revisit the conversation later. It's impossible to remember everything; moreover, information may strike you as more important later, as you grow to understand the project's context. Try to jot down key phrases verbatim so that you retain the original meaning. While taking notes is especially useful for you, it also shows others that you value what they are saying, according it sufficient importance that you're recording it for later use.

Of course, it's just as easy to signal a *lack* of attentiveness. Be especially careful not to appear distracted by mobile phones or tablets, which tend to create the appearance that you value another conversation or diversion more highly. Don't try to text covertly. If you're using the device to take notes during a conversation, tell the other person what you're doing.

Seeking understanding

Active listening occurs when you aren't merely taking in information passively but are working to understand it fully and to engage its implications. Try the following approaches to active listening:

- Paraphrase back what someone has told you. Putting their message in your own words not only helps you think through the content but also ensures that you're representing it accurately.
- Ask sincere questions, soliciting opinions or explanations. Rhetorical questions and thinly veiled opinions ("Wouldn't it be better if we . . .?") are insincere because they either suggest that the other person is wrong or attempt to sway him or her to your position.
- Avoid interrupting the speaker. Even if you want to agree, to paraphrase, or ask clarification, wait for a pause in the conversation.
- When you disagree with someone else's statement, ask a question. Instead of telling them why they are wrong, try to figure out *why* they think what they do. Understanding what has informed their perceptions is important, particularly when they are exhibiting fear, anxiety, or distrust of a technical solution.

Inviting dialogue

Engineers obviously have to ensure that clients are regularly briefed on project developments, and that the technical team's own questions are answered. Exchanging information in this way, however, does not guarantee that those clients—or other stakeholders—feel that their input and concerns are truly valued. To engage in a productive conversation with meaningful impact, it often takes some work to issue a sincere *invitation* for everyone in the conversation to exchange ideas frankly.

- Everyone is used to hearing "Any questions?" at the end of a presentation or an email, but this is seldom interpreted as a sincere request for input. It's better to tell your listeners how much you value their suggestions and questions; you may also want to prompt them with specific topics on which you'd like to hear their ideas.

Sustainable City

Share

Home » ... » Public Works » Environmental Services » Sustainable City

The goal of the city's sustainability efforts is to provide a high quality of life for generations to come by striking a long-term balance between the social, economic and environmental factors that contribute to a sustainable community. The City Council has adopted Sustainability Guiding Principles and Environmental Guiding Principles that articulate the framework for a sustainable community. Click here to view the city's video on sustainability.

The Community Vision, adopted in 2010, calls for the city to build on existing sustainability initiatives to emerge as a leader in green development and sustainability. Read the Envision Carlsbad working paper on sustainability.

Working in partnership with the community, the city has implemented several cost effective, efficient programs.

- Alga Norte Community Park
- Energy and clean air
- Solar power for commercial facilities
- Clean creeks, lagoons and the ocean
- Water conservation
- Sustainable drinking water supply
- Recycled water
- Open space
- Trash and recycling
- Transportation

Carlsbad's potential leadership in sustainability was a theme that emerged from the stakeholder interviews. The framing of this website with the title of "Sustainable City," combined with the Community Vision that was adopted the year following the report, suggests that the stakeholder interviews are informing the direction of the City.

The programs listed directly respond to comments by stakeholders, particularly the programs in solar power, water conservation, recycled water, and trash and recycling.

The City of Carlsbad's web page on sustainability directly responds to the stakeholder interviews. Such direct responses advance the relationship with community members, helping them to know that their perspectives are valued.

- Do not mistake silence for agreement. Some people are uncomfortable expressing explicit disagreement or may feel that you will not take their disagreement seriously. When you want honest input or feedback, ask for it, and be ready to listen respectfully and thoughtfully.
- If you are planning a meeting, or gathering notes for your report, your first inclination may be to brief the other attendees on decisions and ongoing actions. Ask whether you can instead host an open discussion, perhaps listing participating stakeholders instead of scheduled speakers. If they expect nothing more than high-ranking individuals delivering reports, attendees may lack motivation to do more than to listen passively—or they might be tempted to skip the meeting altogether.
- Reflect on the conversation afterward, and follow up if needed. In particular, consider any moments in which you experienced or perceived discomfort or doubt. This might tell you which topics or ideas require further discussion to reach a consensus or to clear up ambiguities.

Summary

Analyzing stakeholder audiences

- To think strategically about how to serve audience needs, first consider the broad constituencies within an organization—groups of stakeholders likely to be affected by your work or its results.
- While a good audience profile begins with the most important stakeholder concerns, it goes on to address a number of other characteristics that should guide your strategy for audience accommodation, such as professional credentials and experience, technical expertise, and personal and cultural background.

Listening to stakeholders

- At the lowest level of listening, engineers' attention is dominated by specifications, computer models, and quantitative analysis—potentially excluding people's perspectives and social contexts.
- More advanced listening competence arises from *observant* and *open-minded* attention to users and stakeholders, and allows engineers to design more effectively with their needs and experiences in mind.
- At the highest levels of listening, stakeholders can become full partners in design, and engineers can go beyond merely

collecting user input to developing more personal connections with others' experiences.

- *Human-centered* design and *empathic* design aim to maximize meaningful participation by users and stakeholders in design projects.

Techniques for listening

- To avoid *ignoring* others, or merely *selective listening*, listeners can work on showing attentiveness. Take written notes on spoken statements, and avoid electronic distractions and other forms of multitasking.
- Misunderstandings can be avoided when partners in a conversation ask *sincere* questions—not leading or rhetorical ones—and *paraphrase* others' statements back to them.
- Eliciting the greatest contributions from clients and stakeholders requires that they feel *valued* and perceive an authentic *invitation* to participate in a meaningful conversation. Engineers can facilitate this degree of involvement by framing their own ideas carefully and structuring interactions in an open, equitable way.

Meeting your ethical obligations

3

Engineers in the United States are generally rewarded for maximizing profit for their employers. However, failure to consider more than the "bottom line" can have disastrous consequences for the engineer and the public. Examples in this module come from the public record documenting General Motors' recalls of cars after reports of faulty ignition switches. (Insufficient torque on a plunger in the ignition assembly allowed the ignition suddenly to switch off—an event that caused a number of accidents and as of July, 2015, at least 119 deaths.) The design flaw affected nine vehicle models across the company's brands. Within the company, many engineers and managers studied the phenomenon and analyzed possible solutions, but they reported their findings in ways that allowed the risks to be treated as minimal. The legal and economic consequences for GM have been dire.

Student engineers should be aware of these pressures. They will have to make real decisions and take real actions in cases where staying within the law and not making waves are inadequate guides to action. Hence the need for the *discipline* of ethics.

Much more than a list of rules for behavior, ethics comprises a fundamental set of intellectual tools for choosing action. Like engineers, ethicists propose models, define their assumptions, and describe the constraints under which models are expected to work and conditions for which they fail. Ethical problems in engineering, like design problems, are complex and ambiguous.

Objectives

- To describe the fundamental canons of engineering ethics as they apply to everyday engineering work and engineering communication
- To anticipate typical pressures that come with work in industry, such as the pressure to avoid calling attention to safety concerns
- To recognize actions that violate rights and duties of engineers, their employers, their clients, and the public
- To weigh courses of possible action according to likely consequences and according to the demands of ethical rules such as those prohibiting deception
- To identify communication—in language or in visuals—that constitutes a "deceptive act"

3.1 Ethics in engineering

People stand to benefit from an engineer's professional activities, but people also bear risk or may be harmed if something goes wrong. Taken all together, these people represent the engineer's *stakeholders*, people on whose behalf engineers have obligations. Identifying these obligations starts with the fundamental canons.

The fundamental canons of the engineering profession

The major engineering professional societies, such as IEEE, ASME, ASCE, and AIChe, all have codes of ethics specifying the moral obligations that they expect their members to follow in professional practice. In general, the codes present minor variations of shared fundamental canons. (These have special importance for the licensed Professional Engineer [PE], as the license to practice is contingent on compliance with the canons.) The National Society of Professional Engineers (NSPE) formulates these canons in the following way:

Engineers, in the fulfillment of their professional duties, shall:

1. Hold paramount the safety, health, and welfare of the public.
2. Perform services only in areas of their competence.
3. Issue public statements only in an objective and truthful manner.
4. Act for each employer or client as faithful agents or trustees.
5. Avoid deceptive acts.
6. Conduct themselves honorably, responsibly, ethically, and lawfully so as to enhance the honor, reputation, and usefulness of the profession.

(Reprinted by Permission of the National Society of Professional Engineers [NSPE] www.nspe.org.)

NSPE and the other professional associations elaborate on these directives in their codes of ethics and update them as engineering practice evolves. For instance, the American Society of Civil Engineering decided in 2009 that "the safety, health, and welfare of the public" require engineers to follow "principles of sustainable development," and so ASCE added language to that effect (along with a definition of "sustainable development" to govern decision making).

Applying codes of ethics

Engineers take these obligations seriously, and they form a crucial part of a new engineer's education. Ideally, you should be able to discuss the canons in the same knowledgeable, intelligent way that you can discuss basic physics, chemistry, computing, or design. Like other basic principles, the canons can help you to anticipate concerns with a proposed course of action.

Sometimes, the canons may prohibit or mandate an action so clearly that little deliberation is needed. For instance, you might be studying a proposed manufacturing process for your employer when you determine that the process would release a hazardous effluent into the air. In such a case, the first fundamental canon requires you to take action of some kind—finding an alternative process, or perhaps installing emission controls.

When first studying the fundamental canons, engineers sometimes experience discomfort at the ambiguities and uncertainties of ethical reasoning:

- Are EPA regulations and the Clean Air Act sufficient to define "the safety, health, and welfare of the public"? Are all emissions assumed to be safe if not prohibited by such regulations?

- If an engineer crafts a public statement favorable to an employer, has he or she violated the canon to report in an "objective and truthful manner"?
- If safety risks affect employees rather than "the public," should they be analyzed differently?

Questions like these are admittedly different than technical questions on which engineers spend much of their time. In many ways, though, they parallel important steps of the design process, when competing priorities force difficult decisions, often without complete information. For instance, choosing materials for a car body requires us to consider weight, strength, cost, availability, manufacturability, the state of our relationships with suppliers, and the demands of marketing attempting to satisfy the whims of potential customers. Similarly, ethical decisions under the fundamental canons can be regarded as "optimizing" among obligations to the public, clients, employer, and one's own career subject to ambiguous information from contractors, contradictory messages from corporate ("reduce costs" and "safety is our number one priority"), and—sometimes—one's conscience.

Working in a corporate culture

While there are some engineering jobs in the government and the nonprofit sector, most engineers work in industry—and most of those in large corporations. While all engineering jobs will likely present difficult decisions, the corporate drive for profit puts additional pressure on engineers and executives. In recent decades, corporations have become increasingly focused on profitability. Competing values, such as health and safety, are often compromised by the imperative to maximize financial returns to investors or owners. Short-term quarterly profit can dominate most decisions, even to the point of overriding the long-term interests of the business itself (as happened at Enron).

Executives are often under enormous pressure to increase profits, and engineers sometimes encounter resistance when their technical judgment contradicts or questions a desired business decision. Engineers can even be unfairly asked to take off their "engineering hat" and put on a "management hat." Conversely, an engineer might be told to stick solely to technical concerns with a reassurance that the corporation's legal and marketing departments will manage adverse consequences. It is far from easy to know when to speak out against a business decision that you believe to be wrong. Fortunately, the discipline of ethics provides analytical frameworks for making decisions in such situations.

Others agreed that GM is sensitive to using the word "stall" in service bulletins and closely scrutinizes any bulletin that does include "stalls" as a symptom.[391] Others at GM confirmed that there was concern about the use of "stall" in a TSB because such language might draw the attention of NHTSA.[392] Oakley said that at times, he included "hot" words to draw attention to an issue from Product Investigations personnel who review Service Bulletins before release.[393] Oakley also noted, however, that he was reluctant to push hard on safety issues because of his perception that his predecessor had been pushed out of the job for doing just that.[394] As discussed above, discussions with Altman and other engineers alleviated Oakley's initial concern that the Ignition Switch presented a safety issue.[395]

The vocabulary of "hot" words is specific to GM but is typical of corporate culture that prohibits use of words that might expose the company to liability.

Oakley's predecessor (Courtland Kelley) sued GM in 2003 about moving too slowly to address vehicle safety concerns. The case was dismissed, and Kelley was relocated within the company.

This excerpt from the Valukas report reveals two troubling aspects of corporate culture that contributed to the GM ignition switch fiasco. Courtesy: National Highway Traffic Safety Administration (NHTSA). (Valukas, 2014.)

At some point in their careers, many engineers will feel, as GM safety inspector Steven Oakley did, "reluctant to push hard on safety issues." This aspect of U.S. corporate culture has been with us since at least the 1930s. Sociologist Elliott Krause states his findings about the profession bluntly:

> Professions, in theory, are supposed to have codes of ethics. Not so in engineering. One thing that engineers almost never do, given their values, is to complain when they work on projects that maximize profits through cutting back on safety. Whether the area is nuclear engineering or the O-ring seals on the space shuttle, whistleblowing on the company will lead to being fired, and usually also to being ostracized by other companies working in the same field. The moral is not lost on U.S. engineers: do not question the safety aspects of your work if you want to remain employed. The codes of ethics of engineering societies are mere pieces of paper, and the officers of the associations that have drafted the codes are practically all in corporate management. (Krause, 1996) © Yale University, used with permission.

In counterpoint, failures caused by "cutting back on safety" are usually the only engineering events that make the news. Engineering that "works" is generally not newsworthy (even the Apollo program moon landings lost television viewership as the novelty wore off). Practicing engineers make countless decisions day after day that support product safety and protect the "safety, health, and welfare of the public." Are there exceptions? Certainly—too many, in our opinion. Fortunately, knowing your professional rights and responsibilities and knowing that ethical conflicts can be managed effectively and without heroics gives you a framework for making decisions consistent with your values and the ethical canons expressed by the engineering professional societies.

3.2 The engineer's rights and duties

Ethical perspectives can be roughly divided into two varieties. *Rights* theories stress human moral rights and define obligations around the need to preserve or respect those rights. *Duty* ethics reverses the order of priority, formulating rules and duties that in turn create and explain the moral rights that we perceive.

Engineers' rights

Engineers have the standard rights of employees, such as the right to privacy, to safe working conditions, to freedom from harassment and mistreatment, and to object to company policies without fear of retaliation. For practical purposes, federal and state laws define the degree to which—and the cases in which—these rights can be enforced.

Engineers also have more specific professional rights. To fulfill the responsibilities to the public that in part define the profession, an engineer has the right to form and express his or her own professional judgment, including the right to disagree with employers or colleagues. In addition, engineers can refuse to participate in unethical activities, such as forging documents and engaging in fraud.

Engineers' obligations to the employer or client

The canons require the engineer to act as a "faithful agent" of the employer—applying one's professional skills to advance an organization's objectives and economic success. Being a "faithful agent" also means protecting proprietary information; even talking in generalities can divulge confidential information. In addition, the engineer's "faithful agency" can be endangered by conflicts of interest: When you perceive such a conflict, notify a supervisor and discuss whether you need to remove yourself from the project.

In addition, employers have a right to expect that engineers will make decisions with a firm's liability exposure in mind. From this perspective, imagine your working documents being made public in the course of a lawsuit. Can you publicly defend the work? Could it be construed as negligent? Unethical?

One important practice to manage your employer's exposure to liability is to document all work in writing. In particular, *never give verbal orders*, because your organization will rely on the written record for vindication in case of litigation. A comprehensive written record has other practical benefits:

- Writing encourages extra attention to issues that prompt disputes—such as specifications, deadlines, or responsibilities.
- Written communication is less prone than verbal communication to misinterpretation.
- Written records help new personnel pick up where a previous engineer left off.

Along the way, the investigators were misled by the GM engineer who approved the below-specification switch in the first place; he had actually changed the ignition switch to solve the problem in later model years of the Cobalt, but failed to document it, told no one, and claimed to remember nothing about the change.

A failure of documentation delayed discovering the cause of the ignition switch deaths.

By failing to document a design change, GM engineer Ray DeGiorgio made the ignition switch failure investigators' job harder—they couldn't easily understand why later models did not have the ignition switch problem. Courtesy: NHTSA. (Valukas, 2014).

A concern for limiting employer's liability does not include failing to document design changes.

Obligations to the public

The canons obligate engineers to "hold paramount the health and safety of the public." In daily practice, then, engineers must explore and mitigate the potential ill effects of their work. The public expects engineers' products to do no harm when used properly. To conduct their work lawfully, engineers must also know and comply with all laws and regulations affecting their specialty. Since regulations have generally arisen in response to previous engineering failures, adhering to them also demonstrates an engineer's concern for public safety.

The engineer is also obligated to report concerns about the safety of a product or process to his or her supervisor, initiating a process to address the concern within the organization. In the event that this process fails, however, engineers have the duty of whistleblowing—disclosing information to an outside source who can take action—if serious harm to the public could be prevented or remedied. Whistle-blowing is never an easy decision and comes with a great personal cost—reporting on colleagues and possibly friends, facing possible retaliation from an employer, and jeopardizing your career and the well-being of your family. Before blowing the whistle, consult a trusted colleague, an ethics committee of a professional society, and a lawyer, who can advise you on legal protections for whistleblowers and their applicability to your situation.

3.3 Analyzing consequences of actions and ethical principles

Philosophers design criteria, systems, and methodologies for judging the morality of actions or decisions. Such systems might be regarded as decision-making tools comparable to industry methodologies like Six Sigma or Total Quality Management (TQM). Ethical positions are not merely subjective opinions: Like other decisions, a position's merits must be judged within a well-defined system of intellectual practices.

Judging morality by predicting the consequences of actions

Engineers, who are problem-solvers by temperament, often appreciate ethical methodologies that frame problems in the most familiar way. *Utilitarianism* defines morally correct decisions by *maximizing utility*—that is, acting to produce the greatest possible benefit for all of those affected by the action. This is a variety of *cost-benefit analysis*, resembling the type of thinking that engineers often perform with the assistance of decision matrices and predictive models.

The nature of engineering practice will almost certainly create some deliberations that need to be conducted in this way. Building a chemical plant in a small town can reasonably be expected to create both *benefits* (jobs, profits, economic growth) and *costs* (emissions, traffic, noise), and responsible decision making needs to account for them all.

Naturally, not all costs and benefits can be quantified easily—how does one weigh a single accidental death? Engineers using utilitarian reasoning should take care that they don't give undue weight to the factors that are most easily quantified. (During past safety crises in the American auto industry, the public was scandalized when they learned that carmakers' analysis of potential recalls had reduced the loss of human lives to a simple dollar value.)

From:	Raymond DeGiorgio
To:	Brian Stouffer
CC:	Brian Thompson; David DeFrain
BCC:	
Sent Date:	2012-10-05 14:10:27:000
Received Date:	2012-10-05 14:10:28:000
Subject:	Re: 2005-7 Cobalt and Ignition Switch Effort
Attachments:	

Brain,
Due to the low volume (I am thinking we would only replaced switches on vehicles which the customer brought into the dealer), I would expect the price of the switch to cost approx. $150.00.

If we replaced switches on ALL the model years i.e. 2005, 2006, 2007 the piece price would be about $10.00 per switch. This cost is based on volume of 1.5 Million unites total.

Best Regards,

Raymond DeGiorgio
Switch Lead Engineer

DeGiorgio suggests replacing only the switches brought into the dealer ($150) as opposed to a recall, which would require replacing all of the switches at a total cost of $15 million.

DeGiorgio's email includes a typo ("Brain" instead of "Brian"), which suggests a hurried response, if not a careless one.

In this email, DeGiorgio's response implies a utilitarian logic, though he fails to account for the potential settlement cost if the unrecalled switches cause fatalities. Courtesy: US Congress. (Energy and Commerce Committee, 2014).

Judging morality by formulating rules for action

Everyone thinks in utilitarian terms at least some of the time. However, many ethicists see problems in determining the moral rightness of actions through their consequences. The most important critics of the utilitarian approach are the followers of the philosopher Immanuel Kant, who believe that objective rules of right conduct can be discovered through rigorous logical inquiry. An action is therefore judged to be right or wrong based not on its estimated consequences, but on its adherence to those universal rules.

For Kant, specific ethical rules are derived from a central principle called the *categorical imperative*. The categorical imperative requires that, to be morally justified, a course of action must work effectively as a *universal* rule for *everyone*'s actions. You can select that course with a good conscience only if you believe it to be the right one for anyone else facing the same deliberation.

Followers of Kant often focus on lying and other forms of deception. From a utilitarian perspective, deception might not always seem to be absolutely wrong: for instance, you could create an obvious benefit for your teammates, making them happy, by praising their work even when you actually believe it to be poorly done. (A rigorous utilitarian analysis would, of course, note potential negative consequences: The project might receive a poor grade, and your teammates would be losing a learning opportunity.) Kant, though, addresses this from an absolute, universal perspective. We can't possibly wish for a rule that advocates lying, because a world of pervasive lying would destroy all trust.

Kant's approach is useful to adopt when you consider some of the fundamental canons, such as the prohibition on "deceptive acts" in your engineering work. For instance, you may need to pursue contracts by delivering the lowest bid that you can. In crafting these budgets, you will obviously want to avoid making your proposal's budget so optimistic that it becomes deceptive, or takes advantage of a client's limited knowledge of technical matters.

3.4 Types of unethical communication

As a practicing engineer, you may feel pressured to act in an unethical way—by your supervisor, your clients or colleagues, or even your internal drive to succeed professionally. Most of these pressures involve communication—whether and how you ought to disclose information to someone. To respond in an ethical way when you feel trapped by conflicting rights and duties, learn to recognize the commonly encountered types of unethical communication.

Concealing information

Engineers tend to value open communication in their work: perform an analysis or run a test and report the results; develop a design meeting stakeholders' needs and present a prototype; be asked to explain the history of project decisions (the project "decision tree") and answer factually. However, conflicting rights and duties can create pressures on engineers to violate this norm and conceal information:

Concealing information (from others in your firm, clients or customers, and/or regulators)	An oil service company destroys engineering documents after a disaster so that the documents cannot be subpoenaed or used by their technical client in their defense.
Concealing a conflict of interest	A professor of petroleum engineering who is also a senior consultant with a petroleum engineering company publishes a study with findings that are favorable to the practice of fracking. The author acknowledges that the study is funded by petroleum firms but cites a university affiliation only, failing to disclose his affiliation with a petroleum firm that does consulting work related to fracking.
Preventing acknowledgment of unpleasant information	To test the effluent of a manufacturing process, a company hires a consultant, who finds that the process fails to meet safety standards and phones the company to give a verbal preview of her report. An executive tells her that she's fulfilled her contract: A written report is unnecessary, but the company will pay her in full.

Unlike the Preliminary Information that preceded this Service Bulletin[388] or an earlier Service Bulletin request drafted by Steve Oakley, the Technical Service Bulletin did not describe the problem as involving a "stall."[389] According to Oakley, the term "stall" is a "hot" word that GM generally does not use in bulletins because it may raise a concern about vehicle safety, which suggests GM should recall the vehicle, not issue a bulletin.[390]

Choosing words to downplay safety concerns—rather than calling attention to them—is a common concern for communication ethics.

Others agreed that GM is sensitive to using the word "stall" in service bulletins and closely scrutinizes any bulletin that does include "stalls" as a symptom.[391] Others at GM confirmed that there was concern about the use of "stall" in a TSB because such language might draw the attention of NHTSA.[392] Oakley said that at times, he included "hot" words to draw attention to an issue. [...]

All technical service bulletins (TSBs) must be sent to NHTSA—the National Highway Traffic Safety Administration—and are posted on its website.

As a result, those looking at the stalls caused by the inadvertent turning of the Ignition Switch rated the problem a "convenience" issue rather than one of "safety"—an error that led to enormous consequences for many. GM personnel further compounded this error by providing information to its dealers that obscured the problem—removing the word "stall" from its TSB precisely because that word might trigger customers' concerns about safety.

The technical service bulletin to dealers in 2005 did not include the word "stall," precisely because that word might trigger concerns about safety. Courtesy: NHTSA. (Valukas, 2014).

Doublespeak

The term *doublespeak* comes from George Orwell's novel *1984*; William Lutz defines it as unethical "language that pretends to communicate but really does not . . . language which makes the bad seem good, something negative appear positive, something unpleasant appear attractive, or at least tolerable. It is language which avoids or shifts responsibility." In addition to *jargon*, Lutz identifies three types of doublespeak:

Euphemism	Avoids unpleasantness with unduly innocuous words. The U.S. Department of Defense used to measure hazardous radioactivity in "sunshine units" (now "strontium units"). "Department of Defense" is itself a euphemism; the agency was historically called the Department of War.
Gobbledygook/ bureaucratese	Muddles explanations with nebulous buzzwords—as when "working together" is recast as "creating win-win synergy"—and unneeded modifiers and generalizations ("at this particular point in time").
Inflated language	Imposes complicated or unduly abstract words on an idea to sound more impressive. *Dilbert* creator Scott Adams mocked inflated language in engineering workplaces with the expression "I utilized a multi-tined tool to process a starch resource." Translation: "I used my fork to eat a potato."

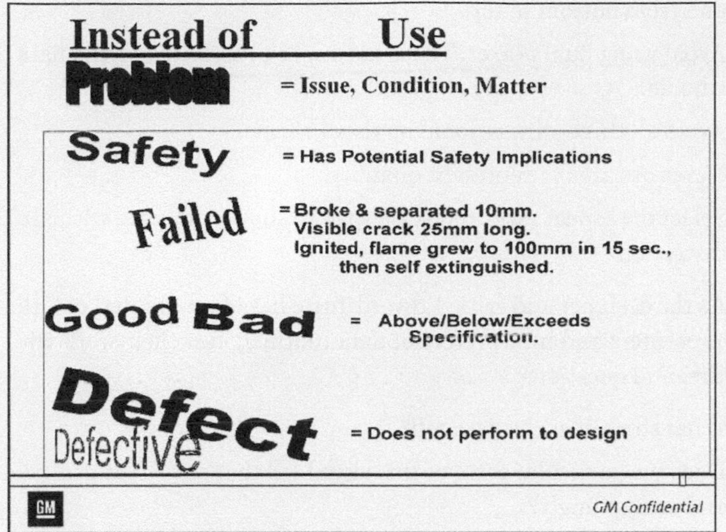

As part of their liability and risk management practices, many companies suggest using euphemisms in place of direct assessment of safety concerns, as GM does in this slide. (NHTSA, 2014).

The alternatives suggested for "failed" push the writer to be more specific in describing the failure, which can make technical writing more effective. More effective would be to say it failed and why/how/to what degree.

For "problem," "safety," and "defect," on the other hand, the presenters suggest euphemisms. More direct, frank assessments may have prompted more concentrated efforts to solve the ignition module problems—or might have prompted decision-makers to allocate financial resources for replacement parts.

Misleading visuals

Like language, visual displays can present exactly the kinds of "deceptive acts" prohibited by the fundamental canons of engineering ethics. A visual that misleads purposefully is usually the result of the designer attempting to deemphasize bad news—visual doublespeak. More commonly, a visual may mislead viewers *inadvertently*, for any one of several reasons:

- Omitting data the designer deems unimportant
- Using conventional designs without considering whether the design obscures important findings
- Oversimplifying the story for the sake of clarity
- Using visual effects or embellishments that alter the audience's perception of the story in the data

Some decisive standards can be applied to prevent the most frequent causes of unintended deception:

- Never omit inconvenient data.
- Always start a numerical scale at zero for bar charts.

- Always assign numerical scales that increase from left to right and from bottom to top.
- Avoid using dual y-axes. These are a special favorite of lying liars who lie.
- Avoid all three-dimensional or visual effects.
- Never use area to represent quantity.
- Select the aspect ratio and scale span to show difference without distortion.

As the designer and critic Edward Tufte has often pointed out, the best presenters and interpreters of data routinely test their work with fundamental questions:

- What story does the data tell?
- Does the particular form of this visual tell that story truthfully and compellingly?

Finding a nuanced story, and figuring out how to tell it and to explain its context, is essential in meeting one's ethical obligation to the audience. Although some audiences may prefer simplicity, make sure that the simpler message does not conceal unpleasant truths, and avoid telling them only what you think they want to hear. Be prepared to fully and thoughtfully explain complexity.

Summary

Ethics in engineering

- The fundamental canons are guidelines *written by engineers* to provide a basic framework to guide engineers' decision making.
- Ethical decisions often involve ambiguity and conflicting loyalties in much the same way that engineering design involves ambiguity and competing priorities.

The engineer's rights and duties

- Employers are obliged to respect engineers' rights to fair pay, safe working conditions, and other labor safeguards. Companies should also be as loyal to workers as they are to shareholders.
- Engineers are obliged to act as competent, faithful agents of the employer.

- Engineers are obliged to hold paramount the health and safety of the public. If their employer's actions risk public safety, engineers may have no recourse but whistleblowing—jeopardizing their career.

Analyzing consequences of actions and ethical principles

- Acting to produce the greatest benefit for everyone affected—utilitarianism—focuses on specific consequences rather than universal rules of behavior.
- Kant's categorical imperative requires that, to be morally justified, a course of action must work effectively as a universal rule for everyone's actions.
- Kantian ethics prohibits practically all forms of lying and deception.

Types of unethical communication

- Concealing information is a characteristic shared by several high-profile cases in which people came to harm.
- Doublespeak makes something negative appear positive to avoid or shift responsibility.
- Doublespeak can occur in visual design as well as in written or spoken language.

Accommodating global and cultural differences

4

As an engineer, you will work with people from a large variety of cultures, whether you travel extensively or not. Members of your engineering team, co-workers, executives, technicians, clients, suppliers, vendors, regulators, and the public will have cultural values, assumptions, and communication styles that differ from your own. Being aware of these differences and their origins can help you avoid embarrassment and misunderstandings and facilitate more effective communication.

Objectives

- To describe major cultural dimensions affecting communication styles
- To recognize how the cultural dimensions of communication styles can give rise to unintentional friction, miscommunication, or misunderstanding between members of an engineering team or between engineers and their many audiences
- To begin developing the ability to avoid giving or taking unintended offense in communicating with people whose cultural values or assumptions differ from yours

4.1 Recognizing cultural values and assumptions

Everyone is born into national and/or regional *cultures*—shared systems of beliefs and customs that are passed along generationally, encompassing "the full range of learned human behaviors." As children, we learn our own cultures' norms; as adults, we learn further cultures of our occupation and our workplace. Insofar as we are immersed in our particular groups' ways of seeing the world and acting in it, those ways can appear not as one option among others but as correct, necessary, and inevitable.

The ways in which we think, feel, and act seem natural and sensible to us because they are sanctioned and supported by our cultures. When we encounter a different culture, its beliefs and customs will likely seem exotic or odd—a perspective that anthropologists call *ethnocentrism*. The work of cultural anthropologists such as Florence Kluckhohn and Fred Strodtbeck, Geert Hofstede, and Edward T. Hall can begin to help us recognize our own ethnocentrism.

Of course, truthful generalizations about culture are based on large population statistics. Every engineering student knows that statistical results have no predictive value applied to an individual. Thus, cultural generalizations can say very little about how a specific person thinks, feels, or acts. Still, by learning about the dimensions of our own cultures, we take the first step in developing the ability to effectively communicate with people from different national, regional, gender, occupational, and organizational cultures.

The major dimensions of culture treated here encompass distinct spectra of values and assumptions:

- Primacy of an individual or a group
- Primacy of equality or hierarchy
- Scheduling for time or for people
- Orienting goals for the short term or the long term
- Primacy of the writer or the reader for the meaning communicated

4.2 Emphasis on the individual or the group

Some cultures, like those in the U.S. Great Plains and Mountain West, emphasize individual identity *over* group membership. In other cultures—including many in East Asia and Latin America—the individual's sense of self is established primarily *through* group membership.

Individualistic	Focuses on achievement, personal success, and competition with others for recognition and monetary reward: "the self-made man," and "pulling oneself up by one's bootstraps"
Group-oriented	Values group loyalty and harmony, with status determined through association with a group such as family, occupation, or organization

Cultures within one of these categories might resemble each other in some ways and differ strongly in others. In China, a strong allegiance to one's own group tends to imply a competitive attitude toward other groups. In Honduras, on the other hand, competitiveness is rarer in general, and a group might experience little desire to promote itself or to establish status over others.

Preferential treatment that is considered unethical in an individualistic culture may be accepted in a collectivist culture: in the United States, hiring a close relative might be perceived as unfair, and even as an abuse of one's power; in many group-oriented cultures, on the other hand, such close personal connections are likely to be seen as desirable assets to group unity.

4.3 Preference for equality or hierarchy

Interpersonal interactions force us to predict and try to meet the expectations of others, and those expectations are influenced by cultural norms: rules or codes of behavior, written and unwritten. The seriousness of these rules often correlates with a culture's tendency toward hierarchy and with the extent to which it accepts uncertainty:

Egalitarian	Treat people as equally valuable contributors to the group goals without regard to differences in economic or social status
Hierarchical	Treat substantial differences in status as inevitable, defining features of any social group: authority figures are likely to be revered, and any perceived disrespect from subordinates is likely to create discomfort

Egalitarian cultures, such as the dominant cultures in the United Kingdom, Australia, and the United States, promote relative equality and tend to have a high tolerance for uncertainty. There may be enormous differences in wealth and status among members of a group, but there will likely be an emphasis on even low-ranking members having influence and importance within the group: Everyone can expect to have his or her voice heard and to be treated as part of "the same team."

Hierarchical cultures, as in China or India, tend to emphasize status, place less value on individual achievement, and have a lower tolerance for uncertainty. Large differences in income at different levels of an organization are likely to appear natural and inevitable, regardless of one's own position in the hierarchy.

Other than laws and some policies that are especially important to an organization's core business or identity, cultures in the United States place relatively little importance on rules. More than this, though, senior and junior members of organizations are likely to be treated similarly—often in casual workplace environments in which colleagues work closely together.

In contrast, in hierarchical cultures, input or advice from junior colleagues is seldom sought or given, and questioning a decision made at a higher level would appear highly inappropriate. In many parts of the world, including the United States, organizations may show variability, ambiguity, and even tension when it comes to acknowledging inequities in power and status; even within one national culture, such factors may vary by industry or region. A Midwestern agricultural equipment manufacturer will seldom resemble an engineering consulting firm in New York City.

4.4 Experiences of time

While hours and minutes are standard measurements of time, the perception of time across cultures varies significantly.

Scheduling for time or for people

Monochronic	Emphasize time management and punctuality, doing things promptly at their set times
Polychronic	Focus on people more than schedules, performing multiple activities at once (such as taking a phone call mid-meeting) and using loose scheduling with little interest in being “on time”

In the typical U.S. workplace, time is a commodity with value, and therefore being late for an appointment or a deadline threatens others' productivity. American working life is governed by concerns with productivity and efficiency: Maximizing output over time is a defining priority in some monochronic cultures.

In a polychronic culture, like in India or most Latin American countries, meetings and appointments might never start “on time.” U.S. engineers who travel to these countries should not be surprised when someone shows up an hour late for an appointment and offers no apology or explanation for the delay. It would be a mistake to apply monochronic cultural expectations and become upset.

Orienting goals for the short term or the long term

Americans are likely to be deeply concerned with the results of a present project or a proximate milestone—hence the importance of quarterly earnings reports to many American companies. In contrast, China has a long-term orientation, which can be seen in its dedication to the Three Gorges Dam. Although controversial in China and globally, the Three Gorges is undeniably a massive project—the world's largest hydropower dam, with a reservoir surface area of about 1000 square km (400 square miles)—that took nearly a century of planning. (The project was initially envisioned in 1919 by Sun Yat-sen, the founding father of the Republic of China.) Partnering with a company in a country with a long-term orientation is not easily done and requires significant investments of time and resources to build the relationship from the beginning.

Long-term orientation	Focus on investment in the distant future (company's health in ten years)
Short-term orientation	Focus on present events and immediate future prospects (quarterly reports)

4.5 Role of the writer or the reader in conveying meaning

Our cultural values and intellectual histories inform how we communicate and the expectations for who is responsible for stating or extracting the meaning of the work:

Writer responsible	Communicates explicitly, using deductive reasoning and logically stated arguments where the writer must clearly state the meaning.
Reader responsible	Communicates implicitly, deeply exploring a topic through nonlinear arguments where the audience is expected to extract meanings that are implied and indirect.

In most cultures that espouse equality, it is the responsibility of the writer to provide access to the ideas being conveyed: Such democratic ideals emphasize frankness, with everyone being empowered to state and access arguments directly. Reader-responsible practices, on the other hand, often emerge in countries with long histories with which everyone is expected to be familiar. References to historical figures or events, seldom considered relevant in writer-responsible cultures, may be prominent.

Writer-responsible cultures prefer explicit communication. Audiences in these cultures expect:

- Clear arguments with a thesis statement, support, and a consideration or rebuttal of a counterargument
- Deductive reasoning and logical evidence in support of an argument
- Coherent documents with clear transitions not just at the larger organizational level, but also at the sentence level; the topic of one sentence leads into the topic of the next
- A writing style that is concise, practical, and efficient

Reader-responsible cultures tend to be more implicit in their communication, relying on a shared understanding of the situation—which is easier to expect in cultures that privilege group membership. Audiences in these cultures expect:

- An exploration of the topic in great detail and grounded in context
- Inductive reasoning (to maintain social harmony) and appeals to ethos and pathos
- Non-linear organization; audiences make connections between topics and extract meaning
- A writing style that is detailed, complex, and repetitive

From: society-presidents15@ieee.org [mailto:society-presidents15@ieee.org] **On Behalf Of** Nazih Khaddaj Mallat
Sent: Thursday, January 29, 2015 10:28 PM
To: SOCIETY-PRESIDENTS15@LISTSERV.IEEE.ORG
Subject: IEEE Society/Council Coordinator in Region 8
Importance: High

Dear all Society/Council Presidents,

My name is "Nazih Khaddaj Mallat": 2015 IEEE Region 8 Chapter Coordination Subcommittee Chair.

As you know, Region 8 is one of the fastest growing IEEE Regions in the world with respect to Chapter activity, and it places much emphasis and effort in Chapter activity. Region 8 believes in strong coordination and cooperation with the IEEE Societies and Councils and has been fortunate to have representatives/coordinators appointed from most societies/councils as liaisons to the Chapter Coordination Committee. I attach the list of society/council liaisons as of April 2014.

As you may have made changes to your appointed liaisons, I kindly request that you review the list and advise whether the currently listed liaison is still active, or if not, kindly consider the appointment (or reappointment) of a Region 8 Chapter and Membership liaison.

The first paragraph reminds the recipients of their shared values and interests, helping to set the stage for an appeal for action.

In a writer-responsible culture, this request would likely be formulated to emphasize directness rather than elaborate courtesy. Here, the author accommodates this desire to "get to the point" by calling attention to the action item with bold type and highlighting.

The timing of the meeting is mentioned here tangentially, but astute readers will understand that this meeting date is actually a deadline for a reply.

The closing re-emphasizes the appeal to pathos, and the urgency of the request is only now stated.

Kindly note that Region 8 holds an annual **Region 8 Committee meeting in March, 2015**. As part of the meeting, we traditionally hold a Chapter Coordination face-to-face meeting with representatives/liaisons from the different Societies. An announcement of the 2015 meeting will be announced once more information is available. The meeting is intended to discuss issues of interest to societies/councils and Region 8 chapter activities. In particular, it is intended to "square the chapter-region-society triangle" and also provide some information to the coordinators as well as hear from them on items related to the interface between the three entities. In the 2014 Chapter Coordination Committee there were representatives of over 20 Societies/councils and from the feedback we had received that was a very successful meeting.

We wish to repeat and improve on that successful and effective meeting, and therefore, **kindly provide the necessary contact information urgently. Your kind help is much appreciated.**

Best,

Nazih Khaddaj Mallat, Ph.D.
2015 IEEE Region 8 Chapter Coordination Subcommittee, Chair

This email to a group of professional society leaders asks for updates to the contact information of an IEEE society. It is clear from the content of the email that the author is from a reader-responsible culture, attempting to accommodate a writer-responsible audience. ©2015 Nazih Khaddaj Mallat. Reprinted with permission.

The differences in communication styles can result in miscommunication. As intercultural communication scholar Matthew McCool explains, those of us from writer-responsible cultures often see reader-responsible writing as disorganized, needlessly detailed, and prone to tangents. Likewise, reader-responsible cultures often perceive writer-responsible writing as "obvious, simplistic, practical and narrow." The value differences extend to what kind of research is privileged: Qualitative reasoning may be favored in reader-responsible cultures on matters where writer-responsible expectations would demand quantitative data. In addition, handshakes or verbal agreements may be more important than written contracts in some reader-responsible cultures. (The spoken word may be privileged as connected to one's honor, while it can seem distasteful to spell out all details explicitly in writing.)

4.6 Considerations for face-to-face communication

Simply knowing that there are differences in communication styles should help you to think twice before making a quick—and potentially incorrect—assumption. In a face-to-face conversation, you will not want to alter your conversational style dramatically, which might be stilted and come off badly. However, you will certainly want to receive other ways of communicating gracefully and without judgment.

Immersing yourself in another culture

You may elect to study in another country while in college, or you may be expected to travel abroad or host international guests as part of your professional duties. Finding out in advance about the culture and customs of the country can make you a gracious guest or welcoming host. You can build a positive rapport with someone by genuinely expressing an interest in his or her culture and being open to new cultural experiences. For example, learning at least a few phrases in his or her native language—greetings, pleasantries, courtesies, basic questions—is generally appreciated. Even if you mangle a phrase with a good-faith attempt, you've shown interest in reaching out to your international counterparts, a gesture not made by all native English speakers.

Since the general "rules" or expectations about public conduct vary across cultures, familiarize yourself with the local etiquette. Here are some important areas where the rules differ and which you will want to learn in advance:

Greeting gestures	In some Mediterranean and Latin American countries, greetings include cheek kisses; a bow instead of a handshake is customary in Japan.
Eye contact	Americans expect and respect eye contact. In Asia, Africa, and Latin America, extended eye contact is disrespectful and would be seen as a challenge to authority. Between men and women in some Middle Eastern countries, it might be interpreted as a sexual advance.
Displays of deference	Some countries show greater respect for the elderly than is customary in the United States (which privileges youth), and older people may even have separate lines in customs or in government offices. Explicit deference to government officials or senior business leaders might be expected.

Customs of dress	Generally, conservative dress is a safe default assumption. Many countries' dress is more formal than that of the United States, even for recreational activities (American tourists are easily identified abroad when wearing shorts, baseball caps, and sneakers).
Personal space in public spaces	In densely populated areas, on public transport, in elevators, and in lines, expect less personal space than you are used to. Backing away to regain your personal space may also seem rude.
Contracts	While binding in writer-responsible cultures, they are not always binding in reader-responsible cultures.

Learning some of these expectations in advance may save you from having an awkward moment. Since it will not be possible to know fully what to anticipate, be aware that you may also encounter different standards of behavior in regards to political discussions, sexuality, and alcohol consumption that may make you uncomfortable (depending on where you are traveling, they may be more visibly in your face or restricted). Observe how others act to figure out the unstated rules of behavior.

Working on a cross-cultural team

Even if you never leave the United States, you will still need to communicate with individuals from other cultural backgrounds—whether with engineering peers or with clients, executives, technicians, suppliers, vendors, government agents, or the general public. Before even entering their profession, most engineering college students will experience working on a team with a student from a different cultural background. Recognizing some of the ways that people communicate differently can help avoid miscommunications. Potential places for misunderstandings include:

- *Punctuality.* Since expectations vary across cultures, it is important to establish as a group what counts as early and what counts as late.
- *Greetings.* When U.S. students ask a colleague "how are you?" as we pass in the halls, we expect a brief reply, but in people-oriented (polychronic or group-oriented) cultures it would be rude to say "Great" and to keep walking.

- *"Small talk."* If someone asks personal questions at the beginning of a meeting, don't assume that they are slacking or not interested in the project. They may be trying to build team rapport, which is seen in many cultures as necessary for working together.
- *Circular talking.* In contrast to the American style of direct, linear communication, much of the world avoids upfront statements of a message's "bottom line," instead suggesting it with anecdotes and painstaking establishment of context. Saying to someone "just tell me what you mean" will likely be perceived as rude.

While U.S. students are accustomed to actively engaging in their education—contributing to class discussions, asking questions, and working in small groups and teams in classes—other educational systems are more hierarchical. A student team member who seems unengaged may in fact be unfamiliar or uncomfortable with the egalitarianism of a student team and may be waiting for a project leader to give instructions. Someone from a country that values group harmony will be less likely to express an opinion and may be embarrassed if asked a direct question that puts him or her on the spot.

When working with team members from East Asia, it is important to be sensitive to the concept of "losing face." While no one wants to be humiliated, in East Asia it is not uncommon for people to alter their behavior because of a concern for someone else's self-worth. In a way, people are responsible for preserving the dignity of others. For example, if an American brought the wrong slides to a business presentation, he or she might "save face" by making a self-deprecating joke and giving the presentation without the visual aids. In contrast, in China the audience would never point out that someone brought the wrong slides to a business presentation because doing so would cause the presenter to "lose face."

Summary

Recognizing cultural values and assumptions

- How people perceive the world and how they communicate is profoundly shaped by culture.
- Cultures vary significantly by country, region, gender, occupation, and organization.
- The culture in which you have lived or worked for a long time becomes deeply

ingrained, becoming difficult to notice until the expectations set by your culture are not met.

- Experts who study cultural differences have identified specific dimensions that help compare cultures across the world.
- Cultural generalizations can say very little about how a specific person thinks, feels, or acts but are useful in reflecting on our own cultures.

Emphasis on individual or the group

- Individualistic cultures emphasize personal identity while group-oriented cultures emphasize group loyalty and harmony.

Preference for equality or hierarchy

- Egalitarian cultures treat others as equals, despite their status, while hierarchical cultures emphasize such status differences.

Experiences of time

- Monochronic cultures prioritize schedules and efficiency while polychronic cultures emphasize people and multitasking.
- Cultures with a long-term orientation tend to make short-term sacrifices for a long-term gain, while cultures with a short-term orientation will focus on immediate results like quarterly reports.

Role of the writer or the reader in conveying meaning

- Writer-responsible cultures emphasize explicit communication, where meanings are clearly stated and quantitative evidence is valued, while reader-responsible cultures emphasize elaborate, suggestive context, inductive reasoning, and qualitative reasoning.

Considerations for face-to-face communication

- Greeting gestures, expectations for making eye contact, and personal proximity are important elements to consider when having a conversation with someone of a different culture.
- Business etiquette involves knowing when to arrive at important meetings and events, what to wear, and to whom you should show deference.
- When communicating face to face in a different culture, notice how others act and try to glean the unstated rules—although it's always better to learn them in advance if possible.
- Group-oriented cultures spend more time building team rapport (which may be perceived as "small talk" by Americans) before working.
- Much of the world talks around points, providing context for their point to be understood (which may be perceived as "circular talking" by Americans).
- Students from hierarchical cultures may be unfamiliar with the egalitarianism of a student team and may be waiting for a project leader to give them instructions.
- Someone from a country that values group harmony will be less likely to express an opinion and may be embarrassed if asked a direct question that puts him or her on the spot.
- In East Asian cultures it is not uncommon for people to alter their behavior to prevent someone from "losing face."

Designing documents for users

5

Good design aims for *usability*. Donald A. Norman, author of *The Design of Everyday Things*, writes:

> Consider the objects—books, radios, kitchen appliances, office machines, and light switches—that make up our everyday lives. Well-designed objects are easy to interpret and understand. They contain visible clues to their operation. Poorly designed objects can be difficult or frustrating to use. They provide no clues—or sometimes false clues. They trap the user and thwart the normal process of interpretation and understanding. Alas, poor design predominates. (2)

Not every engineer considers documents—books, reports, handouts, technical journals, or email messages—as "objects" to be designed in this way. As a result, many give little thought to the decisions involved in preparing a document. If we regard margins, colors, and typefaces as secondary considerations, undeserving of much attention, we're likely to end up with uninformed choices based on what looks good to us at the moment—or we may just go along with the default settings of our word processor or spreadsheet.

Fortunately, the fields of document design and typography can teach us techniques and principles that deliver superior results. Documents that follow the fundamental rules of layout and typography aren't just better-looking than the standard-issue output of Microsoft Office: The differences are practical, helping audiences to locate and comprehend content more easily. And if the author looks like a more polished professional along the way, so much the better.

Objectives

- To design documents with an eye toward the needs of audiences, making deliberate decisions about layout, typography, and color
- To use document design to create visible "chunks" of information that aid audience comprehension and usability
- To divide an informational display space—a page or a screen—into meaningful design elements by applying white space
- To use design elements purposefully to serve the content and the user, not merely the author's aesthetic preferences

5.1 Professional audiences as users

At first glance, purposeful document design may seem like an extra demand on the time of the writer or presenter. This may be true in some cases, especially as you first learn how to create a document layout. Even those first attempts, though, are likely to be worthwhile, making your message far easier for audiences to understand. Moreover, you will reap the benefits of that accessibility as well: a well-designed document is far easier to revise and edit.

Thus, one way to think about design techniques is that they make it easier for audiences to take in more information in less time, increasing the "bandwidth" of your communication. This can be somewhat misleading, though, since it suggests a linear, continuous path of information starting in your brain, continuing through your document, and finally arriving in the brains of your audience.

We know enough about professional documents, though, to know that they are seldom used in this way, with readers starting at the top of page 1 and proceeding sentence by sentence until the end of the last appendix.

Instead, it may be best to think of your document as a kind of "software," with your audience composed of *users*, each concerned with the tools that he or she needs at the moment, for a variety of purposes:

- Using data to perform new calculations
- Deciding whether to authorize proposed actions
- Performing steps of a process
- Planning new research
- Prioritizing tasks and allocating resources

As in our engineering work, then, we should be designing for usability. Doing so requires purposeful arrangement of a print page, an email message, or a presentation slide for the behaviors that we know to expect from users:

- Skipping information not immediately needed for the action at hand
- Attempting to locate key content without reading fully
- Turning pages or scrolling only when necessary
- Moving back and forth between information in the message and other work

At times, we may be able to fight some of these tendencies with excellent writing, artfully composing sentences and paragraphs that keep our readers engaged. Even this effort, though, benefits from design: your beautifully written paragraphs demand a typeface and margins known to promote sustained reading.

5.2 Chunking: Dividing content into manageable units

The basic principle of document layout is that users are able to navigate, comprehend, and retain content more easily when that content is divided into discrete units. Most layout rules and techniques, then, amount to guidelines for the size and spatial placement of those units. Information designers often cite the title of a famous study—"The Magical Number Seven, Plus or Minus Two"—in which Princeton psychologist George Miller measured the number of items that his test subjects could generally retain in "working memory." Our brains work around such limits by aggregating lower-level items into meaningful "chunks"; when speaking or writing, we're seldom conscious of individual letters or sounds, focusing instead on the larger units of words and phrases. The key is to divide content into just enough coherent "chunks" to make it easy for a reader to take in and to retain.

Résumés create obvious "chunks" of content around each experience or credential.

Boldface text and small-caps text create subtle emphasis, drawing the reader's attention to especially impressive accomplishments.

A résumé can use smaller margins than most other documents because of the additional white space around the page.

EDUCATION

Rose-Hulman Institute of Technology — MAY 2015
Bachelor of Science in MECHANICAL ENGINEERING
Concentration in AEROSPACE ENGINEERING, Minor in SPANISH
GPA: 3.47/4.0
Related Coursework: Light Weight Structures, Introduction to Aerospace, Aerospace Design, Spanish Language & Culture IV - VI

WORK EXPERIENCE

CROWN EQUIPMENT CORPORATION, Greencastle, IN — JUNE 2013 - PRESENT
Product Engineer
- Headed project to eliminate overheating of FC4500 lift truck traction motor
- Designed and prototyped custom fan brackets to optimize airflow through chassis
- Fans are scheduled to be installed on all future trucks and save over **$155,000** per year on warranty costs

JUNE 2012 - AUG 2012
- Assisted engineers with construction and modification of a variety of lift trucks
- Drafted comprehensive drawings of new and existing parts for many projects
- Redesigned overhead guards for battery-powered lift trucks to meet OSHA standards

LEADERSHIP EXPERIENCE

ROSE-HULMAN FLOOR HOCKEY CLUB — SEPT 2011 - PRESENT
Founder and Vice-President
- Organized interest and presented club idea to Rose-Hulman representatives
- Established an intramural sport, in addition to club, due to popularity on campus
- Collaborate with athletic directors to receive funding and reserve courts for games
- Working to establish a four-team winter league for competitive play

PI KAPPA ALPHA FRATERNITY — DEC 2012 - JULY 2013
Project Leader
- Headed 5-figure kitchen renovation project under Financial Distribution Committee
- Incorporated tear-down with associate class work day to save funds
- Communicated regularly with contractors to finish project on time and under budget

Design principles are especially important for documents, like résumés, that readers are unlikely to read in a linear order from beginning to end. © 2014 Josh Rychtarczyk. Reprinted with permission.

Balancing positive and negative space

Negative space (or "white space," generalizing from a white background) refers to the empty areas separating a page's components—paragraphs, headings, visuals, and captions. However, negative space should not merely be space left over where these "positive" elements are absent; rather, it should be consciously constructed to cue the reader to relationships among textual and visual components. The most common uses of negative space are:

- Limiting the area of content by setting margins at the top, bottom, and sides
- Identifying paragraphs by indenting a first line
- Identifying paragraphs by increasing line spacing between paragraphs
- Separating elements of a list
- Increasing line spacing above and below headings
- Separating a visual from surrounding text with horizontal and vertical gaps

- Separating elements of a gridded layout with horizontal and vertical gaps

In deciding how much white space to use, be mindful that a reader can be distracted by a crowded display caused by the absence of white space; nearby components clash with one another and create confusion, as readers are unsure where to direct their attention. Too much white space can also be distracting: The reader wonders, "Why is there so much blank space?" This is less of a risk, though: Adding quite a bit of white space will often create a clean, modern look. (Apple documents and software displays, for instance, often seem to echo the sleek, minimalist lines of the company's devices.)

Margins and text columns

Like salt, white space is sometimes applied "to taste." However, there are some rules to keep in mind. In particular, any block of text requires enough white space that the line width doesn't exceed about 80 or 90 characters; more than that taxes our ability to read effectively. Text can also be arranged in multiple columns, each buffered with white space: The shorter line widths created in this way make for faster reading, useful for documents like brochures or posters.

5.3 Relationships among content: Proximity, alignment, repetition, contrast

Robin Williams (*The Non-Designer's Design Book*) identifies four principles by which designers establish relationships among components on a page:

Proximity	Signals *relatedness*. When components are close to one other, readers infer a connection between them. For example, a caption should be closer to its visual than it is to the surrounding text. White space is an important tool in establishing proximity.
Alignment	Creates a sense of *order*. For example, left-aligning a section heading and the following paragraphs signals a purposeful ordering of the argument.

Repetition	Signals *parallelism*. For example, setting two subheadings in the same typeface and size indicates that they occupy the same level of logical hierarchy within an argument.
Contrast	Signals *differences*. For example, using a heading typeface that contrasts with its following paragraph typeface signals that a new component of the argument is beginning.

Proximity and alignment match the figures to their captions, and also suggest relationships with the accompanying body text. This graphic, for instance, suggests likely decisions for "people who live in areas with . . . EV charging infrastructure," discussed in the text at right.

The most familiar contrasts in type separate serif body text from sans serif fonts used in captions, headings, and other types of text.

Repetition allows readers to distinguish components of the page easily: Based on color and typeface, one can quickly recognize section headings, body text, and data graphics.

FIGURE 2. Lifetime Gasoline Consumption and Fuel Costs
Electric vehicles slash oil consumption and cost thousands of dollars less to fuel than gasoline vehicles.

ELECTRIC VEHICLES	0* GALLONS $5,200	SAVINGS OF NEARLY $13,000
GAS HYBRID VEHICLES (50 MPG)	3,300 GALLONS $9,800	
GAS-POWERED VEHICLES (27 MPG)	6,100 GALLONS $18,000	

*Electric vehicles consume no gasoline and contribute very little to oil consumption, since less than 1 percent of the electricity in the United States is generated with petroleum.

Note: Fueling/charging costs are based on $3.50-per-gallon gasoline, an electricity price of 11 cents/kWh, a discount rate of 3 percent, 166,000 lifetime miles, and an EV efficiency rating of 0.34 kWh/mile.

Savings through Electric Vehicles

Reaching a Half the Oil future means having more freedom to choose how our vehicles are powered and what type of vehicle suits our individual needs. While higher fuel efficiency standards save consumers money by giving drivers and businesses options for using less gasoline, technology now provides the option of using none at all.

Electric vehicles (EVs)—such as the plug-in hybrid Chevrolet Volt, fuel-cell cars that run on hydrogen gas, and the battery-electric Nissan Leaf—can cut U.S. oil use by nearly 1.5 million barrels a day in 2035 and save drivers thousands of dollars in fuel costs (UCS 2012d). Compared with a compact gasoline-powered vehicle with average fuel economy, a typical midsize battery EV could save its owner nearly $13,000 on fuel costs over its lifetime (Figure 2) (UCS 2012e). Even compared with the cost of fueling a gasoline hybrid vehicle that achieves 50 miles per gallon (mpg), a battery EV could save drivers more than $4,500 after accounting for the cost of "filling up" with electricity.[1] Such savings go a long way toward offsetting the additional cost of the vehicle and any home-based charging equipment, and prices are expected to decline as investments in EV technology and manufacturing continue (UCS 2012e).

Expanding EVs' share of the automobile market is critical not only to halving our oil use in 20 years, but also to developing a twenty-first-century U.S. auto industry. EVs were just reintroduced by auto manufacturers in 2011, but in the following two years sales grew rapidly, reaching more than 120,000 units worldwide by the end of 2012 (Figure 3) (IEA 2012). It will take time for the market to develop further, but state and federal policies can meanwhile encourage the manufacturing of EVs and their components (such as batteries and fuel cells) and stimulate investment in charging and fueling infrastructure. Such policies could accelerate the market to a point where 30 to 40 percent of new vehicles sold in 2035 will run primarily on electricity or hydrogen instead of oil.

Different drivers need different options, and there is no one-size-fits-all electric vehicle technology.

For example, people who live in areas with more developed EV charging infrastructure might opt for battery-electric vehicles, others may pack their families into fuel-efficient minivans powered by hybrid gasoline/plug-in-electric powertrains, and others may pull thousands of pounds of cargo with a fuel-cell-electric truck. Regardless of the type of vehicle one owns, in a Half the Oil future the prevalence and diversity of vehicles will give consumers more ways to use less oil, reduce their environmental footprint, and guard against volatile gasoline prices.

FIGURE 3. Plug-in Electric Vehicle Sales in Model Year 2012
Despite some bumps in the road, EV sales are charging forward. Automakers are making large investments that will bring dozens of new models to the market, thereby creating more competition and making EVs accessible to more drivers.

Sources: UCS 2012e; WardsAuto 2012.

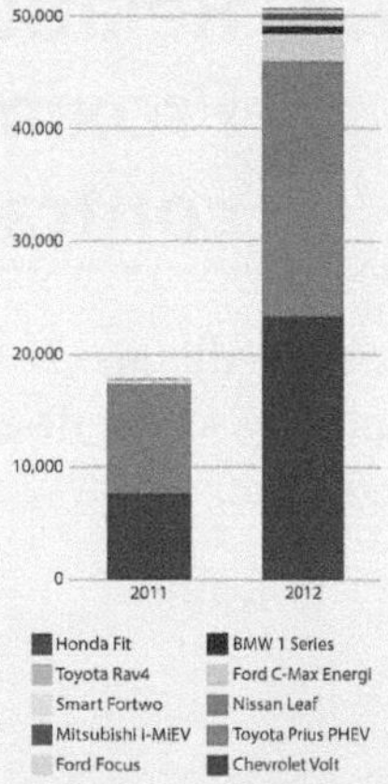

1 Electricity prices vary across the country. A closer look at the costs of charging an EV at home in 50 major U.S. cities shows that decisions on rate plans and on the time of day when you charge can significantly alter the amount you will pay to power your EV.

4 Union of Concerned Scientists

Effective page design—as in this report on reduced petroleum consumption by the Union of Concerned Scientists—is a result of relatively simple principles (proximity, alignment, repetition, and contrast) applied in a consistent, thoughtful way.

Designing a layout with a grid

A grid is an invisible table used as a layout tool, particularly useful for achieving proximity and alignment. When creating a table or arranging multiple elements on a page, you can use the grid lines to adjust the placement of page elements and the spaces between them, then adjust the settings in your word processor to hide them.

Gridding is more efficient and more reliable than entering spaces with the Tab key and space bar, and allows more precise control. Elements in a grid are easily aligned, alignment is easily revised, and elements are easily moved from one area of the layout to another.

WBS	Description	DDT&E Total (FY12$M)	Flight HW Total (FY12$M)	DD&FH Total (FY12$M)
06.1.1	Payloads	65.0	28.0	93.0
06.1.2	Command & Data Handling	50.1	18.3	68.5
06.1.3	Communications and Tracking	29.7	13.7	43.4
06.1.4	Guidance, Navigation, and Control (GN&C)	17.2	12.7	29.9
06.1.5	Electrical Power Subsystem	190.3	62.1	252.4
06.1.6	Thermal Control (Non-Propellant)	26.0	13.2	39.3
06.1.7	Structures and Mechanisms	52.1	26.0	78.0
06.1.8	Propulsion System	156.0	67.5	223.5
06.1.9	Propellant	0.0	0.0	0.0
	Subtotal	**586.4**	**241.6**	**828.0**
	IACO	41.6	12.6	54.1
	STO	37.7		37.7
	GSE Hardware	77.0		77.0
	SE&I	109.9	35.6	145.5
	PM	42.5	18.3	60.8
	LOOS	40.6		40.6
	Spacecraft Total (with Integration)	**935.7**	**308.0**	**1243.7**
Prime Contractor Fee (10% less payload)		87.1	28.0	115.1
Spacecraft Total with Fee		**1022.7**	**336.0**	**1358.7**

Figure 16. Cost estimate for the Prime Contractor (including fee) in FY'12 $.

The data in this display is more prominent because grid lines have been hidden.

The locations of the hidden grid lines can be deduced by looking at the proximity and alignment of the entries—but readers can easily focus on content alone.

Bold type is used to highlight the primary information presented.

Gridding principles make a number-filled table easier for readers to comprehend. © 2012 Keck Institute for Space Studies, California Institute of Technology. Reprinted with permission.

5.4 Setting type for ease of reading

Writers often begin work without thinking much about type. The choices available in word-processing and document layout software can appear bewildering if one simply tries out fonts to see what looks good; on the other hand, sticking to a handful of default fonts like

Calibri or Times New Roman can prevent a document from standing out among all the others that look more or less the same.

Selecting typefaces

The typographic categories that matter most for workplace documents are *serif* and *sans serif* typefaces. Serifs are the small strokes adorning the stems of letters in such typefaces as Times New Roman, Palatino, Georgia, and Garamond. Sans serif typefaces—such as Helvetica, Arial, Calibri, Futura, and Verdana—lack those strokes. Other categories exist but seldom have a place in professional documents: In general, avoid typefaces that are whimsical, overly complicated, or highly stylized.

Serifs are visible as "feet" at the bottom of letters that end in a straight line.

Engineering Communication (Times New Roman)
Engineering Communication (Palatino Linotype)
Engineering Communication (Georgia)
Engineering Communication (Garamond)

These same locations have lines that end abruptly in sans serif typefaces.

Engineering Communication (Helvetica-Narrow)
Engineering Communication (Arial)
Engineering Communication (Calibri)
Engineering Communication (Verdana)

To determine whether a typeface has serifs, inspect the bottom of letters that end in straight lines. The top group shows serif typefaces while the bottom group shows sans serif typefaces, all in the same font size.

Serif typefaces are the default standard for body text in print documents. Readers often perceive serif typefaces as traditional, formal, and professional, perhaps based on their history in publishing. (The original Times Roman, for example, was commissioned for the *Times* of London.) Essentially all long, formal print publications use serif typefaces for their body text, and many readers find them most readable for such documents. Generally, you should use serif fonts for professional documents.

Sans serif typefaces, on the other hand, are often preferred for on-screen reading; the absence of extra flourishes makes them easier to read when a display's resolution is lower than paper's. Web browsers and operating systems almost always display text in sans serif typefaces; Microsoft commissioned Verdana specifically for readability at small sizes on screens, and Apple used versions of Helvetica for its operating systems before switching to its own sans serif typeface, San Francisco.

In print documents, sans serif fonts are most frequently used for headings or sidebars but can also be suitable for short documents, such as résumés, designed for quick reading. Because they evoke computers rather than books or newspapers, many sans serif typefaces strike readers as more modern and "high-tech" than those with serifs.

Adjusting type style, size, and spacing

The size of "points" varies among typefaces, but 10 or 12 points will work for the main body text within most documents. Headings are likely to be slightly larger (and may use a contrasting typeface); footnotes, bibliography entries, block quotations, and other supplementary text may be slightly smaller.

Typefaces include a variety of formats for emphasis in document headings and within sentences:

italic and **bold**	Italic type suffices for emphasis in most cases. (It has other uses, too—mathematical symbols and the titles of books and journals are all set in italics). Bold type conveys heavier emphasis, allowing key words, phrases or sentences to "jump out" at readers even at first glance.
Underline	Underlined text was important for typewritten text that could not use italics. It is no longer used for emphasis.
CAPITALIZING	Text in ALL CAPS will be sufficiently emphatic but impedes readability and may convey anger or aggression that you do not intend. SMALL CAPS, though, can be effective within headings or for especially heavy emphasis—for instance, if you wish to note key terms defined in a glossary.

In general, use "sentence case" even in headings, capitalizing only the first word and proper nouns.

5.5 Using color

Where to use color

Default settings in many common software packages generate documents that use color capriciously and arbitrarily, so that data graphics, text headings, and even page backgrounds are made to appear in vivid, striking colors. A deliberate design approach, though, will limit color to the uses that create or enhance meaning and clarity. The safest rule when producing a print document is to keep it black and white.

Text almost never needs color. (A single color *can* be used for emphasis, or to shade text boxes.) Color in photographs may be aesthetically pleasing but is seldom important in conveying data (except in some cases when the photograph is capturing experimental results). Data graphics, on the other hand, can often achieve greater clarity and capture more content by using color.

Choosing a color palette

Additionally, one might well encounter audiences whose eyes do not detect the differences among particular parts of the color spectrum; red–green color blindness is the most common limitation, but other parts of the color spectrum may also be affected.

Data graphics in particular should use a carefully constructed color palette. Smaller objects, such as lines and points, should use bright or dark colors with strong hues and intensities; lighter, pastel colors are appropriate for larger areas, such as those on maps or pie charts. Sample palettes (including RGB values) are shown in Appendix I of Stephen Few's *Show Me the Numbers*.

Users are likely to photocopy the document at some point, so colors in your document should in most cases translate effectively to grayscale. Moreover, color blindness and other perceptual factors can prevent readers from understanding your information even when they are using the original color document. Even audience members with excellent color vision often struggle with color-coded graphs, charts, or maps when colors (or grayscale values) are insufficiently distinct. Such problems can be avoided by using a limited, carefully chosen palette, such as the ones available at the website of the cartographer Cynthia Brewer, *colorbrewer2.org*.

Brewer distinguishes among three central types of color palettes:

- *Sequential* data, spanning a range from small to large values, is best rendered in a set of shades of a single color, with lighter shades indicating smaller values and darker shades representing larger values.

- Diverging schemes resemble sequential ones but emphasize an important or decisive break within the range of values (warming vs. cooling, profit vs. loss) by using two colors—one for the lower end and one for the upper end of the data range.
- *Qualitative* data is sorted among categories, without registering magnitude. Each category is represented in a distinct hue.

In selecting your palette, you should also pay attention to colors' typical connotations and the ways that they are used in conventional visual language. For instance, a diverging scheme for temperature will almost always represent warmer temperatures in red and cooler ones in blue.

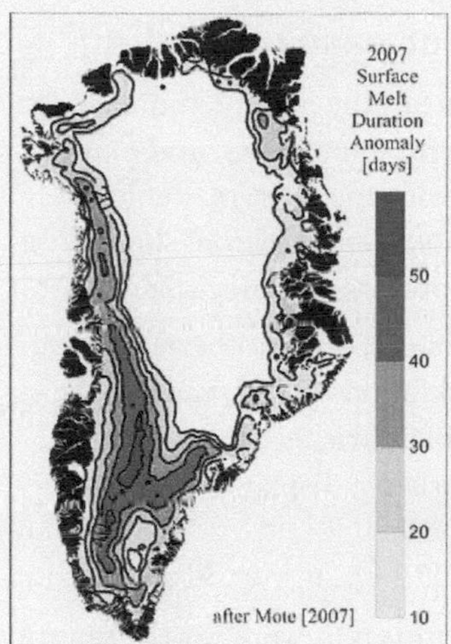

Figure G1. Surface melt duration departure from average for summer (Jun–Aug) 2007 from SSM /I; units are days. The average is based on the summers from 1973 to 2000 (excluding 1975, 1977, and 1978). Only departures >10 days are included. (Figure after Mote 2007)

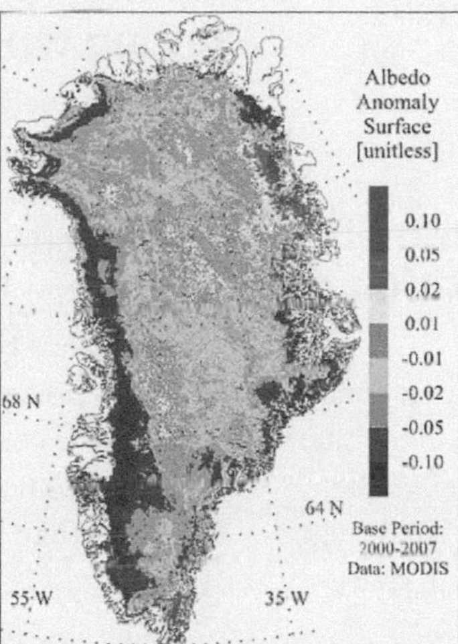

Figure G2. Albedo anomaly (unitless) for 8–23 Aug (days 220–235) 2007 vs the 2000–07 average (from algorithm based on Liang et al. 2005).

The map on the left uses a sequential color scheme: Increasingly saturated shades of red indicate more days of melting due to above-average temperatures.

The diverging scale on the right shows increased reflectivity in red and decreased reflectivity in blue.

Red often means "higher" and blue "lower." Here, though, the designer may want to reverse the scale, since the lowered reflectivity actually comes from higher temperatures. For a non-expert audience, the dominant blue color may create the impression that Greenland is cooling rather than warming.

These two maps show climate-related measurements in Greenland. The map on the left measures the number of days when snow and ice melted. The one on the right shows changes in the reflectivity of the surface caused by that melting (because snow and ice are more reflective than grass and soil). Courtesy NOAA (Richter-Menge et al., 2008).

Summary

Professional audience as users

- Purposeful document design makes it easier for users to locate information needed for a variety of uses.

Chunking: Dividing content into manageable units

- The basic principle of document layout is that users are able to navigate, comprehend, and retain content more easily when that content is divided into discrete units.
- The key is to divide content into just enough coherent "chunks" to make it easy for a reader to take in and to retain. Negative space should be consciously constructed to cue the reader to relationships among textual and visual components.

Relationships among content: Proximity, alignment, repetition, contrast

- Proximity signals *relatedness*. When components are close to one other, readers infer a connection between them. For example, a caption should be closer to its visual than it is to the surrounding text. White space is an important tool in establishing proximity.
- Alignment creates a sense of *order*. For example, left-aligning a section heading and the following paragraphs signals a purposeful ordering of the argument.
- Repetition signals *parallelism*. For example, setting two subheadings in the same typeface and size indicates that they occupy the same level of logical hierarchy within an argument.
- Contrast signals *differences*. For example, using a heading typeface that contrasts with its following paragraph typeface signals that a new component of the argument is beginning.
- A grid is an invisible table used as a layout tool, particularly useful for achieving proximity and alignment.

Setting type for ease of reading

- Serif typefaces are the default standard for body text in print documents, and many readers find serif type most readable for such documents. Generally, you should use serif fonts for professional documents.
- Sans serif typefaces, on the other hand, are often preferred for on-screen reading; the absence of extra flourishes makes them easier to read when a display's resolution is lower than paper's.
- The size of "points" varies among typefaces, but 10 or 12 points will work for the main body text within most documents. Headings are likely to be slightly larger (and may use a contrasting typeface); footnotes, bibliography entries, block quotations, and other supplementary text may be slightly smaller.

Using color

- A deliberate design approach will limit color to the uses that create or enhance meaning and clarity.
- The safest rule when producing a print document is to keep it black and white. Data graphics, on the other hand, can often achieve greater clarity and capture more content by using color.
- Users are likely to photocopy the document at some point, so colors in your document should in most cases translate effectively to grayscale. Moreover, color blindness and other perceptual factors can prevent readers from understanding your content, so use a limited, carefully chosen palette.

Audiences

...shape your communication more than any other factor. How much technical detail should be present? Which design criteria or business considerations will others see as most important? Analyzing your readers or listeners accurately and insightfully will help you to make correct decisions. Fortunately, you need not make guesses about the individual preferences of engineers, executives, clients, or others for whom you are writing. The professional roles played by these groups determine the information they will ordinarily seek, the type of relationship that an engineer can expect to establish with them, and general rules for how to start accommodating their needs.

Audiences

Engineers

6

Engineers most frequently communicate with other engineers. Engineers interact with engineers in their own working group, engineers outside their working group but in the same company or organization, and engineers outside their organization.

Objectives

- To explain how engineering training impacts expectations for communicating with engineers
- To identify reasons for communicating with other engineers and what is likely to be considered important information to include
- To explain how the underlying relationship is likely to impact the tone of communication between engineers

6.1 Who they are and how they think: Credible arguments required

Professionally, engineers apply scientific and mathematical knowledge to develop solutions for technical problems. Therefore, engineers are particularly attuned to the methodological soundness of an experiment and whether an argument is supported by credible evidence. When communicating with other engineers, it is especially important to clearly outline your methodology and to provide credible evidence in support of your argument. If you are working on a project with another engineer, it is also important to be clear how your work affects his or her work. If these communication expectations are not met, engineers are likely to be skeptical.

Meeting these expectations requires a level of detail tailored to the particular audience's needs and expertise. For example, an audience unfamiliar with the background of your work expects to be given sufficient information to understand its context and how it relates to their own responsibilities. An audience unfamiliar with your methods expects sufficient detail to be able to follow your argument. All engineering audiences expect supporting evidence to be correct and credible, especially when the material is unexpected, outside the audience member's expertise, or contrary to desired outcomes.

6.2 Why you communicate with your engineering peers

The usual purpose of communicating with this audience is day-to-day project management to coordinate work, allocate resources, and identify and resolve problems. Informal correspondence—phone calls, emails, instant messages, and face-to-face conversations—is usually the default method of communicating to other engineers.

Work coordination	Coordination of effort required when working with colleagues
Status update	Information on progress towards a project milestone
Issue notification	A technical or administrative problem that needs attention
Decision	A resolution has been found to a previous issue, roadblock, or concern—often involving choosing between alternative solutions
Professional advice	Guidance on navigating career choices, conflict resolution, or understanding unwritten rules that define an organization's culture

Work coordination

Engineering work almost always occurs in a team or group setting. To make progress towards project goals, frequent communication with other engineers is required. The back-and-forth discussion of results, next steps, and answering questions could occur in person when your colleague is in the cubicle next to yours, or by phone call or email.

Setting off the meeting time in a paragraph of its own makes it stand out.

Indicating that you have a question before providing technical data prepares the reader as to how to process it.

Using technical jargon between engineering colleagues is appropriate.

Replies can be brief while still providing the needed information.

From: Brackin, Patricia
Sent: Thursday, February 13, 2014 9:13 AM
To: Bob H.; Maldonado, Ron
Subject: FW: Bond test Equipment Update

Hi Bob,

Everything is set for testing tomorrow afternoon (Friday, 14-Feb at 3:00 PM).

We have the ability to use the slow speed that you need. We can also test the base. Ron Maldonado want to make sure that you realize that we will be applying a compressive force to the base. We will support the legs in our tensile testing machine and then apply a compressive load in the spot that you have designated. We can also record deflection.

I do have a question about the base. I have that it is 3/4 "thick 6061 Aluminum. Is there any additional designation? In the material handbook, if the designation is 6061-0; the yield is 8 ksi. If the designation is 6061-T6, the yield is 40 ksi. I am calculating a stress of about 9.1 ksi (worst case). The ultimate strength for both is materials is well over the 9.1 ksi so we shouldn't get a failure.

Looking forward to tomorrow afternoon.

Patsy

From: Rob H.
Sent: Thursday, February 13, 2014 11:49 PM
To: Brackin, Patricia
Subject: RE: Bond test Equipment Update

Patsy,

The base is machined from Alcoa MIC 6® cast and ground tooling plate. Here is a link with a description and a PDF link with specifications. Probably should have given this to you earlier, but hopefully you can at least compare the specs with the numbers you used for your calculations. We use this material for a lot of our parts.

http://www.alcoa.com/mold/en/product.asp?prod_id=619

I will also bring a short piece of aluminum pipe with a shaft collar attached so we can test it separately.

Bob

From: Brackin, Patricia
Sent: Friday, February 14, 2014 9:27 AM
To: Rob H.
Subject: RE: Bond test Equipment Update

Hi Bob,

Thanks for the info. The yield for Alcoa MIC 6 is quoted at 15 ksi, so we are fine at 2000 lbs.

See you this afternoon!

Patsy

An email exchange between engineers demonstrates close coordination of effort required to make progress towards project goals. © 2014 Brackin. Reprinted with permission.

Status updates

A status update keeps colleagues up to date on progress towards a project milestone. The audience is interested in what you have accomplished recently, how it relates to upcoming milestones, and its relevance to their responsibilities. How far ahead or behind schedule you are is an important consideration. Don't overwhelm colleagues with constant status updates—decide when a reasonable chunk of progress has been made that is worth communicating.

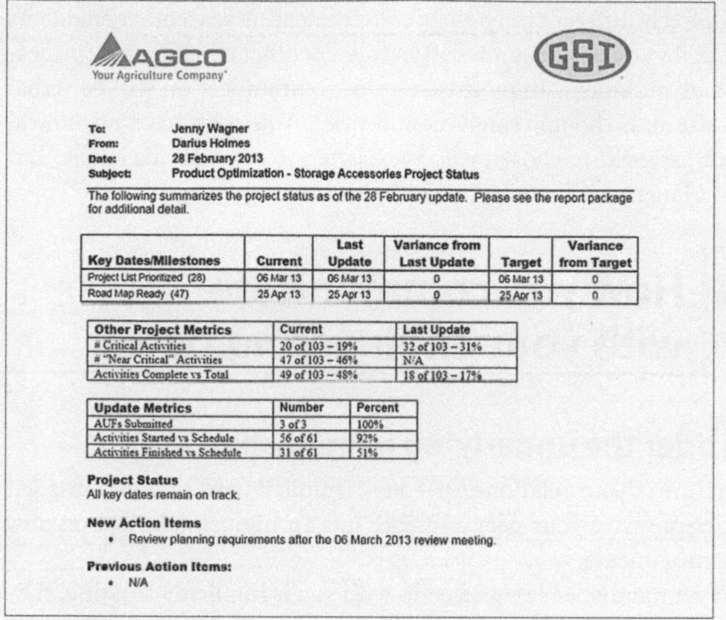

To: Jenny Wagner
From: Darius Holmes
Date: 28 February 2013
Subject: Product Optimization - Storage Accessories Project Status

The following summarizes the project status as of the 28 February update. Please see the report package for additional detail.

Key Dates/Milestones	Current	Last Update	Variance from Last Update	Target	Variance from Target
Project List Prioritized (28)	06 Mar 13	06 Mar 13	0	06 Mar 13	0
Road Map Ready (47)	25 Apr 13	25 Apr 13	0	25 Apr 13	0

Other Project Metrics	Current	Last Update
# Critical Activities	20 of 103 – 19%	32 of 103 – 31%
# "Near Critical" Activities	47 of 103 – 46%	N/A
Activities Complete vs Total	49 of 103 – 48%	18 of 103 – 17%

Update Metrics	Number	Percent
AUFs Submitted	3 of 3	100%
Activities Started vs Schedule	56 of 61	92%
Activities Finished vs Schedule	31 of 61	51%

Project Status
All key dates remain on track.

New Action Items
- Review planning requirements after the 06 March 2013 review meeting.

Previous Action Items:
- N/A

A memo between engineers keeps colleagues aware of progress towards project milestones. © 2015 The GSI Group, LLC. Reprinted with permission.

This memo communicates the status of a project to colleagues in a company-standard format.

At quick glance, other engineers will know how this project is progressing.

More information is presented elsewhere—"Please see the report package for additional detail."

Issue notifications

Problems—either technical or administrative—often need to be brought to your audience's attention. Engineers expect you to describe the issue, its causes, possible solutions, actions you intend to make, and actions you require of them to resolve the problem. The cost of investigating and resolving an issue is usually an important consideration. If another engineering team is waiting for your test results before they can finalize the design details, you should let them know if the test machine breaks down and it won't be repaired for two weeks.

Decisions

A decision is a notification of an issue resolution, particularly a choice between different possible solutions that impact the audience in different ways. The audience expects you to summarize the issue, the decision, and its rationale. If the decision affects the scope of work, due dates, or cost, the audience will expect a statement of authorization by appropriate supervisors.

Professional advice

A completely different purpose of communicating with other engineers is to ask for advice about unwritten rules, conflict resolution, or professional advancement. In most cases, this communication will be verbal and informal, though rarely confidential. You may have an official mentor assigned or chosen who is expected to provide this advice, but informal mentoring can be just as important.

6.3 How you communicate with your engineering peers

Consider the underlying relationship

The nature of your relationship—one of implicit trust or one of implicit skepticism—with your peer audience has an important effect on how you communicate.

If your audience's relationship with you is implicitly trusting, they tend to be comfortable with an informal style and to respond supportively to your assertions. Such relationships develop with peers with whom you have had successful past collaborations inside or outside your working group. Such collaborators are generally aware of one another's professional strengths and weaknesses.

If your audience's relationship with you is implicitly skeptical, they tend to object to an informal style and to need proof of your assertions. Such relationships might be due to the newness of the relationship (you've not worked together before) or to unsuccessful past collaborations inside or outside your working group. At their worst, such audiences view any communication as an opportunity to find fault with your work and to assign liability to your organization.

Engineers as audiences for formal documentation

Despite the mostly informal communication that occurs between engineers, some communication requires formal documentation. Engineers have to create a record of their decisions—what was done and why—as part of a company's record. These decisions are often captured in *memos* or *short reports* that are not final reports. This record might be revisited by other engineers in the company when it is relevant to a future project, but it would also be used to determine liability if something went wrong.

Engineers also serve as secondary audiences on formal documents such as *final reports*. Although engineers are not the primary audience for these documents, often an organization will have an engineer review the document. Therefore, these documents need to include a technically sound argument with a level of detail appropriate for an engineer. Some of the details might appear as appendices instead of in the main body of the document, but they should be available.

Summary

Who they are and how they think: Credible arguments required

- Engineering training forces engineers to be responsive to the appearance of methodological soundness and credible evidence. Strive to communicate both.
- When results, conclusions, or decisions are contrary to expectations or experience, expect engineering audiences to be more doubtful.

Why you communicate with your engineering peers

- The usual reason for communicating with other engineers is to coordinate the day-to-day project management through informal correspondence.
- Status updates communicate recent progress towards project completion.
- Notification of issues informs other engineers of problems that might impact their work.
- Decisions resolve an issue and direct continued work.
- Seeking professional advice from other engineers is usually conducted verbally and informally.

How you communicate with your engineering peers

- Consider the underlying relationship—trust or skepticism influences the engineer's comfort with informality and likelihood of questioning the work of others.
- When formal documentation is needed to record decisions and work that was completed, memos or short reports are often used. This documentation may be important for future projects or to determine liability.
- When engineers are expected to be secondary audiences for concluding documentation like final reports, make sure to include appropriate details that establish a technically sound argument.

Technicians and technical staff

7

Engineers communicate regularly with technicians and technical staff. Technicians may work "in-house"—that is, employed in the same company as the engineer—or be externally contracted.

Objectives

- To explain how technicians' training and job responsibilities differ from engineers and impact their knowledge, skills, and abilities
- To identify reasons for communicating with technicians and what is likely to be considered important information to include
- To explain how to build relationships with technicians by showing respect for their understanding and significant practical experience

7.1 Who they are and how they think: Implementers

Technicians operate equipment, oversee technical procedures, and manage many of the day-to-day details of laboratories, machine shops, and other industrial facilities. Technicians' backgrounds and professional preparation are more varied than those of engineers: They may have studied engineering or engineering technology in universities, in the military, or simply in the workplace. The technician often has the most "hands-on" practical experience, which can complement the engineer's greater familiarity with the theory underlying a technical problem or task. A technician may be your most valuable ally when you need a part machined, a circuit board fabricated, or a server updated: The skill necessary to perform such jobs makes much engineering work possible.

Technicians' thinking usually concerns a task's practical possibility and ease, taking an instruction and moving as quickly as possible to the efficient, correct, safe way to execute it. They often know when equipment limitations create practical challenges for executing a design or specification that an engineer has created on paper (or in modeling software). Therefore, technicians may be skeptical if your directions seem impracticable—but if you explain how and why something will work, the technician will trust that it will. (If it doesn't, of course, you should take responsibility.) Engineers typically occupy a higher position than technicians in an organization's hierarchy, influencing the tone of communications between engineers and technicians. The best guideline for communication with technicians is for both parties to show respect for one another's knowledge, skills, and abilities. Technicians expect you to respect their training and experience. In return, they respect the effort and ability your engineering degree represents—unless you give them reason to think otherwise.

The hydraulic pressure on the annular preventer was increased to stop the suspected leak.[87] The riser was refilled. The drill pipe and kill lines were closed. The drill pipe pressure unexpectedly rose from 273 psi to 1,250 psi in six minutes.[87] Testimony indicates the Transocean *toolpusher* and BP *company man* deliberated their different interpretations of the negative test results.[91] The tool pusher asserted the evidence was indicative of a well leak problem. The BP company man asserted that the anomalous results were caused by the riser leak. It was decided to conduct a second negative test.[92]

A technician correctly identified the meaning of the test result, but his interpretation was ignored by the BP manager (the "company man").

Excerpt from a report on the Deepwater Horizon disaster. Courtesy: UC Berkeley Center for Catastrophic Risk Management. (Deepwater Horizon Study Group, 2011).

7.2 Why you communicate with technicians

Experienced engineers know that technicians' expert advice—for example, in manufacturing, maintenance, operations, testing, or troubleshooting—is invaluable in foreseeing problems, developing practical alternatives, and saving time and money. When approached for advice, technicians need and expect a clear statement of the problem, the desired result, and the relevance to their responsibilities.

You will often communicate with technicians to direct their work. They expect clear directives that are practicable and consistent with similar work they've done before. Important considerations include the budget to which the work is to be charged, and your deadline's potential disruption of other ongoing projects.

The engineer's specification is based on calculations—but John asks the technician for advice, recognizing that Shane may raise practical considerations that merit changes to the design.

John doesn't simply issue an order: He prefaces a request to Shane by continuing to brief him on the design—enabling Shane to suggest improvements and to ask careful questions.

From: John Gassler
Sent: Thursday, August 25, 2011 8:02 AM
To: Shane Ellis
Cc: Paul LaRose
Subject: RE: Pneumatic requirements for vibe system

Shane,

The requirements are the same. I had looked at the speed requirement and didn't think that we needed a 1 CV valve. Calculating the pressure drop at 5 PSI and 80 PSI line, it would require .352 CV. If there is a reason that 5 PSI would be detrimental, let me know. The force required is very minimal for the size of the cylinder; I foresee us turning the pressure down to 40 PSI.

That brings me to the next question: are you quoting regulators and flow controls? I think we would prefer the regulators to be on the manifold and the flow controls mounted directly on the cylinders (both front and back). A couple of places would need to be mounted close by (but in line) because of space—so, if you would, please include both styles. I will decide how many of each are required at the time that I order.

John Gassler

John (the engineer) responds in this email to questions posed by Shane (a technician), who is preparing a price quote and a bill of materials for equipment needed to implement a vibration system that John is designing.

7.3 How you communicate with technicians

Foster trust

Your communications should be respectful of technicians' expertise and of the demands on their time. If your audience's relationship with you is implicitly trusting, they tend to improve your work by volunteering suggestions based on their experience and training. If your audience's relationship with you is implicitly skeptical, they tend to "work to spec"—that is, to work exactly as directed without volunteering suggestions for improvement, even if you have erred.

Avoid verbal work orders

You may verbally discuss problems and potential solutions, but written instructions, technical drawings, and specifications reduce ambiguity and establish responsibility if something goes wrong—another aspect of the hierarchy of employment roles and responsibilities. Engineers are responsible for design; technicians are responsible for implementation.

From: Shane Ellis Lay [mailto:SEllis@EngineeringGroup.com]
Sent: Tuesday, August 30, 2011 10:24 PM
To: John Gassler
Cc: Paul LaRose; Gary D Fuchs; Gregg Elrod; Chris Delia; Brent Atkinson
Subject: RE: Pneumatic requirements for vibe system

John and Paul,

Please see the breakdown of your machine list below. Please provide answers to any questions, specifically those in red, and review all comments I made due to changes in your part numbers, as well as confirming quantities. I've added switches (all PNP with 5M cables, as we discussed) for the cylinders, and am sensing extend and retract on all. Please note also the PIAB smart pump vacuum generator info.

If you're OK, we'll confirm the remaining part numbers and quote everything below. Please advise and we'll get you a quote very quickly. I've attached the spreadsheet you sent me initially.

Thanks!
Shane

CONVEYOR/INSERTER ASSY
Table Lift Cylinder:

- Bimba M-041-DT, %" bore x 1" stroke, switch track, 1/8"npt ports. Qty: 1 (you spec'd a single acting spring return cylinder (T-041-M) ... don't you need double acting? I've added the *"D"* for double acting.) *Yes: double acting.*

- MSCQCX: Bimba, PNP switch for track mount with M8 quick connect and 5M cable. Qty: 2

Card Inserter Extend/Retract Cylinder:

- UGS-0611-A, Bimba ultran slide, magnetically coupled, 7/8" bore x 11 stroke with stroke adjustment, 1/8" input ports. Qty: 1 (this is worst-case move of 11" in .2 seconds at 80 psi) (removed "T" option on slide as will have adjustability with end sensing with prox switch) *Yes!*

PCQCX: Bimba, PNP inductive prox, M8 connection and 5M cable. Qty: 2

Here, the engineer and technician have discussed a change over the phone, but included the details in email to provide a record of all decisions. Such records help to protect all parties.

Shane's suggestions to John (in red) show the benefits of the collegial relationship that the two have developed: Had Shane simply executed the order as written, important parts would not have worked for John's design.

John (the engineer) responds in this email to questions posed by Shane (a technician), who is preparing a price quote and a bill of materials for equipment needed to implement a vibration system that John is designing. © 2011 Numerical Concepts. Reprinted with permission.

Finally, remain aware that technicians work differently from engineers, practically and legally. According to the U.S. Fair Labor Standards Act (FLSA), engineers are "learned professional employees," are paid a salary, and are therefore exempt from minimum wage and overtime pay guarantees. Technicians, on the other hand, are subject to those guarantees. (Depending on their credentials, responsibilities, and the industry in which they work, some technicians are salaried employees instead; they may be designated as "members of technical staff.") You may be willing to work all night or all weekend to finish a project, and likely have the autonomy to do so. But adding extra hours to technicians' schedules is more complicated and must be coordinated with the technicians themselves, their supervisors, and the managers of the budgets covering their overtime pay.

Summary

Who they are and how they think: Implementers

- Technicians operate equipment, oversee technical procedures, and manage many of the day-to-day details of laboratories, machine shops, and other industrial facilities.
- Technicians' thinking usually concerns a task's practical possibility and ease, taking an instruction and moving as quickly as possible to the efficient, correct, safe way to execute it.
- Engineers typically occupy a higher position than technicians in an organization's hierarchy.

Why you communicate with technicians

- Technicians' expert advice is invaluable in foreseeing problems, developing practical alternatives, and saving time and money.
- Engineers need to coordinate with technicians to manage schedules, deadlines, and budgets.

How you communicate with technicians

- Foster trust, creating productive partnerships that take advantage of technicians' expertise.
- Avoid verbal work orders; maintain a written record of decisions and specifications.

8

Executives

Executives occupy a broad range of positions of authority in an organization. Supervisors and engineering managers may oversee a specific project or a small technical staff; a senior vice president in a large corporation, on the other hand, might be responsible for a huge division with tens of thousands of employees.

Objectives

- To select content for the most common executive functions—authorizing actions and allocating resources
- To accommodate busy readers by foregrounding the most important conclusions (as in the "answers first" format)
- To subordinate technical data and detailed findings to main ideas important to the organization's goals
- To observe the organization's customs for communicating with executives at different levels of hierarchy

8.1 Who they are and how they think: Authorizers

The title "executive" refers to the task of *executing* or carrying out decisions on behalf of the organization and its stakeholders (such as employees, investors, clients or customers, and the public). Engineers, programmers, and other technical professionals are often in a position of having to supply executives with accurate, reliable technical information in order to enable those decisions. For instance, you might need to report to a senior manager or a vice-president on a prototype's performance, leaving it to the executive to decide whether to move ahead with further testing or with a production model.

Executives may be trained engineers themselves, have training in business, or some combination of the two (perhaps a master's degree in business administration or engineering management). Either way, though, executives' decision-making role generally requires them to focus on business considerations (market strategy, client relationships, return on investment) and to rely heavily on engineers to be right on technical matters.

8.2 Why you communicate with executives

Technical findings

Managers and organizational leaders often need to make decisions—especially to initiate or evaluate projects—that require findings from experiments or design exercises. These decisions can be the most important outcomes of engineering reports, which need to be clear enough that executives get an immediate sense of the technical findings' "big picture" implications.

Technical findings may sometimes constitute bad news—indicating that a desired specification hasn't been met, or that a project requires an expensive change in materials or procedures. Nevertheless, executives expect, and engineers should deliver, honest, frank communication about such matters: Such honesty can be vital to avoid disaster. For instance, a concern with company finances may initiate an effort to make a system cheaper, a process faster, or a part lighter. If such results would compromise safety or environmental quality, engineers should take special care to interpret and explain findings to decision-makers in such a way that the potential consequences are clearly understood.

Status updates

Executives are typically keeping track of many different projects or subprojects at any one time. The engineers working on one of those projects might need to report their progress to an immediate superior—an engineering manager or supervisor—as well as a higher-level executive. Managers close to the project will need fairly extensive detail about ongoing technical developments, individual work assignments, budgets, and calendars. Managers will also have a direct interest in the ongoing relationships with external stakeholders such as clients. Higher-ups are probably checking to see whether major milestones have been met and whether unexpected challenges or delays are changing the long-term outlook of the project.

Requesting resources

When you discover a need that can't be met within an engineering project's existing budget—for instance, you need to hire a technician, pay overtime wages to meet a deadline, or purchase expensive new equipment—you will need to submit requests for executive approval. This may be as simple as filling out a standard form, but large requests will typically require you to explain why the project team needs the new resources—the positive impacts they will have, and the negative consequences that may occur in their absence.

Seeking approval for new ideas

Many projects and workplace practices originate in a "top-down" fashion, with employees receiving instructions from their superiors. "Bottom-up" communication can be just as important, though, in generating some of the most valuable ideas. Any employee may have an idea that leads to a new project, a new product, or a way to make an existing process safer or cheaper; when you have such an idea, the authority to move it forward will probably lie with executives. For that reason, executives represent the primary audience for most internal proposals (see Module 14: *Proposing*).

8.3 How you communicate with executives

Avoiding "data dump"

Engineers sometimes accidentally alienate or frustrate executives by overwhelming them with technical detail—often the results of an experiment conducted to inform an important business decision. Whether they come from experimental data or from the predictions of a theoretical model, engineers need numbers to understand problems and to evaluate possible solutions. It's understandable that engineers are enthusiastic about the results of painstakingly designed experiments, and about the advanced mathematics that provide engineering with much of its subtlety and beauty.

Executives' decisions can hinge on those numbers, too, but the executive is typically looking for a bottom-line answer as to *what the data means* rather than an exhaustive report of *what data was generated.* Naturally, those exhaustive details remain important in substantiating the engineer's position, but they don't usually need to be part of the report to the executive: It's enough to have the detailed findings available when they're needed for further work. When you're sure that you do need to include more detailed data, placing it in attachments or appendices will usually be most convenient for executive audiences.

Answers first

In addition to *selecting* the practical, meaningful content that executive audiences need, you will want to *structure* your communication so that those audiences can access that content quickly. This typically means stating major conclusions up front—near the beginning of a memo, email, or presentation, and in a dedicated *executive summary* placed first in a longer report. (See Module 12: *Reporting in an industrial organization* for a guide to the Answers First report format, and Module 26: *Summaries* for more specific instructions on writing an executive summary.)

Negotiating hierarchy

One defining aspect of an executive's position is the responsibility to oversee subordinates, usually following a formal structure specified in an *organizational chart*. Thus, one executive might have direct, formal authority over your work, while another might work in a completely different unit. In most organizations, though, hierarchy matters regardless of whether you report directly to the person with whom you're communicating: If an executive's level in the organization exceeds your own, it may be safest to treat him or her with some deference—or at least to use the same care with which you would treat your own boss.

In deciding on the appropriate degree of formality in dealing with executives—for instance, in whether to address them by their first names—you should consider the customs of your organization; if you know individuals' preferences, account for those as well. You will rarely go wrong, however, with just a bit of extra care in professional courtesy—for instance, replying promptly to an email that requires some follow-up, and including salutations and signatures in those replies.

Table 1. Demand for new market-rate rental housing in Central Pittsburgh, 2011-2014.[6]

Zero Bedrooms		One Bedroom		Two Bedrooms		Three or More Bedrooms	
Monthly Gross Rent ($)	Units of Demand	Monthly Gross Rent ($)	Units of Demand	Monthly Gross Rent ($)	Units of Demand	Monthly Gross Rent ($)	Units of Demand
900 to 1,099	55	1,100 to 1,200	300	1,300 to 1,499	570	1,500 to 1,799	140
1,100 to 1,200	20	1,300 to 1,499	110	1,500 to 16,99	200	1,800 to 1,999	50
1,300 to 1,499	5	1,500 to 16,99	20	1,700 to 1,899	40	2,000 to 2,199	10
Total	80	Total	430	Total	810	Total	200

While it's important not to overwhelm decision-makers with data, this doesn't mean that data should be absent. Huge spreadsheets and pages of complex calculations should rarely be at the center of your communication strategy—but a simple table analyzing a target market may be very persuasive.

The proposal's authors have anticipated the business mindset that readers will bring to the document: Information is carefully selected to situate the venture in the existing market, establish its value proposition, and estimate its potential for growth.

While Urbanize Pittsburgh's development in Garfield will only create sixteen apartments, this data shows that Urbanize Pittsburgh has an opportunity to further expand in different neighborhoods with larger apartment buildings, filling the demand for urban living spaces.

2.4 Unique Approach

Housing developments and community organizations already exist in Garfield: the Glass Lofts provide upscale housing, and the Bloomfield-Garfield Corporation aims to engage and revitalize the neighborhood community.[9, 10] However, these two businesses are separate from each other, and their clients are unlikely to interact. The Glass Lofts are also outside of the price range of Garfield residents: condo prices start at $269,000, while the median household income in zip code 15224, which includes Garfield, is $34,867.[9, 11]

Urbanize Pittsburgh housing will be affordable for a wider range of clients than profit-driven real estate due to our cross-subsidization approach. We will also be reclaiming and renovating existing buildings rather than building new ones. Finally, we plan to engage Urbanize Pittsburgh residents in the larger Garfield community, linking new residents with those who have been living in the area for a long time.

Urbanize Pittsburgh is a social enterprise proposed by students affiliated with Engineers for a Sustainable World at the University of Pittsburgh. To convince executive readers to invest in their project, the authors must make a strong business case for their proposal to construct LEED-certified apartments in the Garfield area of central Pittsburgh.

On average, U.S. workplaces expect employees to display less deference to senior executives than is practiced in many other parts of the world. When travel and international correspondence bring you into contact with executives from overseas offices, plan to adjust your approach, usually to a higher degree of formality and courtesy than you use in the United States.

Summary

Who they are and how they think: Authorizers

- Executives make decisions that define an organization's or unit's strategic direction and practices; those decisions often involve authorizing a project.
- Executives' education and training may have focused primarily on business, on a technical field, or both. In general, design your communication with executives to accommodate a non-expert, non-specialist audience.

Why you communicate with executives

- Engineers must frequently brief executives on experimental or other technical findings, enabling strategic decisions.
- Engineers must frequently update executives on the status of ongoing projects, using more technical detail for immediate supervisors and broader overviews for higher-level officials.
- When a project requires additional resources—equipment, personnel, time, or money—you will need to seek authorization from executives.
- Whether formal or informal, internal proposals (see Module 14: *Proposing*) target executives as their primary audience.

How you communicate with executives

- Avoid "data dump": Emphasize not raw findings or detailed quantitative data but their overall meaning and practical implications.
- Most correspondence with executives should deliver Answers First, with conclusions and major findings frontloaded at the beginning of the document; comprehensive data and detailed analysis are supplied later, often in attachments or appendices.
- Observe your organization's customs for addressing executives at a higher level of authority or seniority. When in doubt, opt for a slightly greater level of formality.
- Consciously adjust the formality of your communication when interacting with executives from outside of the United States; this will most often mean increasing your level of formality.

9

Clients

Clients are the individuals, firms, institutions, or government agencies to which engineering firms provide professional services. Because clients' projects nearly always require professional services in a wide range of specialties, the engineer's firm is usually just one of several *contractors* or *consultants* working on a project, each providing services in their areas of expertise.

Objectives

- To explain how a contract between a client and an engineering firm reflects their relationship and the way that each party perceives it
- To identify reasons for communicating with clients that are likely to be most important at different stages in a contractual relationship
- To manage communication according to a client's overriding interest—meeting project goals on time and under budget—while responsibly apprising clients of events that upset any of their expectations

9.1 Who they are and how they think: It's in the contract

Clients want a fair exchange—quality work at a reasonable cost. The contract formally defines the terms of this exchange by describing the engineering work and outcomes, deadlines for project milestones, budget, and payment schedule.

Clients may or may not have engineering expertise, so an engineering firm might consider some of the clients' expectations impractical. Thus before a contract is signed, clients generally negotiate these terms with engineering firms to reduce ambiguity and increase the likelihood of success. Engineers can expect different procedures for negotiating contracts with government agencies and with corporations; regardless of the details, though, expect your firm and the client to be similarly eager to avoid ambiguity that might lead to litigation.

Clients generally think they've described their intentions clearly, in the contract and in communication during the course of a project. To avoid misunderstandings, engineers should practice effective listening by asking sincere questions—not leading or rhetorical ones—and paraphrasing clients' statements back to them. (See Module 2: *Understanding your audience*.) Engineers should frame their own ideas carefully and structure interactions with clients in an open, equitable way.

During a project, clients generally require periodic reports on progress, which reassure them that the terms of the contract are being met. At a minimum, engineering firms will report progress as specified in a contract; a firm with a longstanding relationship with a client might have more frequent and less formal interactions. (Such informality doesn't make your promises to the client any less binding, though.)

9.2 Why you communicate with clients

In almost any client project, the successive stages of a contractual relationship define a few major goals:

Prior to signing a contract	Negotiate terms of the contract.
While under contractual obligation	Provide periodic updates on progress to reassure client that the terms of the contract are being met.
At the conclusion of the contract	Provide a final report or product as outlined in the contract.

Negotiating professional services

Clients are interested in matching your services to their needs at the lowest cost. Clients expect a compelling argument that your firm's expertise and experience meet their needs—including evidence that your engineering firm can do what you claim you can do in the time and budget allowed.

A client expects the engineer to have carefully read all materials—such as a request for proposal (RFP)—in which clients state their needs. An RFP usually presents not only a description of the work to be completed but also some budget expectations, the format and process for submitting a proposal, and resources for obtaining further information about the project and its requirements.

The client will also likely involve lawyers to make sure that the contract contains no unfavorable terms; your firm will probably do the same.

The engineer's work is expected to meet an "ordinary care and skill" standard of one's peers.

The engineer agrees to be contractually responsible for the accuracy of his or her work.

Ordinary and customary design elements supplied by manufacturers and suppliers are acceptable to the client as long as they meet the "standard of care" requirement.

ARTICLE 6 – GENERAL CONSIDERATIONS

6.01 *Standards of Performance*

A. *Standard of Care:* The standard of care for all professional engineering and related services performed or furnished by Engineer under this Agreement will be the care and skill ordinarily used by members of the subject profession practicing under similar circumstances at the same time and in the same locality. Engineer makes no warranties, express or implied, under this Agreement or otherwise, in connection with Engineer's services.

B. *Technical Accuracy:* Owner shall not be responsible for discovering deficiencies in the technical accuracy of Engineer's services. Engineer shall correct deficiencies in technical accuracy without additional compensation, unless such corrective action is directly attributable to deficiencies in Owner-furnished information.

C. *Subconsultants:* Engineer may employ such Subconsultants as Engineer deems necessary to assist in the performance or furnishing of the services, subject to reasonable, timely, and substantive objections by Owner.

D. *Reliance on Others:* Subject to the standard of care set forth in Paragraph 6.01.A, Engineer and its Subconsultants may use or rely upon design elements and information ordinarily or customarily furnished by others, including, but not limited to, specialty contractors, manufacturers, suppliers, and the publishers of technical standards.

Excerpt from the EJCDC E-500 Standard Form of Agreement Between Owner and Engineer for Professional Services. Reprinted by permission of the Engineers Joint Contract Documents Committee (EJCDC). For more information about the EJCDC, please visit www.nspe.org.

Providing professional services

Having entered into a contractual relationship, the engineer's task is to fulfill its terms. During the course of a project, the engineer may communicate with a client to:

- Discuss specific problems and possible solutions
- Clarify a client's needs
- Coordinate details such as meetings, reports, samples, or other milestone deliverables
- Coordinate with other firms working on the project
- Request changes in the contract outcomes, budget, or deadlines
- Respond to errors or omissions
- Negotiate disputes (usually with the help of counsel on all sides)

Application Engineering Bulletin

Title: Automotive and Bus Aftertreatment (DPF and SCR) 2013 Installation Requirements		This AEB is for the following applications: ☒ Automotive ☐ Industrial ☐ Marine ☐ G-Drive ☐ Genset ☐ Filtration ☐ Emission Solutions
Date: 12 May 2014	Refer to AEB 5.31 for Safety Practices, Guidelines and Procedures	AEB Number: 33.442
Engine models included: 2013 Automotive ISX15, ISX12, ISL & ISB		
Owner: John Doe	Approver: per Procedure VPI-GAE-8921	Page 1 of 144

This AEB supersedes AEB 33.442 dated 4 February, 2014.

The AEB outlines the main SCR system components and the known Cummins Inc. requirements for application in on-highway truck and bus for 2013, 2014, 2015, and 2016 models. The AEB contain information that Cummins Inc. has learned during development to date. Revisions to this bulletin are anticipated as experience is gained.

Copyright © 2014 Cummins Inc., Reprinted by permission.

The title states the specific issue addressed in the bulletin.

The summary (in bold) serves clients' needs by noting that revisions are possible as "experience is gained."

The cover page of a bulletin to Cummins' truck and bus clients notifying them of installation requirements for emissions "aftertreatment" hardware. Printed with licensed permission from Cummins, Inc. © 2014 Cummins, Inc., all rights reserved.

Concluding professional services

On the successful conclusion of a project, the client expects a final report, product, or other deliverable as set out by the contract, as well as a final invoice. If disputes over errors, omissions, changes, or other disagreements arise, the communication may involve counsel.

C. *Effective Date of Termination*: The terminating party under Paragraph 6.05.B may set the effective date of termination at a time up to 30 days later than otherwise provided to allow Engineer to demobilize personnel and equipment from the Site, to complete tasks whose value would otherwise be lost, to prepare notes as to the status of completed and uncompleted tasks, and to assemble Project materials in orderly files.

D. Payments Upon Termination:

1. In the event of any termination under Paragraph 6.05, Engineer will be entitled to invoice Owner and to receive full payment for all services performed or furnished in accordance with this Agreement and all Reimbursable Expenses incurred through the effective date of termination. Upon making such payment, Owner shall have the

Contracts state specifically how the project is to conclude, including the conditions for payment upon termination of the agreement.

limited right to the use of Documents, at Owner's sole risk, subject to the provisions of Paragraph 6.03.E.

2. In the event of termination by Owner for convenience or by Engineer for cause, Engineer shall be entitled, in addition to invoicing for those items identified in Paragraph 6.05.D.1, to invoice Owner and to payment of a reasonable amount for services and expenses directly attributable to termination, both before and after the effective date of termination, such as reassignment of personnel, costs of terminating contracts with Engineer's Subconsultants, and other related close-out costs, using methods and rates for Additional Services as set forth in Exhibit C.

Excerpt from the EJCDC E-500 Standard Form of Agreement Between Owner and Engineer for Professional Services. Reprinted by permission of the Engineers Joint Contract Documents Committee (EJCDC). For more information about the EJCDC, please visit www.nspe.org.

9.3 How you communicate with clients

Focusing on the client's needs

Clients want their projects to run on time, to remain under budget, and to do what they're supposed to do. Clients value conciseness (high information density) and clarity (communicating with them in terms they can understand). Above all, don't waste a client's time. Use as much detail as the client needs, but omit unnecessary detail, as you would do for executives in your own organization.

Clients also need to know as soon as possible when something happens to upset any of their expectations. No client *wants* bad news, but identifying a problem early gives everyone more time to devise a solution. However, individual engineers should always discuss problems with their own supervisors or executives before reporting to a client. Managing errors, omissions, and disputes often involves extra cost that will have to be negotiated with a client.

Being a faithful agent of your firm and the client

The fundamental canons of engineering ethics require an engineer to serve as a "faithful agent" of both one's own employer and the client. (See Module 3: *Meeting your ethical obligations.*) Thus, you will want to take special care in weighing your actions in any case in which your

Automotive and Bus Aftertreatment (DPF and SCR) 2013 Installation Requirements AEB 33.442

31. The NOx sensor cable between the NOx sensor control module and the NOx sensor probe **must** not be modified when using a SCR with the "CES to route" strategy. If the SCR uses an "OEM to route" strategy for the NOx sensor cable (opt out of CES routing), OEM **must** route the NOx sensor cable according to the requirements in Sensors (SCR Device) Section of this AEB.
32. The maximum cable temperature on the PM sensor must not exceed XXX ºC (XXX ºF).
33. The maximum temperature on the PM sensor control module must not exceed XXX ºC (XXX ºF).
34. The PM sensor cable between the PM sensor control module and the PM sensor probe **must** not be modified when using a SCR with the "CES to route" strategy. If the SCR uses an "OEM to route" strategy for the PM sensor cable (opt out of CES routing), OEM **must** route the PM sensor cable according to the requirements in Sensors (SCR Device) Section of this AEB.

Requirements 32, 33 and 34 required for models ISB, ISL, ISX12 and ISX15 starting 1 January XXXX.
Requirements 32, 33 and 34 will be required for vehicles <14k lbs GVW. See Cummins Application Engineering for specific production implementation timing.

Decomposition Reactor

1. The decomposition reactor **must** be supplied by Cummins Emission Solutions.
2. The decomposition reactor **must** be mounted in the horizontal plane within ± 3 degrees even when SCR device is mounted vertically.
3. All connections between DPF and SCR catalysts **must** use gaskets and full marmon joints.
4. The decomposition reactor **must** be installed such that the injector in the diesel exhaust fluid dosing module is directed towards the SCR device inlet. Reference Figure 35.

An excerpt from an Application Engineering Bulletin. Printed with licensed permission from Cummins, Inc. © 2015 Cummins, Inc., all rights reserved.

A section of installation instructions is boxed to focus the client's attention on instructions that have changed since the previous bulletin.

Important requirements are highlighted.

firm's interests and the client's do not perfectly coincide (as in negotiating fees). Fortunately, most contractual relationships align the parties' interests in a way that avoids too many situations of this type. Some general guidelines for managing these relationships can help to prevent problems:

- Brainstorm or speculate about project issues inside your firm only. Clients might get nervous about your firm if they are unfamiliar with the brainstorming process.
- Keep proprietary information inside your firm.
- Communicate with a client in your area of technical expertise only.
- Make statements consistent with your level of authority in your firm.

CUMMINS PROPRIETARY: This information is confidential and classified PROPRIETARY per CORP-99-99-99-99, and shall not be disclosed to others in hard copy or electronic form, reproduced by any means, or used for any purpose without the written consent of Cummins Inc.

Revision 13, 12 May 2014 Page 2 of 157 © Copyright 2014 Cummins Inc.

The document footer from the Cummins Application Engineering Bulletin. Printed with licensed permission from Cummins, Inc. © 2014 Cummins, Inc., all rights reserved.

The document footer reminds the reader that the installation instructions are proprietary.

Document everything

During the negotiation phase of a project—before a contract is signed—the client typically specifies a process for disseminating proposal requirements and answering questions. Posing and answering questions in writing reduces the likelihood of misunderstandings and ensures that all parties receive the same information. Every engineering firm bidding on a project is trying to submit the proposal that has the lowest cost; information from the client can be critical in determining whether a less expensive method, design, component, or process will meet the client's needs.

Once a project is under way, unambiguous documentation of every key decision limits the risk of litigation. A firm's liability insurers and legal staff rely on documentation that engineers produce to establish a clear record of decisions supported by technically sound evidence. Therefore, one reason to be especially cautious in communicating with a client verbally is that such communication is part of the project record. Many firms exhort their engineers to "never give verbal orders" and to follow up conversations with written memos.

A firm that is having difficulty meeting its contractual obligations—for instance, remaining within budget or on schedule—must issue a written request and eventually come to a written agreement that amends the contract.

Revisions to the aftertreatment installation instructions are documented in a "Change Log" at the end of the document.

Change Log

Rev	Date	Author	Description	Page(s)
13	12May14	John Doe	Added PM (Particulate Matter) Sensor to Table 1.	13
			Changed text from "indicates diameter of outlet connection for DPF and the inlet connection for SCR system" to "gives the inlet and outlet diameters for different configurations of DPFs and SCRs".	65
			Added text "If there is relative motion between the DPF and SCR there should be a flex section installed between the DPF and SCR."	66
			Changed text from "The DM includes a 40 µm non serviceable" to "The Dosing Module (DM) includes a non-serviceable".	69
			Changed text from "two filters: main filter (10 microns) and an inlet filter (100 microns). Only the main filter is serviceable" to "3 filters/filter screens: a main serviceable filter, a non-serviceable inlet connector filter screen, and a non-serviceable backflow connector filter screen".	78
			Added text "The DEF lines should be secured to a rigid mounting bracket and not to adjacent plumbing."	81
12	4Feb14	John Doe	Changed text from "electronics package" to "control module".	8
			Changed text from "on the" to "of the NOx on the sensor control module".	8
			Added text "the surface temperature".	8
				8

An excerpt from an Application Engineering Bulletin. Printed with licensed permission from Cummins, Inc.

Ultimately, though, good documentation is more than a hedge against legal risks: It's a legacy of work accomplished and a central part of sound engineering practice. In future projects, past work can be an important resource, saving engineers and their employer time and money.

Summary

Who they are and how they think: It's in the contract

- The terms of the professional exchange between an engineer (or engineering firm) and a client are usually defined in a contract.
- Engineers and clients negotiate the terms of the contract before signing, trying their best to avoid misunderstandings that might lead to future litigation.
- During a project, clients generally require periodic reports on progress to be reassured that the terms of the contract are being met.

Why you communicate with clients

- Negotiating professional services. Clients are interested in matching your services to their needs at the lowest cost. A client expects the engineer to have carefully read all materials—such as a request for proposal (RFP)—in which clients state their needs.
- Providing professional services. After having entered into a contractual relationship, the engineer's task is to fulfill its terms.
- Concluding professional services. On the successful conclusion of a project, the client expects a final report, product, or other deliverable as set out by the contract as well as a final invoice.

How you communicate with clients

- Focus on the client's needs. Clients want their projects to run on time, to remain under budget, and to do what they're supposed to do. Clients also need to know as soon as possible when something happens to upset any of their expectations.
- Be a faithful agent of your firm and the client alike. This ethical obligation is best followed by keeping proprietary information inside the firm and communicating with a client only within your area of expertise and level of authority.
- Document everything. In the negotiation phase, written communication helps the client and the bidders. Once a project is under way, unambiguous documentation of every key decision is vital to help a firm manage the risk of litigation. Lastly, good documentation serves an engineering firm as a legacy of work accomplished—potentially saving time and money on future projects.

[illegible] though, good documentation [illegible] than a hedge [illegible] al risks. Like a legacy of work [illegible] and a central [illegible] of sound engineering practice. In [illegible] past work can [illegible] resources, saving [illegible] employer time [illegible]

The public and the public sector

10

Most audiences with whom engineers communicate day to day—other engineers, technicians, executives, or clients—are directly involved with the engineering work in a way that is relatively easy to understand. Outside of his or her own organization, an engineer can think of everyone else affected by the work—in success or failure, individually and in groups—as the *public*. Public interest is also served by the *public sector*, those who provide basic government services such as public roads or education.

Engineers' ethical obligations to the public and the public sector are explicit: Hold paramount the safety, health, and welfare of the public; conduct their engineering practice lawfully; and issue public statements only in an objective and truthful manner. Meeting these obligations requires communication with the public, individually and in groups, and with representatives of the local, regional, national, and international public sectors.

Objectives

- To define the public and the public sector and explain how they are stakeholders in engineers' work
- To list engineers' ethical obligations in communicating with the public and the public sector
- To list reasons that engineers communicate with the public and the public sector

10.1 Who they are and how they think: Health and safety are first priority

The public are those individuals and groups with a stake in engineering outcomes but only an indirect voice in decisions. The voice of the public is thus heard through hearings, media campaigns, marketing surveys, and appeals to public institutions such as regulatory agencies and courts.

One can imagine the public as a vast, undifferentiated citizenry, or as a single citizen or consumer voicing concerns. While political rhetoric typically distinguishes the public from "special interests," engineers will often encounter public audiences in the form of communities defined by an interest of some sort:

- Geographical communities: the Park West Community Association
- Advocacy groups: AARP (formerly the American Association of Retired Persons)
- Charitable foundations: the Gates Foundation or the American Cancer Society
- Nongovernmental organizations (NGOs): Save the Children
- Business groups: the U.S. Chamber of Commerce

The public *sector* includes public institutions such as school districts or public university systems and cultural and leisure services such as local park districts or a national park system. It encompasses government at all levels: a city council, a local building department, a regional health board, national regulatory agencies such as the Food and Drug Administration (FDA), and international agencies such as the World Health Organization (WHO). Members of the public sector may have a technical background and be familiar with engineering jargon.

Public and public sector audiences vary, but most share three key characteristics:

- Desire for attention to and action on their concerns
- Respect for engineers' contribution to society—but mistrust for corporate assurances
- Dependence on technology and expectation that it will be safe

Wants its concerns to be heard

A basic principle of engineering design is to solicit input from stakeholders in the early stages of a project, bringing their life experiences and social context to bear on engineering decisions. Members of the public, as individuals and groups, want engineering decision-makers to hear their concerns and act on them to protect and support their interests as they see them, in the short term and in the long term. The concerns of the public sector, of course, are often enforced by law.

Different groups' interests often conflict. Engineers are not expected to reconcile irreconcilable differences but are ethically obliged to determine all stakeholders' needs and address them to the best of their ability. For example, a proposal to build a wind farm 5 miles off the coast of Cape Cod in Massachusetts is supported by the Natural Resources Defense Council but opposed by the International Wildlife Coalition. Should the project go forward, engineering decisions should account for the opposition's concerns. For example, the number of birds killed by the moving turbine blades could be reduced if the turbine rotates more slowly, increasing the visibility of the blades to birds in flight—accomplished by making the blades longer, allowing them to rotate more slowly but with greater torque, generating the same amount of electrical power.

One of the audiences targeted in the report is the Nuclear Regulatory Commission (NRC)—the agency responsible for overseeing nuclear power production to be sure that public concerns are addressed.

It is important to recognize that even if Diablo Canyon were designed to withstand ground motions of 0.75g—which has not been shown using a robust, rigorous, and legally acceptable way—Californians would be at risk since larger earthquake can occur. Extending the vehicle safety/nuclear plant seismic protection analogy one final time, designing the plant to withstand the seismic acceleration equal to the SSE does not protect a nuclear plant from all earthquakes any more than seatbelts, airbags, and other safety features protect occupants during every crash. Diablo Canyon and other nuclear power reactors are vulnerable when faced with hazard levels greater than the design basis they are protected against.

In particular, the chance of an earthquake causing ground motion at Diablo Canyon greater than 0.75g is $3.9\text{x}10^{-3}$ per year (NRC 2011).[2] Put another way, such an earthquake is likely to happen once every 256 years. To put this value in context, the Diablo Canyon reactors are more than 10 times more likely to experience an earthquake larger than they are designed to withstand than the average U.S. reactor.[3] Of the 100 reactors currently operating in the U.S., the two at Diablo Canyon top the NRC's list as being most likely to experience an earthquake larger than they are designed to withstand.[4]

Several analogies ("a toss of the die," "seatbelts, airbags, and other safety features") accommodate readers without specialist knowledge of nuclear safety or of probability estimates.

Another way to look at this risk is that the chance such a large earthquake will occur at Diablo Canyon over the 40-year lifetime of the plant is 40 divided by 256, or about 1 in 6—which is a toss of a die.

As shown in Figure 2, dozens of earthquakes have occurred at or near the Diablo Canyon site. These past earthquake do not mean that Diablo Canyon will experience an equal number of earthquakes in the future. They also do not mean that Diablo Canyon will avoid earthquakes of greater magnitude and/or proximity in the future. They mean that Diablo Canyon sits on ground that frequently shakes a lot and its seismic risks should be evaluated very carefully.

The report includes a careful disclaimer about the certainty of future predictions. It reaches audiences effectively by using much less precise language—"ground that frequently shakes a lot"—than one might expect from a technical report.

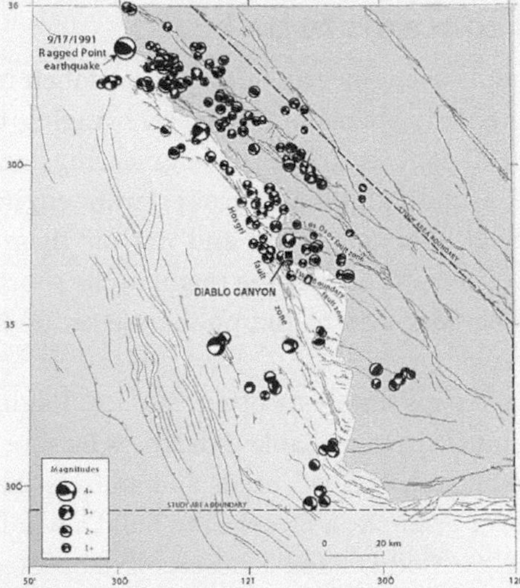

In this report, the Union of Concerned Scientists (UCS) reports on the hazards that earthquakes pose to the Diablo Canyon Power Plant in California. Because of its proximity to the Hosgri and San Andreas fault lines, UCS is concerned about the likelihood of earthquakes.

Depends on technology and expects it to be safe

The public generally takes for granted that established technology is safe to use. Drivers do not expect to die because of an ignition switch design flaw; airline passengers expect aircraft to be flight-worthy; bridges, dams, and buildings are expected to stand; boilers shouldn't explode, pipelines shouldn't leak, and elevators shouldn't fall.

The public sector, on the other hand, generally does not take safety for granted and often imposes regulations based on past engineering failures. For example, following numerous fatal boiler explosions in the early 1900s, Massachusetts passed the first boiler construction codes. Following the Deepwater Horizon disaster in 2010, the U.S. Bureau of Ocean, Energy, Management, Regulation, and Enforcement (BOEMRE) required deep well operators to "obtain certification by a professional engineer of their drilling, casing, and cementing program to assure well integrity."

However, failures do not always result in new regulations. The 2014 Vehicle Safety Improvement Act, introduced in Congress following the publicity of the GM ignition switch failure, was opposed by members of the auto industry such as the National Automobile Dealers Association and the Tire Industry Association and was not enacted. Instances such as these erode the public's trust.

Respects engineers' contribution but mistrusts corporations

Scientists and engineers constitute about 5% of the U.S. labor force, so most public audiences will not have an engineer's technical background and training. Nevertheless, the majority of Americans, between 56% and 68% depending on age group, think that engineers contribute positively to society's well-being.

However, the majority of U.S. scientists and engineers (60% in 2012) are employed by for-profit corporations. Many communities are suspicious of corporate assertions because companies have financial incentives to overstate benefits and understate costs. Mistrust can be based on fear and misinformation—such as the persistent (and groundless) fear that high-voltage power lines are a health hazard—or on the facts of such engineering failures as the Deepwater Horizon, the Upper Big Branch Mine, or the GM ignition switch.

A panel of scholars that produced a 1985 National Research Council study, *Support Organizations for the Engineering Community*, found a large part of the responsibility for this mistrust to reside within the engineering profession itself, which is often reluctant to communicate with the public. The study attributes this reluctance to a belief that the public is incapable of understanding engineers' work, a fear that talking to the press (the public's major source of information) violates their obligations to their employers, and a hesitancy to enter discussions of public policy, either publicly or within their firm. Overcoming such attitudes to communicate effectively with the public, the NRC panelists argue, not only improves the profession, but also helps to combat "a widely held stereotype of engineers as inarticulate".

10.2 Why you communicate with the public

Engineers communicate with the public and the public sector to:

Plan a project	Throughout an engineering project, public input should inform decisions about what to do (or what not to do) and how to do it.
Report on progress	A contract or company policy may compel engineers to notify the public about progress through annual reports or other periodic methods.
Shape opinion	Engineers may provide technical information to sales and marketing or public relations efforts on behalf of their firm.
Explain failure	Members of the public, like all stakeholders, have a right to know when failures occur and what can be done about them.
Be a whistleblower	At the risk of their careers, engineers may feel compelled to report their safety concerns to regulators, to lawmakers, or to the public.

Planning a project and reporting on progress

People are bound to have strong opinions concerning their own safety, health, and welfare. Communication with the public can uncover contextual issues and community concerns that are important to project success. These issues might appear during dedicated stages of a project's planning phase—during a town hall meeting, for instance—or through regular conversations with stakeholders throughout the project. (See Module 2: *Understanding your audience.*)

Planning also requires communication with public sector agencies to acquire information, to interpret regulations, to obtain necessary waivers or exemptions, and to obtain permits and approvals. The format and the content of these communications are often prescribed by the agency; their goal is usually documented evidence such as plans, specifications, or reports that the engineering work complies with the agency's requirements.

When working for the public sector, progress reports are usually contractually required. These notifications take the form of written reports or public hearings to keep the public informed of project progress, explain delays, and predict time to completion. With these communications, members of the public are most concerned with how the project affects their group through expected benefits, temporary inconveniences, and long-term costs.

Shaping opinion

Engineering firms' marketing and sales departments or public relations departments will call on engineers to provide technical information to support their efforts. For example, marketing teams may require technical specifications to promote the latest product's features and operation. Public relations executives might need test results to address public safety concerns. A challenge for the conscientious engineer is to make sure that technical information edited for public consumption is true, without understated costs or overstated benefits.

Explaining failure and being a whistleblower

Communicating with the public in the event of a failure has unique challenges because of the inevitable loss of trust when engineering projects go wrong. Expect the public's view of the engineers, managers, and others involved in the project to be skeptical and possibly hostile because of the damage and inconvenience caused. There may even be a government hearing or legal deposition where you testify about what happened. While the primary audience of such a hearing is the public sector agency, the public nature of the hearing means you are also communicating with the public at large.

Communicating with the public or public sector to blow the whistle on your firm, or a firm with which yours is contracted, is one of the

The needs and concerns of the potential customer are highlighted by summarizing relevant regulations.

The features and function of a diesel particulate filter are explained in jargon-free terms that potential customers can understand.

PM Reduction.

The 2007 emissions standards for particulate emissions were 90% lower than 2002 engines, from 0.1 to .01-g/hp-hr. While previous reductions in particulate matter (PM) emissions have been achieved through engine combustion improvements and oxidation catalysts, the stringent 2007 particulate standards required very effective particulate aftertreatment.

The active diesel particulate filter (DPF) is the technology used for meeting the U.S. 2007 PM emissions standards.

Active Diesel Particulate Filters (DPF).

Filtration of exhaust gas to remove microscopic-sized PM is accomplished using porous ceramic media generally made of cordierite or silicon carbide. A typical filter consists of an array of small channels that the exhaust gas flows through. Adjacent channels are plugged at opposite ends, forcing the exhaust gas to flow through the porous wall, capturing the PM on the surface and inside pores of the media. Particulate matter accumulates in the filter, and when sufficient heat is present a "regeneration" event occurs, oxidizing the PM and cleaning the filter.

14

As part of a marketing release describing Cummins on-highway diesel engines, this page describes for potential customers the technology that reduces particulate matter emissions. Printed with licensed permission from Cummins, Inc. © 2015 Cummins, Inc., all rights reserved.

most difficult professional decisions you can face. Engineers in nearly every well-known whistleblowing case were correct in their fears for public safety, exhausted all possible avenues before going public, and lost their careers nevertheless. If it happens to you, obtain your own legal counsel first to improve the likelihood of a good outcome for the public and yourself. (See Module 3: *Meeting your ethical obligations.*)

10.3 How you communicate with the public

When communicating with the public and the public sector, engineers strive to make technical matters clear while balancing their ethical obligations to faithfully serve their employer and to support public safety, health, and welfare. Engineering codes of ethics also have something to say about *how* you communicate with the public: You are expected to report "all relevant and pertinent information," to express technical opinions only in your area of competence, and to reveal potential conflicts of interest.

To make technical matters clear, especially for audiences without a technical background, avoid technical jargon. Communicate in a way that is technically correct, yet easy to understand. The words you use to discuss a technical topic with other engineers draw on a deep shared experience and mutual understanding of what the terms of the art mean and how they are used—a product of your specific educational background. When communicating with someone with a different educational background, however, find ways to make the matter clear without using jargon.

July 25, 2012

Jada Marshall
Fulton County School Corporation
3145 Peachtree St.
Atlanta, GA 30079

Re: Fulton County, GA
Magnolia Blvd. Road Expansion
Des. No. 2011-R45-M29

Dear Ms. Marshall:

Thank you for your response regarding the proposed Magnolia Blvd. road expansion project. Your feedback is appreciated as the purpose of the early coordination process is to identify concerns so that they can be properly addressed before the project is finalized.

The opening paragraph begins by acknowledging the feedback that was received from the recipient, emphasizing that the feedback is important and being treated with respect.

The letter responds to each concern in separate paragraphs by giving details about how potential negative impacts will be minimized—without using jargon.

The paragraphs begin by paraphrasing the concern so that the recipient knows that the concern is understood (and to prevent misunderstandings).

Negative impacts are not left out, but clear reasoning is given and possible mitigating alternatives are mentioned.

The letter closes by re-emphasizing the importance of the concerns and keeping the lines of communication open.

Your first concern is the accommodation of traffic flow during construction and the need to provide access to both parking lots for school buses and other vehicles. Traffic will be accommodated at parking lot entrances during construction since there are no other access points to the schools. Access will be through the construction site. The Georgia Department of Transportation specifications state the contractor must maintain access to all drives for property owners. Pine Bluff Road needs to be closed during construction in the vicinity north of the venetian blind company as extensive grade work is needed in this area. A detour route will be posted for this closure. At this time, construction sequencing has not been fully identified. However, it is anticipated the work in front of the school, as well as the grade correction, will be completed during the summer months as much as possible to minimize impacts.

Your second concern is the issue of right-of-way acquisition and potential loss of parking for the middle and/or elementary schools. Right-of-way is needed a few feet into the parking areas of the schools. The current plans are preliminary and can change based on feedback from the property owners. There may be options available such as a small curb or wall to mitigate and avoid the lots.

Your concerns are understood and your willingness to provide feedback is appreciated. These concerns will be taken into consideration when finalizing the plans.

If you have any further questions regarding th s matter, please contact me at 404-555-7301 or lshaheen@strandassociates.com. Thank you for your input.

Sincerely,

STRAND ASSOCIATES, INC.

LINDA SHAHEEN

This justification letter responds to the concerns of the stakeholders represented by the school corporation. © 2015 Strand Associates, Inc. Reprinted with permission.

To keep communication open, never belittle the concerns of the public. Instead, listen and respond empathetically, even if the concern is based on what you think is misinformation, fear, or a misunderstanding. Take their concerns seriously and act on them in good faith.

Express technical opinions only in your area of competence: If you don't know something, admit it. Don't guess. Find out and get back to the questioner after you have it figured out. Making incorrect public statements not only is unethical but also erodes public trust in your competence and in the good faith of your employer.

Summary

Who they are and how they think: Health and safety are first priority

- Members of the public are stakeholders with only an indirect voice in engineering decisions, while members of the public sector are those who provide basic government services.
- Members of the public, as individuals and groups, want engineering decision-makers to hear their concerns and act on them to protect and support their interests as they see them, in the short term and in the long term. The concerns of the public sector are often enforced by law.
- Members of the public expect technology to be safe, and the public sector often imposes regulations based on past engineering failures to increase the public's safety.
- Both the public and the public sector respect engineers' contribution to society but mistrust corporate assurances.

Why you communicate with the public

- Engineers obtain input from public stakeholders during project planning and communicate with public sector agencies to acquire information, to interpret regulations, to obtain necessary waivers or exemptions, and to obtain permits and approvals.
- If called upon to supply technical information for a firm's marketing and sales departments or the public relations department, the challenge is to make sure that technical information edited for public consumption is true, without understated costs or overstated benefits.
- Communicating with the public in the event of a failure has unique challenges because of the inevitable loss of trust when engineering projects go wrong. Expect the public's view of the engineers, managers, and others involved in the project to be skeptical and possibly hostile.

How you communicate with the public

- Make technical matters clear while balancing your ethical obligations to your employer, the public, and yourself.
- Avoid jargon, communicating in a way that is technically correct yet easy to understand.
- Listen empathetically to the concerns of the public.
- Express technical opinions only in your area of competence; issue public statements only in an objective and truthful manner.

Genres

...define the categories into which writing and speaking tasks most often fall. Are you reporting or proposing? Making an inquiry or a request? To audiences inside of your organization, or outside of it? The better you understand the purpose and context of your message, the more precisely you can identify the rules that you need to follow and the expectations that your readers will hold. Often, your job will be to meet those expectations as completely as you can, maximizing the chances that a scientific journal will publish your experimental report or that a prospective client will accept your bid rather than your competitor's. At other times, you will want to adapt the genre to meet your needs and your readers', as when formatting a memo to be easy to read in an email message. In both cases, you can communicate most effectively by understanding the most important rules, the ones that might be negotiable, and the purposes that all of them are supposed to serve.

Genres

Reporting in a research community

11

An effective report will persuade the reader that the writer is competent, perceptive, and careful, and that the report's content is therefore trustworthy. Achieving these results requires report-writers to attend to concerns that are fundamentally ethical:

- Reporting the full context of the work
- Showing the results honestly
- Discussing expectations and where results deviate from them
- Conceding unpleasant truths rather than concealing or omitting them
- Exploring the potential for negative consequences (which may include social or environmental impacts as well as economic ones)

With such aims in mind, technical audiences of research reports expect careful, detailed exposition of methodology and analysis. Their needs are served by a standard report structure that presents a defined sequence of *introduction, methods, results, and discussion*—for short, *IMRaD*. Not every engineer or scientist knows this label, but the structure itself almost certainly shapes their writing, as you will see if you read a journal article written by one of your engineering professors. The IMRaD format is thus a convenient standard, and the first option to consider when writing for readers with an interest in the details of an experiment or study.

Objectives

- To appeal to the three main concerns of a research report's typical audience—methodological soundness, credible evidence, and well-supported conclusions
- To use the IMRaD format to present complex, technical work in a standard format that the audience understands
- To modify the IMRaD format as needed for purposes and content where its standard sections are insufficient

11.1 Writing for a technical audience

Most engineering research reports target an expected audience of other engineers and technicians. This expectation allows authors to assume basic background knowledge and to use standard jargon without many explicit definitions. (You will likely recognize the needs of audiences and contexts that require exceptions to this rule. In civil engineering, for example, a geotechnical engineer might need to define terms and concepts that wouldn't be familiar to structural engineers reading the report.)

It is helpful to think of the audience for a research report as a well-meaning but skeptical one, ultimately interested in uncovering the truth even if it means pointing out the flaws in your research report. Such an audience has three main concerns:

Methodological soundness	Research reports must provide enough detail for a qualified reader to reproduce the reported procedure. The limitations of your equipment, data collection, or analysis might be well known to the audience.
Credible evidence	The exact nature of credible evidence depends on the claims being made—but you can safely assume that the evidence will be

dissected thoroughly by readers with an active interest in interpreting its implications.

Well supported conclusions — Conclusions must be clearly supported by everything else in the research report, since the audience is equipped to notice—and to object to—any unsupported leaps of logic or unwarranted claims of importance.

Point-of-Care Technologies

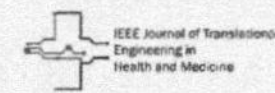

Received 2 October 2013; accepted 28 March 2014. Date of publication 30 May 2014; date of current version 12 June 2014.

Digital Object Identifier 10.1109/JTEHM.2014.2327612

Effective CPR Procedure With Real Time Evaluation and Feedback Using Smartphones

NEERAJ K. GUPTA[1], VISHNU DANTU[2], AND RAM DANTU[1]

[1]University of North Texas, Denton, TX 76203, USA
[2]Clark High School, Plano, TX 75075, USA

CORRESPONDING AUTHOR: N. K. GUPTA (neerajgupta@my.unt.edu)

This work was supported by the National Science Foundation under Grant CNS-0751205 and Grant CNS-0821736.

ABSTRACT Timely cardio pulmonary resuscitation (CPR) can mean the difference between life and death. A trained person may not be available at emergency sites to give CPR. Normally, a 9-1-1 operator gives verbal instructions over the phone to a person giving CPR. In this paper, we discuss the use of smartphones to assist in administering CPR more efficiently and accurately. The two important CPR parameters are the frequency and depth of compressions. In this paper, we used smartphones to calculate these factors and to give real-time guidance to improve CPR. In addition, we used an application to measure oxygen saturation in blood. If blood oxygen saturation falls below an acceptable threshold, the person giving CPR can be asked to do mouth-to-mouth breathing. The 9-1-1 operator receives this information real time and can further guide the person giving CPR. Our experiments show accuracy >90% for compression frequency, depth, and oxygen saturation.

INDEX TERMS CPR, oxygen saturation, frequency of compression, depth of compression, smartphones.

Heading section of a research paper on using smartphones to improve the effectiveness of administering CPR. © 2014 IEEE, reprinted with permission, from (Gupta, Dantu, & Dantu, 2014).

The title expresses the main idea of the paper.

Author names and affiliations are displayed prominently.

The abstract summarizes the context, goals, methods, results, and conclusion. Publishers usually impose a maximum word count for abstracts.

Index terms help other researchers find this paper in electronic databases. Authors usually select their index terms from lists provided by the publisher.

11.2 Elements of the IMRaD format

Almost all research reports use the IMRaD format:

- **I**ntroduction
- **M**ethods (and materials)
- **R**esults

- **a**nd
- **D**iscussion

The IMRaD structure emphasizes careful, detailed exposition of methodology and analysis. Technical journals use it as a default standard because it meets the needs of researchers who may themselves be conducting experimental work in the same area. Such readers need to understand the nuances of the work in order to evaluate its implications—and at least in principle, they need to be able to reproduce or otherwise verify the findings that you're delivering to them.

Introduction

A research report's introduction explains *why* your work is being done, discussing the context and goals of the research. There may be a small handful of readers who are eagerly awaiting your results or conclusions; perhaps the report was assigned by a manager looking for very specific data. Most readers, though, will take an interest in a study's findings or conclusions only if the introduction persuades them of the importance and relevance of the research being reported. Consider future researchers working on the same project or in the same technical field, and write the introduction in a way that those researchers will recognize the relevance of the report to their work.

Research question	An experiment or study should answer a clear research question. The introduction should state clearly what the experimenter or investigator is attempting to learn—which defines the report's content and purpose.
Goals	Goals address the results that the researcher intends to accomplish. Goals should be explicit, unambiguous, and measurable, and the report should deliver a conclusion for each one.
Review of research literature or prior art	Most research reports summarize previous work in the area and connect the current work to those previous results. A quick connection might be made in a few sentences within the introduction's main body. Making it more substantive might require a section or subsection for a *literature review* or *review of prior art*.

B. USE OF TECHNOLOGY FOR EFFECTIVE CPR

Over the years, awareness amongst the general public that CPR can be a lifesaving procedure has increased. There is a growing use of technology that aids people in performing CPR. Several devices provide CPR training. These devices improve the quality of CPR by providing feedback on proper placement of hands on chest and the correct frequency and depth of compressions [21]–[23]. Mechanical devices which give accurate frequency and depth of chest compression provide automatic CPR. Studies have shown that these automated CPR devices improve the survivability of patients who need out-of-hospital CPR [24]. During an emergency it is likely that a person trained in CPR may not be available. In such situations 9-1-1 operators help the caller to administer CPR by giving instructions over a phone. In such instances, a readily available technology would be useful in ensuring that people untrained could deliver CPR properly. Recently smartphone applications have provided video instructions on how to give CPR [25]. If the application is not available, a 9-1-1 operator can help in downloading that application. However, having to download the application and then watch the video seriously reduces the window of survivability for the injured person. Alternatively, there exist devices that give real-time feedback on the quality of CPR [26]. This paper reports on such a real-time feedback application.

The authors review prior use of technology to aid CPR and summarize and cite each contribution.

The first part of the problem being addressed is that "a person trained in CPR may not be available."

The second part of the problem is that the current use of smartphones to help an untrained person administer CPR takes too much time.

The introduction concludes by asserting the goal of the paper—to "give real-time feedback on the quality of the CPR."

From the introduction section of the research paper on smartphones and CPR

Methods (and materials)

To convince your readers to accept your results as valid, they will need the relevant information about the procedures used to obtain them. This information's prominence and level of detail will vary: Formal experimental reports, which usually assume that readers might need to reproduce the experiment, will cover it most thoroughly. Even in such a report, though, individual steps in operating an apparatus or a software program can be omitted as irrelevant to the design of the experiment.

This section may combine "Methods and Materials." Most often, a paragraph or two will describe the physical specimens, samples, or artifacts on which your experimental work is being done. If such specimens require especially important discussion, "Materials" might be a separate section, but this is quite rare.

In this paper, the authors use the heading "Motivation" to explain the need for the device and how the system works. This section introduces the authors' methods and materials in overview. A subsequent section describes the experimental work in greater detail.

II. MOTIVATION

During an emergency situation, it is highly likely that persons trained in CPR are unavailable. Even though devices can provide automatic CPR, these devices are highly unlikely to be accessible at the time of need. In such cases, an untrained person will need to administer CPR. In these situations, 9-1-1 operators provide CPR instructions over a phone.

A graphic combining images and text helps explain the use of smartphones to improve CPR. The person administering CPR is assumed to be untrained.

The figure shows the flow of information to the 911 operator, who then gives instructions to the untrained person administering CPR.

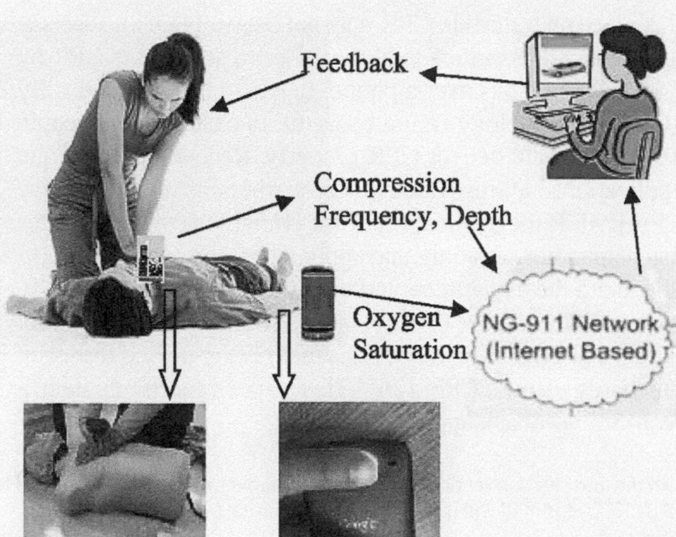

FIGURE 2. CPR and Its Evaluation. The compression frequency, depth and the oxygen saturation levels are reported to the 9-1-1 operator. The person giving CPR has a phone tied to her hands, as shown in the picture below the arrow. A smartphone is also placed near the hand of the person receiving the CPR with finger on the camera lens as shown in the second picture below the arrow.

To understand a graphic, readers often require a detailed caption of two or three sentences—here, providing an overview of the procedure and notes about equipment. This amounts to a condensed version of the paragraph below.

As mentioned in prior sections, continuous chest compressions have been emphasized over cardiopulmonary resuscitation as the most critical CPR procedure to perform in an emergency. That said, mouth-to-mouth breathing remains a viable option in certain cases, especially when trained personnel are present. We present a smartphone application which measurers the blood's oxygen saturation without specialized equipment. In conjunction, one smartphone can be used by the person giving the CPR where it measures the frequency and depth of compressions. A second smartphone can then be used to measure the oxygen saturation level. The data from these smartphones is continuously reported to the 9-1-1 operator who can use the information to guide the CPR giver. A patent has been filed by one of the authors for devices that can measure the vital signs using sensors of the smartphones [32]. Figure 2 summarizes how CPR could be enhanced with these devices.

An outline of methods and materials is located after the introduction and before the experiments and results sections. © 2014 IEEE, reprinted with permission, from (Gupta, Dantu, & Dantu, 2014).

Results

While the report must include the results of the work, findings should not simply be deposited in the report without being adequately explained and contextualized. Engineering managers dread "data dump" in reports—an extensive catalog of findings without an obvious or explicit meaning. A handful of habits can help you to avoid this problem:

- Assess findings with respect to expected values, prior art, or standards.
- Discuss individual findings (usually separately from broader conclusions).
- Note unexpected findings, discussing why they might have appeared and what they might mean.
- Leave no obvious inference unstated.

We conducted experiments using 40 volunteers. Each volunteer performed CPR for about 30 seconds. The number of compressions volunteers performed during these 30 seconds ranged from 30 to 50. Figure 9 shows a scatter plot of each compression for one subject. It shows the accuracy of the depth calculation done by our application as compared to the actual depth as observed in the Mobotix video. Our

One experiment is outlined with sufficient detail that it could be reproduced.

The authors define "accuracy" as the ratio of the smartphone measurement to a separate video measurement. The video measurement is assumed to be the true value.

application's accuracy ranged from a low of 57% to a high of 98%. The other 39 subjects had similar ranges of accuracy.

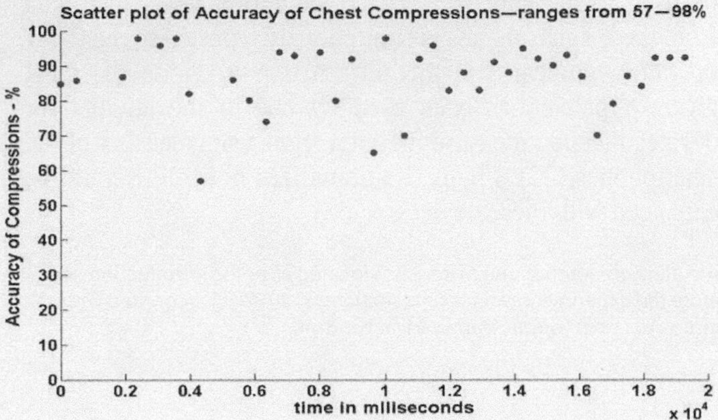

FIGURE 9. Accuracy of Each Compression as Calculated by Our Application for One Volunteer. The range is from a low of 57% to a high of 98%.

An outline of one experiment and its results. © 2014 IEEE, reprinted with permission, from (Gupta, Dantu, & Dantu, 2014).

Discussion (analysis and significance)

At least as important as the findings themselves are the writer's conclusions about them. The report must deliver the broad themes that are (1) directly relevant to the goals of the project and (2) supported by the findings. Analysis of the findings is directed and made meaningful by the goals of the work. Which goals were met (or not met), and by how much? To what extent were the needs met or the authors' hypotheses confirmed?

While analysis relates the work to specific goals, the report must also assess the significance of the results within the overall context—the "big picture" of the project, initiative, or problem with which the authors (and often their readers) are engaged. This discussion is typically future oriented: The report's authors are expected to deliver their authentic professional opinion, based on their findings and experience with the problem, about the next steps needed to improve or build upon this work.

We did an analysis to decide how frequently the alert should be given (Table 1). When our application gave an alert for each compression, out of a total of 38 alerts (for 38 compressions in that session), 21 alerts were less than 90% accurate. This means that for 21 alerts the accuracy of calculated depth as compared to the actual depth was less than 90%. The total of 38 alerts assumes that we give an alert for each compression (assuming no compression is going to be 100% accurate). When our application gave an alert every second, we averaged the calculated depth of all compressions within each second. As Table-1 indicates, the total number of alerts for one second will be 20. Of these 9 alerts will be less than 90% accurate. When our application gave an alert every 6 seconds, then 4 alerts were more than 90% accurate. We continued this analysis through 10 seconds. At 10 seconds, our application gave two alerts, each 100% accurate. We conclude that the accuracy of alerts increases when an alert is given every few seconds rather than for every compression. The accuracy improves because the errors with negative magnitude adjust the errors with positive magnitude with in the time period. So the overall accuracy improves.

The goal of this part of the analysis is explicitly stated.

The meaning of the numbers in the table is explained, leaving no inference to the reader.

The conclusion that follows from the table is stated explicitly.

TABLE 1. Accuracy of alerts over time. The rows show accuracy for alerts given for each compression, and for each second between 1 and 10 seconds. Frequency of alerts shows the time period analyzed. Total alerts shows the total number of alerts that occur during the specified time period. Accuracy Range for Alerts–% shows, in 5% increments, the number alerts with compression accuracy. For example, when an alert is given every second, 6 alerts of the 20 alerts have a compression accuracy of less than 85%. However, when an alert is given every 5 seconds, all the alerts have compression accuracy greater than 85%.

Table captions are sometimes used to summarize the meaning of the data in the table.

Frequency of Alerts	Total Alerts	Accuracy Range for Alerts - % <80	80–85	86–90	91–95	96–100
each comp	38	6	7	8	12	5
1 second	20	3	3	3	8	3
2 seconds	10	0	0	3	4	3
3 seconds	7	0	0	3	2	2
4 seconds	5	0	0	1	3	1
5 seconds	4	0	0	1	2	1
6 seconds	4	0	0	0	4	0
7 seconds	3	0	0	0	1	2
10 seconds	2	0	0	0	0	2

The authors emphasize the main finding of this section of the work by italicizing the sentence.

Within the 6–7 second range, our application's accuracy is reasonable at more than 90%. But, our experiments suggest an alert every 6–7 seconds does not provide persons giving CPR enough time to adjust their CPR compression depth. *Our experiments also suggest that an optimum time for giving alerts is every 10 seconds.* However, we still need to decide the optimum time gap between alerts.

Discussion section to establish the evidence supporting one finding. © 2014 IEEE, reprinted with permission, from (Gupta, Dantu, & Dantu, 2014).

The authors discuss conditions that impair the performance of their system, contributing to their technical credibility.

An experiment provides evidence of the system's limitations.

Even under controlled conditions, the system produces unreliable results when used in a moving vehicle.

E. CPR IN A MOVING VEHICLE

In certain situations, one may have to give CPR as the patient is being transported in a moving vehicle to the hospital. Several factors come into play in a moving vehicle that increases the difficulty of calculating the compression depth accurately when using a smartphone accelerometer. The first, a vehicle moving affects accelerometer readings. If shock absorbers are inadequate, an accelerometer records vehicle movements in the Z axis, skewing the Z axis motion of the chest compressions. Road condition presents another major factor that contributes to increased errors. Bumps in the road, lane changes, and traffic turns also affect readings. Traffic patterns also add to the randomness of readings. The vehicle may have to be slow at times and then accelerate as the traffic moves. It may have to stop at traffic lights and then accelerate. Even if we keep factors constant, such as using the same vehicle, driving on the same road and even driving at the same speed, randomness of a traffic pattern still produces different results each time the CPR is attempted.

We ran an experiment with extremely controlled conditions. We selected a smooth road with no bumps. The road had almost no traffic, required no turns for a few miles, and had no traffic lights. We, then, drove at a constant speed of 30 mph to minimize movements due to vehicle motion. The results of the experiments are shown in the Table 3. The high standard deviation indicates that there was large variation in compression depths. This was caused by driving conditions and the vibrations of the vehicle itself. However the frequency calculations were accurate.

Discussion section includes straightforward exposition of the work's shortcomings. © 2014 IEEE, reprinted with permission, from (Gupta, Dantu, & Dantu, 2014).

Submitting reports for publication

When submitting a report as an article or paper for a professional conference or journal, you *must* follow the expected format. A submission that doesn't meet the expected format—the sections in the wrong order or the wrong fonts—might be rejected, regardless of the quality of its content.

Most engineering journals publish their articles in the IMRaD format. Specific headings and details of the organization might vary slightly from one journal or conference to another; usually, the editors or conference organizers will have specified the correct format—sometimes in a template file that you can edit directly, or a list of required sections and formatting styles. Expect everything from the headings to the typefaces to be specified. Find this format information and follow it exactly.

> REPLACE THIS LINE WITH YOUR PAPER IDENTIFICATION NUMBER (DOUBLE-CLICK HERE TO EDIT) < 1

Preparation of Papers for IEEE JOURNAL OF TRANSLATIONAL ENGINEERING in HEALTH AND MEDICINE

First A. Author, Second B. Author, Jr., and Third C. Author, *Member, IEEE*

Abstract (The abstract should not exceed 250 words. It should briefly summarize the essence of the paper and address the following areas without using specific subsection titles.). Objective: Briefly state the problem or issue addressed, in language accessible to a general scientific audience. Technology or Method: Briefly summarize the technological innovation or method used to address the problem. Results: Provide a brief summary of the results and findings. Conclusions: Give brief concluding remarks on your outcomes. Clinical Impact: Comment on the translational aspect of the work presented in the paper and its potential clinical impact. Detailed discussion of these aspects should be provided in the main body of the paper.

(Note that the organization of the body of the paper is at the authors' discretion; the only required sections are Introduction, Methods and Procedures, Results, Conclusion, and References. Acknowledgements and Appendices are encouraged but optional.)

***Index Terms*—At least four keywords or phrases in alphabetical order, separated by commas. For a list of suggested keywords, send a blank e-mail to keywords@ieee.org or visit http://www.ieee.org/organizations/pubs/ani_prod/keywrd98.txt**

Note: There should no nonstandard abbreviations, acknowledgments of support, references or footnotes in in the abstract.

I. INTRODUCTION

THIS document is a template for Microsoft *Word* versions 6.0 or later. If you are reading a paper or PDF version of this document, please download the electronic file from the IEEE Web site at http://www.ieee.org/web/publications/authors/jtehm/index.html so you can use it to prepare your manuscript.

When you open the template, select "Page Layout" from the "View" menu in the menu bar which allows you to see the footnotes. Then, type over sections of the template or cut and paste from another document and use markup styles. The pull-down style menu is at the left of the Formatting Toolbar at the top of your *Word* window (for example, the style at this point in the document is "Text"). Highlight a section that you want to designate with a certain style, then select the appropriate name on the style menu. The style will adjust your fonts and line spacing. **Do not change the font sizes or line spacing to squeeze more text into a limited number of pages.** Use italics for emphasis; do not underline.

To insert images in *Word,* position the cursor at the insertion point and either use Insert | Picture | From File or copy the image to the Windows clipboard and then Edit | Paste Special | Picture (with "float over text" unchecked).

IEEE will do the final formatting of your paper. If your paper is intended for a conference, please observe the conference page limits.

II. PROCEDURES FOR PAPER SUBMISSION

A. Review Stage

Please check with your editor on whether to submit your manuscript as hard copy or electronically for review. If hard copy, submit photocopies such that only one column appears

This paragraph of the first footnote will contain the date on which you submitted your paper for review. It will also contain support information, including sponsor and financial support acknowledgment. For example, "This work was supported in part by the U.S. Department of Commerce under Grant BS123456".

The next few paragraphs should contain the authors' current affiliations, including current address and e-mail. For example, F. A. Author is with the National Institute of Standards and Technology, Boulder, CO 80305 USA (e-mail: author@ boulder.nist.gov).

S. B. Author, Jr., was with Rice University, Houston, TX 77005 USA. He is now with the Department of Physics, Colorado State University, Fort Collins, CO 80523 USA (e-mail: author@lamar.colostate.edu).

T. C. Author is with the Electrical Engineering Department, University of Colorado, Boulder, CO 80309 USA, on leave from the National Research Institute for Metals, Tsukuba, Japan (e-mail: author@nrim.go.jp).

Partial first page of a template for authors submitting an article to the *IEEE Journal of Translational Engineering in Health and Medicine.* © Copyright 2015 IEEE.

The document template contains some placeholder text that the author replaces without changing the formatting, such as the paper title and author names.

The template includes instructions on abstract length and the expected section headings.

IEEE includes directions on obtaining index terms.

This template includes formatted section headings, subheadings, and main-body text.

Preparation of such a template can also be helpful in collaborative writing—individual members of the team can then place their own content into the style on which the team has agreed.

11.3 Experimental reports

Experimental reports communicate technical *findings*. Having completed your experiments or tests, you know something that you didn't know before, or have compared some finding to an expectation and assessed whether that expectation has been met. For instance, you may be testing a device or component against industry standards, internal benchmarks, regulatory standards, or performance specifications requested by a client.

Experimental reports have evolved to meet the needs of readers who are themselves experimentalists: Their conventions derive from readers' possible need to duplicate the experiment and validate its conclusions. Reports produced in industry or grant-funded research should allow readers to build a test procedure or apparatus based on your description and obtain results that confirm yours—the hallmark of the scientific method. Readers of undergraduate student experiments won't likely be reproducing the work in such a painstaking way; at a minimum, though, they should be able to perform the described analysis on the data reported.

How the sections are specialized for experimental reports

In an experimental report, the crucial *context* explains the need for the experiment. For example:

- A performance specification requires a validation test.
- A company division needs exploratory tests on the performance or capabilities of a competing product.
- A new product or process is being developed, and its limits are being explored.
- Quality control tests need to be performed.
- A problem has been found, but its root cause has not yet been discovered.

Experimental *goals* should be explicit, unambiguous, and measurable:

- Determine the stopping distance of a new antilock brake system.
- Determine the mechanical properties of a new fiber composite.
- Pressure test a well bore.
- Find the peak operating temperature of a processor chip.

- Determine the efficiency of a chemical processing operation.
- Characterize the speed of operation of a software algorithm.

In detailing the *method*, address the material and procedural components of the experiment. The *infrastructure* must be described, often with a diagram or schematic: apparatus, test stands, sensors, lab space. The *sample* or test object may need to be detailed—critical dimensions, materials, configuration, and assembly—so that it can be reproduced. The *procedure* should be explained: test conditions, types of measurements, control groups, test sequence. Demonstrate that the procedure meets the requirements of the experiment's goals and of relevant standards.

The *method of analysis* may have essential elements that require explanation: theoretical principles, representative data, data reduction equations, uncertainty analysis, statistical hypothesis tests. This section can include modeling schematics that help explain the mathematics being used. If one of the report's goals is to present an experimental design, then this design should be justified: Why did it have to be designed? What are the constraints? What are the specific features of the design? What evidence shows that the design is better than the alternatives?

The *results* produced by the analysis include key intermediate findings that conclude with the major finding. These intermediate findings constitute evidence that supports the major finding in response to the experiment's goals. Each finding is described and discussed with the goal of removing any ambiguity. When including a graph, equation, table, or illustration, such maximal clarity is best achieved by explaining what readers are seeing and what relationships or important features they should notice.

1) VALIDATION OF LOW OXYGEN LEVELS

After measuring the Oxygen saturation level under normal circumstance, we wanted to confirm that our system can detect reduced levels of oxygen in the blood. We used occlusion spectroscopy, the method of using over-systolic pressure to temporarily stop the flow of blood to the finger to collect data, (Figure 12). In our study, we temporarily cut off the blood in the root of the finger. This allowed us to measure blood flow in the upper layers of the finger. Later when occlusion is removed, the oxygen saturation level reaches the normal value and our system

The goal of the experiment is emphasized with italics.

First, a reference experiment is described, against which the smartphone results will be compared.

The second experiment uses the same method, but the measurement is made using the smartphone app. The results, shown in the figure, have the same shape and magnitude of the reference data, providing evidence that supports the conclusion.

The conclusion is explicitly stated.

can measure this return to the normal value. The occlusion experiment confirmed that we can indeed measure the depletion of oxygen saturation in the blood using the optical features of a smartphone.

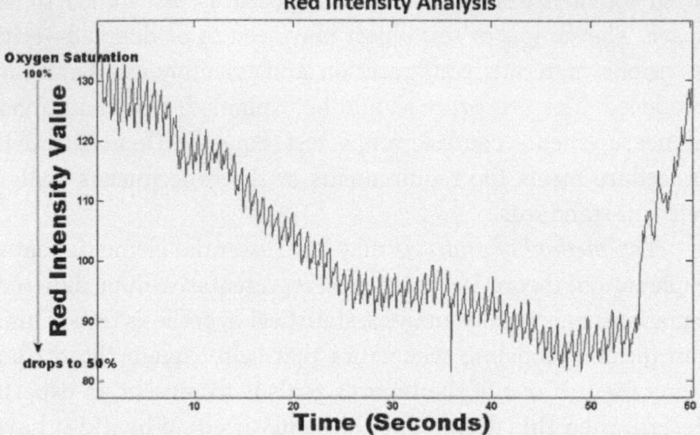

FIGURE 13. Red Intensity of Occlusion. A drop in oxygen saturation level to about 50% occurs. Then the saturation level returns to 100% after occlusion is removed.

Experiment to validate the smartphone measurement of blood oxygen levels. © 2014 IEEE, reprinted with permission, from (Gupta, Dantu, & Dantu, 2014).

11.4 Reports that advance theory

Most engineering research is done in communities that use experimental reports to communicate findings and advance the field. Other reports, though, make such progress by advancing theory. Such reports often approximate the IMRaD organization, but they interpret and often label its major sections differently. Section headings are less likely to read simply "Methods" or "Results"; they will often be more specific to particular issues in the technical content.

How the sections are specialized for theoretical articles

The introductory section should clearly show the need for the theoretical advance being described. Existing work, both theoretical and experimental, must be discussed to connect the current work to the current state of the art. Recent political, environmental, and economic circumstances might be used to clarify the context of the current work.

The methodological section discusses the first principles used, relevant constants and theoretical parameters, as well as necessary assumptions. The general approach taken by the theoretical work is elaborated, showing how the pieces fit together.

The paper then shares the useful theoretical result, referring to the items mentioned in the methods section as needed. For many theoretical reports, a full line-by-line re-creation of the derivation is not needed, but the major steps and intermediate results are listed so that someone else working in the field could follow, and then re-create, the result. General relationships derived might be best presented in graphical form for a few representative situations to assist the audience in visualizing the result.

Finally, results are discussed in a way that ties the major results back to the current state of the art and potential applications. Limitations of the analysis are mentioned, as well as potential next steps to strengthen or generalize the theory to new areas.

I Introduction

II Background

III The thermodynamics of computation

A Modelling computation

B Time-symmetry and reversibility

C The second principle and irreversibility

D From Maxwell's demon to the Landauer principle

IV Physical model of secure computation

A A simple two state register

B The multiple–state case

C Universal factorization of secure computations

D Erasure vs resetting: extracting work from the system

The introduction includes a statement of the new contribution of the work.

The background includes basic definitions and principles applied.

Sections 3, 4, and 5 lay out the connections from previous work to the new theory being advanced.

Implications and applications of the new theory conclude the report.

V The thermodynamics of vulnerability

A Dissipation and intrinsic source code threat

VI Practical implications

VII Future directions

VIII Conclusions

Outline of section headings from a research paper advancing a new theory. from (Malacaria & Smeraldi, 2012).

The introduction outlines the specific contribution of the new theory.

The connections to previous theory are established.

The authors conclude by stating explicitly the novel aspect of the work.

A. Contributions

The overall contribution of this paper can be seen as laying down in a precise sense the thermodynamic foundations of confidentiality.

In Section IV we consider an idealised physical model that is commonly used in the literature of thermodynamics of computation [12], [2]. We generalize it to model an arbitrary number of states with a non-uniform distribution, thus providing a conceptual model for the most general statement of the Landauer principle. We then show that the notion of remaining uncertainty W from Quantitative Information Flow is, up to the multiplicative factor $K_B T \ln 2$, precisely the minimum dissipation associated to secure computation (equation 17 and proposition 3).

Section IV-D demonstrates that net energy expenditure is not required if we allow probabilistic operators into the language. In that case work can actually be extracted by the system (inequality 20), although erasure remains an irreversible operation. Again the energy is bounded by the quantity W which is in this case non-positive.

Section V investigates the thermodynamics of Smith's notion of vulnerability and proves that remaining vulnerability is in general a lower bound on W (proposition 6). The bound becomes an equality if and only if the remaining vulnerability coincides with the difference between the work needed to reset the input and output registers when they are in their maximally disordered state (proposition 4 and 7). To the best of our knowledge this is the first connection between guessability and thermodynamics.

Finally both measures are order related to the magnitude of dissipation in section V-A.

VIII. CONCLUSIONS

The study of thermodynamic aspects of computation dates back to the pioneers of computing starting with Von Neumann. Following works by Landauer and later Friedkin and Toffoli and Bennett illustrated how all computations can be executed reversibly. Thus dissipation, while of great practical importance, seems to have little foundational status in computer science.

Here we established a fundamental relation between dissipation and secure computation by proving that two of the main metrics of confidentiality in computer security, namely information leakage and vulnerability, are essentially measures of dissipation in the thermodynamic sense. These results provide thermodynamic foundations for confidentiality, with Landauer's principle thus implying a fundamental lower bound to the energetic cost of secure computation. Understanding the physics of confidentiality contributes to the debate on the role of irreversibility in other minimally dissipative systems such as nano technologies, molecular and biological computation and quantum computing. Applied fields such as the study of power analysis attacks are also likely to benefit.

In the conclusion, the new fundamental finding is restated.

Applications of the new finding are suggested.

Introductions and conclusion sections from a paper that proposes a new theory.

11.5 Literature reviews

"Don't reinvent the wheel!" says a senior engineer to a junior. In other words, don't waste time and money re-creating someone else's earlier work. In starting a new project, especially one in which you have little prior experience, learn the relevant *prior art*: how others before you defined a problem, their solutions and their consequences, and what challenges remained. When the prior art is gleaned from published sources, the process is called a *literature review*.

The author of a literature review is expected to summarize and discuss published work that establishes the technical, social, and economic context for a problem and summarizes the relevant history of the technology with an emphasis on the state of the art, recent advances, unresolved problems, and promising approaches to solutions.

The review helps the author establish which aspects of their new work are novel, distinguishing it from prior art.

A literature review is a conventional component of research papers and patent applications, with its length and detail determined by the audience's expectations. A literature review can also be published as an independent document—a category of research publication in its own right. Such reviews are the subject of this section.

How the sections are specialized for literature reviews

A typical literature review's *introduction* section will likely resemble those of other research reports, and the last section may be similar to *discussion* sections—especially in emphasizing important opportunities or directions for future research.

Large, renewable energy systems provide the engineering context for the review.

The need for fast circuit breakers is identified as a principal obstacle to further advances in the field.

The goal of the paper is to summarize the current state of technology, recent attempts to advance the technology, and recommendations for the direction of future work.

The conclusion restates the main problem addressed in the paper.

Abstract

The increasing interest in development, operation and integration of large amount of renewable energy resources like offshore wind farms and photovoltaic generation in deserts leads to an emerging demand on development of multi-terminal high voltage direct current (MTDC) systems. Due to preformed studies, voltage source converter based HVDC (VSC-HVDC) system is the best option for realising future multi-terminal HVDC system to integrate bulk amount of energy over long distances to the AC grid. The most important drawback of VSC-HVDC systems is their need for fast HVDC circuit breakers. This paper aims at summarizing HVDC circuit breakers technologies including recent significant attempts in development of modern HVDC circuit breaker. A brief functional analysis of each technology is presented. Additionally, different technologies based on derived information from literatures are compared. Finally, recommendations for improvement of circuit breakers are presented.

6 Conclusion and recommendations

Nowadays, the main obstacle against the realisation of HVDC grids is lack of mature HVDC fault current breaking technologies. In this paper the present technologies of HVDC circuit breakers were summarised and compared. All of presented breaking schemes have limited capabilities in interruption of permanent fault current and need to be significantly improved.

In terms of mechanical circuit breakers as the basic devices for fault current interruption, attempts should be concentrated in optimization of size of resonance circuit's elements. Also the behaviour of arc chamber needs to be improved to reach higher current rating.

The attributes and drawbacks of current technology are summarized.

Since hybrid circuit breakers present more efficiency and acceptable interruption speed, the development of faster mechanical switches with high surge voltage withstand and low concoction losses can lead to more improvements in this area.

Areas of improvement in the technology are described.

In terms of solid-state circuit breakers, application of new wide-band-gap semiconductors like SiC or GaN based switches should be investigated. Also active gate driving technologies can improve the performance of semiconductor switches in pure solid-state circuit breaker. Moreover, accurate dynamic models for semiconductor switches with validity in high voltage and high currents to be used in designs and simulations are necessary to be implemented. In order to provide the possibility of distinguishing the permanent faults from transient grid events applications of DC fault current limiters in HVDC networks can be interesting to study.

Future research directions are summarized.

Abstract and conclusion sections from a literature review paper. The form and content are similar to those in a conventional report of technical findings. © 2014 IEEE, reprinted with permission, from (Mokhberdoran, et al., 2014).

Between the beginning and concluding sections, however, literature reviews differ from a typical engineering report in form and content. They do not chronicle a process (from an experimental method to its results to the analysis of those results). Rather, they divide a field of research into categories based on the kinds of questions being asked, the kinds of data being collected, or the kinds of solutions being proposed. As a taxonomy of research in a carefully limited area, the middle sections in a literature review are often subdivided into numerous subsections.

1 Introduction

2 VSC-HVDC based multi-terminal grid

3 Requirements

The first three sections establish the context and need for the technology being reviewed.

Existing technology is classified and described.

The authors identify important performance characteristics and discuss the advantages and drawbacks of existing technology.

4 HVDC circuit breakers

4.1 Mechanical HVDC circuit breaker

4.2 Hybrid Technologies

4.3 Pure solid-state circuit breaker

5 Comparison of technologies

5.1 Interruption time

5.2 Power losses

5.3 Voltage rating

5.4 Current rating

6 Conclusion and recommendations

Outline of section headings from a literature review paper. © 2014 IEEE, reprinted with permission, from (Mokhberdoran, et al., 2014).

Summary

Writing for a technical audience

- The audience for research reports includes other engineers and technicians.
- The audience expects methodological soundness, credible evidence, and well-supported conclusions.

Elements of the IMRaD format

- The *introduction* section discusses the context and goals of the effort—the need for the work, reviews relevant literature, and specific goals for what is being accomplished.
- The *methods* (or *methods and materials*) section presents information about the procedures used to obtain the results—how the work was performed and how the results were analyzed.
- The *results* section presents the findings with adequate explanation and context.

- The *discussion* section gives the author's conclusions about the results: Which goals were met? What is the next step to move the work forward?
- A particular journal, conference, or organization may have an expected format that must be followed. The headings and structural details may vary from the IMRaD structure, but the sections' basic purposes and sequence will likely be similar.

Experimental reports

- Experimental reports target an audience of other experimenters and should supply enough accurate detail that a reader could reproduce the reported work and check its findings and analysis.
- An experimental report's *introduction* section describes the goals of the experiment and establishes the context that makes it necessary, in the larger project and in the results of other researchers. The report's conclusions should address all goals explicitly.

Reports that advance theory

- Not all technical and scientific papers are experimental; research communities also depend on reports that advance theory.
- Section headings are less likely to read simply "Methods" or "Results"; they will often be more specific to particular issues in the technical content.
- The methods section is used to set up the general approach taken, along with necessary assumptions.
- Useful theoretical results are reported, with major intermediate results shown to aid the audience in following the derivation. General relationships may be presented in graphical form to help the audience understand the result.

Literature reviews

- Literature reviews distill large quantities of information specific to a particular area of research, informing the audience about the state of the art.
- A literature review's introduction and closing discussions resemble those of other reports.
- Middle sections in a literature review survey many studies rather than following a single investigation. Therefore, those middle sections typically divide research into multiple categories.

Reporting in an industrial organization

12

Reports encountered in an industrial setting, particularly progress reports and status reports that inform executive audiences, emphasize the results that follow from the work without dwelling on the details of the procedure or analysis. Readers in industry who will act on the information in research reports may be less skeptical than other investigators in a research community, and they will not likely be reproducing any experiments.

To best serve the needs of this audience, industry reports organize content differently, delivering Answers First. By giving the results early in the report, managers or higher-level executives can make decisions based on clear results and recommendations, referring to the supporting documentation as needed.

Objectives

- To accommodate decision-making readers in industry with conciseness while maintaining credibility, using the "Answers First" format
- To preface central findings with a concise account of the project context, based on analysis of what primary audiences need to know
- To anticipate and accommodate likely *secondary audiences*—readers other than those originally or explicitly addressed—who may need to use the document
- To distinguish between crucial findings that need to be included in the body of the document and work documentation that can be arranged as explanatory supplements such as appendices or attachments
- To apply the Answers First format to reporting genres where it is expected, such as progress reports, design reports, and feasibility studies

12.1 Writing for decision-making audiences in industry

Most industry reports are read by managers or executives at higher levels in the organization. Unlike a research report—which aims at thoroughness, and is based on the assumption of scientific skepticism—a typical industry report is built to enable quick decisions that are nevertheless well informed. To enable such action on the part of your reader, aim for two qualities:

Credibility	Reasoning and evidence should be reported clearly enough to promote confidence in the findings and the effort that produced them.
Conciseness	A truly *concise* document isn't the shortest one possible—it's the one that delivers the most relevant, important information most efficiently, without creating ambiguity or confusion.

Accommodating primary and secondary audiences

The author of an article in a technical journal usually has a good idea of what *type* of reader to target: other researchers working in the same field of study. Engineers in industry have the luxury of knowing—often personally—the exact readers who need to be accommodated in a report: teammates, colleagues in other divisions of the same company, managers, a division vice-president, or a client.

The report's authors will thus know whether or not readers need to be briefed on a project's goals or history, and what level of detail is required. In many cases, such authors can rely on readers' deep familiarity with a project's big-picture objectives and can "cut to the chase" by moving quickly to the new findings, recommendations, or other information that those readers are seeking. Such brevity and directness will often be appreciated by busy readers.

Such brevity requires one important caution, though. Many industry reports find their way to *secondary audiences* who lack the understanding or background of the primary readers who are foremost on the author's mind. These may include:

- Engineers brought in to work on the project or revisit it in the future
- Senior executives who take an unanticipated interest in your work
- Lawyers or regulators

Such readers typically cannot benefit from the quick briefing we design for a well-informed executive; if your document becomes important to them, they may need a more substantial understanding of how your findings or recommendations were reached, or why you approached a task in a certain way.

12.2 Elements of the Answers First format

The Answers First format meets the needs of the most common primary audiences, who are looking for "bottom-line" findings to inform quick but well-informed decisions. At the same time, it supplements those findings with documentation that is sufficiently thorough to back up those findings for any readers who may bring questions about those findings:

- Context
- Results and discussion
- Work documentation

Nearly any genre of report could be presented in this style; it will almost always be required when reporting in a memo or email. (Sometimes, a single element of a report could be presented all by itself in a short report—a review of prior art or a problem description, for instance. As in a full Answers First report, some context is likely needed before delivering such content.)

The title states the fault code number that is the subject of the document. The subtitle describes the relevant hardware (the "dosing unit").

A table immediately summarizes the main points of the document: the fault code number, what it means, and its effect on engine performance.

Liberal use of acronyms, abbreviations, and jargon indicates an intended audience already well acquainted with the system being described.

FAULT CODE 1682 (Air-assisted)
Aftertreatment Diesel Exhaust Fluid Dosing Unit Input Lines—Condition Exists

Overview

CODE	REASON	EFFECT
Fault Code: 1682 PID: SPN: 3362 FMI: 31 LAMP: Amber SRT:	Aftertreatment Diesel Exhaust Fluid Dosing Unit Input Lines - Condition Exists. An error has been detected by the aftertreatment diesel exhaust fluid dosing unit.	Diesel exhaust fluid injection into the SCR aftertreatment system is disabled.

Circuit Description

For an air assisted SCR aftertreatment system, the aftertreatment diesel exhaust fluid dosing unit requires air pressure from the OEM air tanks. The diesel exhaust fluid dosing unit precisely measures the amount of diesel exhaust fluid (DEF or Urea) to be injected into the aftertreatment system. The diesel exhaust fluid dosing unit has three primary cycles. A priming cycle at initial engine start makes sure that diesel exhaust fluid is available at the diesel exhaust fluid dosing unit. During the dosing cycle, the diesel exhaust fluid is being delivered to the aftertreatment nozzle. A purge cycle occurs when the engine is turned off. The purge cycle makes sure that all the diesel exhaust fluid is removed from the diesel exhaust fluid line and aftertreatment nozzle.

Component Location

The aftertreatment diesel exhaust fluid dosing unit location is OEM dependent. Refer to the OEM service manual for more information.

Conditions for Running the Diagnostics

This diagnostic consists of multiple parts, which run when the engine is first started, and make take up to 12 minutes to complete.

Conditions for Setting the Fault Codes

The aftertreatment diesel exhaust fluid dosing unit is **not** able to provide the correct dosing rate to the aftertreatment nozzle.

The specific condition that causes the fault code to appear is described in one sentence.

An example of the Answers First format: a report describing a "fault code" that appears for a type of exhaust aftertreatment failure in a diesel engine. Printed with licensed permission from Cummins, Inc. © 2015 Cummins, Inc., all rights reserved.

Context

Even when you are eager to benefit your reader by arriving quickly at your main ideas—the results and discussion—introduce them with at least basic information about the project or problem and goals. This accommodates secondary audiences and may also provide useful reminders for primary readers who haven't thought about a project's details recently.

Fundamentally, you must address both *what work was done* and *why it was needed*. Such a need may be easy to explain by referring to a project's goals—or you may need to review briefly the results already achieved by other designers and researchers.

Results and discussion

In an Answers First format, results should be explained and contextualized in such a way that their "big-picture" significance is made clear and prominent. Because the Answers First format emphasizes the primary results and discussion, an extensive catalog of findings is unnecessary. To highlight the primary results and discussion:

- Begin with the results that are directly relevant to the goals of the project.
- Explicitly state which goals were met (or not met), and by how much.
- Briefly discuss the significance of the results within the overall context—the "big picture" of the project, initiative, or problem.
- Note unexpected findings, discussing why they might have appeared and what they might mean.
- Leave no obvious inference unstated.

Given the test objectives and the sponsor's interests, it suffices to say that one type of spline geometry is much more stable than the other. Work documentation follows in an appendix.

Conclusions

Axial stability in the spline-foam model in this study was significantly increased by the utilization of the flat spline geometry as opposed to the sharp spline, as well as an increased spline angle. Trends in this data were clear and highly significant. The most axially stable spline geometry is that with flat geometry and a 5 degree spline taper angle.

In contrast to axial stability tests, much less clarity was observed in the difference between spline designs and spline angles as they related to torsional stability. Trends of increased torsional stability were seen in small angles of rotation as a result of sharp spline geometry. However, the overall measure of torsional stiffness resulted in no significant difference between any of the spline geometries or angles tested.

This very brief conclusions section closes a report of tests performed by a biomedical engineering laboratory for a manufacturer of orthopedic implants.

Work documentation

Intermediate findings, accounts of methodology, and other content used in the IMRaD format (See Module 11: *Reporting in a research community*) may not be needed as a primary component of an Answers First report, but such material still needs to be recorded. Think about what evidence could be used to support the results and discussion if the reader were looking for a little more detail.

This kind of work documentation may be required as an attachment or appendix to the report; alternately, it may be stored in a separate project archive. Traditionally, engineers kept logs of their work in notebooks that served such an archival function; today, such information is more likely to be maintained in a wiki or a cloud-based document storage system.

It is useful to relegate work documentation to the parts of the document where you know that it will not interfere with the central message that you need to convey to executives or other readers whose information needs are more basic. Even in an appendix or attachment, though, readers will appreciate explanation that enables them to use the information effectively. Such supplements may contain extensive data or other detailed information, but they are not exempt from the principles and techniques that govern effective communication.

S.No.	Documentation best practices	Status	Reason
D1	standby protocol definition should be properly documented in the SoC guide before LF. standby entry/exit sequence should be captured in the soc-guide , along with the valid mode entries for the SoC (for example always OFF domain will not be ON in any power down mode , which is in theory possible from MC point of view)	DONE	
D2	standby entry/exit and corner cases should be done in the verif plan preparation stage itself by thorough ADD review and peer review with FE/AMS team	DONE	
D3	use case document should be available atleast 10 weeks before LF. ADD should list critical verification scenarios.	DONE	
D4	Send verification plans to IP owners for review. Completion of reviews should happen before the first integration release. Must for new Ips	DONE	Mode Controller and SDIO are the new IP's. Verification Plan review of these modules has been done with IP Owners.
D5	Verification plan should be created ahead of going into execution of actual testcase implementation and debug	DONE	
D6	Verification plan should be updated along with testcase modification	PARTIALLY	Verification Plan to be updated for all modules post LF
D7	The comments in the testcase should be adequate and not out-of-date. Verif lead should get the testcases cross-reviewed within the team to ensure that comments are in place (incl. header)	PARTIALLY	P2P reviews have NOT hapenned. All new patterns have proper headers and comments. Ported patterns have old comments and header, nothing has been changed for these patterns.
D8	Verification team should dissect each and every line of ADD to ensure all scenarios to be covered through verification are identified and translated to testcase	DONE	
D9	For GLS testcase identification, get in touch with STA team to identify the scenarios to be checked in verification based on critical paths and exceptions applied.	DONE	GLS to be targetted after LF. GLS is almost clean before TO.

The documentation column describes the company's expectations.

The status column is updated from "not done" to "done" or "partially done" as the work proceeds.

The descriptions are sprinkled with obscure acronyms (SoC, LF, FE/AMS), indicating the document is intended only for internal audiences.

"Stoplight" color coding is used to emphasize status: red for "not done," yellow for "partially done," green for "done."

Many companies are replacing traditional progress reports with online "dashboards" and other systems for monitoring project status. This spreadsheet helps a project team check its compliance with the company's specified practices for documenting work. © 2015 Freescale Semiconductor Inc. Reprinted with permission.

12.3 Progress and status reports

A progress or status report focuses on recent developments for an audience that already knows the context, prior progress, and goals. Because brevity and clarity are highly prized here, progress and status reports just about always present Answers First, with only a very cursory account of the evidence supporting claims and findings.

The most common progress reports provide periodic *updates to a working group* within an organization, keeping everyone informed of what's being done and letting the working group's supervisor know what resources might be needed. Such reports might be written in a memo or an email; detailed meeting minutes may also serve this function. Alternately, you might deliver a verbal report given during a meeting or on-the-spot updates given when you drop by a supervisor's office. Even in this least formal situation, you will want to be sure not to leave anything out, so it can be helpful to remember the way that you would organize a written progress report.

How the sections are specialized

Because a progress report's typical readers are already familiar with the project's goals and history, a reminder of the *context* is all that's needed. If a particular aspect of the context is important for the results to be discussed, it should be highlighted.

Most progress or status reports focus on *results and discussion*:

- Is the project on schedule?
- What has changed or been accomplished since the last report?
- What new information has been discovered?
- Are we still working on the right problem?
- What obstacles have appeared? How might the obstacles be mitigated?
- What resources and authorization might we need?
- What is the next milestone, and when should it occur?
- If the work is concluded, what are the closure tasks?

Formal progress reports to a project's sponsor will likely be less frequent but more detailed than an internal summary, confirming that progress is being made toward the project goals.

Field reports (or *investigative, test, or trip reports)* are often highly structured because the nature of the reported activity itself is quite predictable. Most writers of field reports rely on a template, whether using a form created by the organization or copying the organization of an earlier document. The final report of a series will likely be more detailed and summarize the entire series.

WTP4 – GMP 3B – Excavation & Construction of Clearwells and Upflow Clarifiers
Monthly Status Report No. 18 – March 2012

COA CIP No. 6683.019
Contractor: MWH Constructors
Scope: Major Excavation for Clearwells, Upflow Clarifiers (UFCs) and Filters; Major Concrete for Clearwells and UFCs.

Start Date: 7/20/10
Original Contract Duration: 714 Days
Approved Time Extensions: 0 Days
Current Contract Completion Date: June, 2012
Percent Complete Time: 87.0%
Percent Complete Dollars: 88.8%

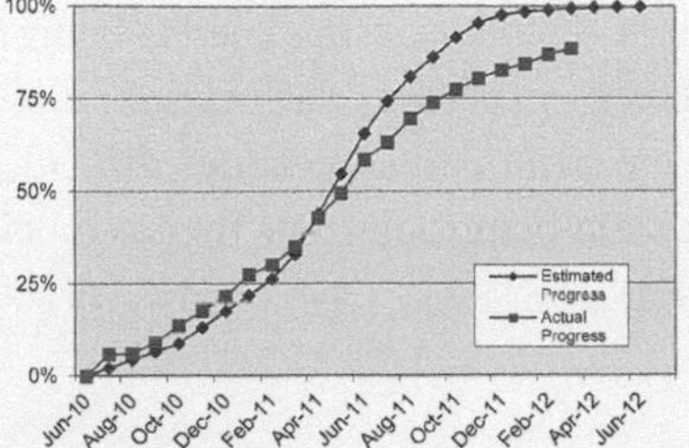

Construction Progress as of 3/31/2012:

S304 Site Excavation and Improvements (Ranger Excavating, LC)

- Excavation for Clearwells, UFCs, and Filters complete.
- Filter area retaining wall and stormwater piping complete.
- Field trailer utilities complete.

S301 UFCs (Austin Engineering Co.)

- All piping beneath UFCs complete.
- UFC 1 and 2 Hydraulic area concrete complete.
- UFC1 and 2 hydraulic leak test complete.
- Gallery wall construction 50% complete.

S302 Clearwells (Laughlin-Thyssen, Inc.)

- Clearwell structural concrete complete.
- CMU velocity walls 95% complete.
- Concrete sealing and finishing underway.

Change Orders:

- CO#1 and #2 completed for UFC over-excavation.

Requests for Information:

- To date, 80 RFIs have been received.

Submittals:

- To date, 183 Submittals and 82 Resubmittals have been received.

Upcoming Milestones:

- Clearwell hydraulic leak test May.

Delivery of UFC Center Column.

City of Austin water treatment progress report for excavation and construction of clearwells and upflow clarifiers (UFC). City of Austin Water Treatment Plant 4 (2012). Courtesy City of Austin, Texas.

The context of this progress report includes the project administrative designations, timeline, and scope.

While not explicitly labeled as "results & discussion," the majority of this progress report is dedicated to bulleted lists of milestones (both completed and ongoing) as well as expected next steps. Most of the information pertains to the questions

Are you on schedule?
Are you on budget?
What has been accomplished?
What's coming next?
What obstacles have appeared?

The photograph in the lower right corner can be considered work documentation—supporting evidence for the claims of UFC progress.

12.4 Design reports

Design reports written in the Answers First format focus mainly on the chosen design, rather than on the process that produced it. A chronological listing of every step taken during the design process is not needed, because the audience is not going to try to reproduce the design process. Information from the design process should serve the report's purpose: making a case that the chosen design satisfies the original need.

How the sections are specialized

The *context* is established with an overview of the original need statement—What existing need is being satisfied by the new design? In many design processes, the original problem statement is vague, so it is necessary (although sometimes difficult) to clarify the exact nature of the problem being solved or the need being met.

This Section Summary identifies the design requirements for a Mars lander.

Because the design problem is a specific part of a (massively) large enterprise, no effort is made to establish its importance in general. Instead, the stakes for specific design choices are investigated. The size of the lander influences many other crucial parameters—perhaps most importantly, the size of the crew.

4.1 Section Summary

To enable delivery of payloads to the Martian surface of the size and mass suitable to support human-scale missions, viable approaches to the critical EDL phases of flight must be devised and investigated. Prior to the Human Architecture Team (HAT) studies described here, a significant amount of analysis on EDL approaches was performed during the EDL Systems Analysis (EDL-SA) study.[3] From the results and recommendations of that study, a number of areas for further work were identified. These included: detailed lander concept definition, lander/aero-assist system integration and packaging, and transition from aerodynamic to powered flight during descent. Two particular aero-assist architectures highlighted for further investigation were the mid-Lift to Drag (L/D) rigid aeroshell and the Hypersonic Inflatable Aerodynamic Decelerator (HIAD).

The definition of detailed lander point designs was undertaken during the HAT Cycle D study phase to provide greater insight into potential integration and packaging issues between the lander and its entry aero-assist system and/or the launch vehicle. Design aspects of the landers were also influenced by the ability to carry specific payloads and by surface operations. The point designs allowed determination of mass properties necessary to assess the suitability of certain aero-assist/lander combinations with regard to aerodynamic stability and control during entry. The mass estimates produced also provided verification data for higher-level, parametric sizing models.

Prior to defining detailed lander concepts, a series of high-level sizing investigations was performed in HAT Cycle C, bounding the size and mass of the Mars lander. These included trade studies and sensitivities on the number of crew, Mars rendezvous orbit, payload delivery capacity, sample return payload capacity, number of lander stages, extent of in situ resource utilization (ISRU) for propellant, propulsion types, and propulsion performance parameters. The investigations initially made use of first-order sizing techniques to examine a wide trade space. This was followed by the use of a more detailed parametric modeling tool (ENVISION) to focus on a narrower trade space and also to set a design point-of-departure for further detailed study.

The Mars lander concepts assessed in the studies were each defined to consist of a Mars Ascent Vehicle (MAV) and a Mars Descent Module (MDM). The MAV design, along with its requirements, has a large influence on the overall lander mass and size. The high-level sizing investigations examined a number of important sizing factors for the MAV, the most important ones being: the number of vehicle phases/stages (one or two), the parking/rendezvous orbit (low or high), and the choice of propulsion system and propellants – Nitrogen Tetroxide/Monomethyl Hydrazine (NTO/MMH), liquid oxygen/liquid methane (LOX/LCH_4), or liquid oxygen/liquid hydrogen (LOX/LH_2).

The scale of the project determines the way that the problem is analyzed and presented. For a mission sending humans to Mars, each NASA report contains many sections, each defining its own problem. Samples in this module have been excerpted from the 596-page "Addendum #2," reflecting 596 pages of updates since version 5 of the major report. Courtesy NASA ("Human exploration of Mars" 2014).

Once the general need has been established, the report states the specific goals of the design from the perspective of the engineering firm. Such statements, often called a "Statement of Work" or a "Scope of Work," begin to establish limits of the firm's design responsibilities.

A design report moves from *goals* into *results* by presenting the criteria that develop out of the need statement—the unambiguous and measurable goals used to determine whether a design is satisfactory.

Horizontal Landers: Instead of vertically-stacked cargo elements, horizontal landers utilize a "side-by-side" arrangement where the various MDM components and cargo elements are spread out horizontally and potentially lower to the ground. Horizontal landers have a rectangular landing gear footprint, and the length of the landing gear is determined by the rectangle's shortest dimension, which is the driver for stability at touchdown. Horizontal landers may offer advantages related to flexible packaging, easy offloading/surface access, and a lower center of gravity at landing leading to reduced landing gear mass. Disadvantages include a greater total length and less efficient use of available launch volume. By nature, the primary load path directions for launch and landing are in different directions, which can result in structural inefficiencies. Because of their greater length, horizontal configurations are likely to have lower lateral stiffness when mounted within the launch vehicle shroud. In some cases this may lead to unacceptably low modal frequencies that are difficult to mitigate. This issue and the limited options for increasing stiffness are discussed in Section 4.6.5.4.

For this work, a decision was made to consider at least one horizontal and one vertical lander configuration in detail. Aside from a desire to explore the advantages and disadvantages just mentioned, this decision was motivated by the fact that the optimum lander configuration for compatibility with HIAD or rigid mid L/D aeroshell systems is currently unknown. Adopting a two-configuration approach guaranteed the generation of one detailed layout with analysis results for each configuration. Design issues and details as well as preliminary mass properties would thus be available regardless of which lander configuration is eventually deemed optimal.

4.6.2 Assumptions

Firm requirements for a human mission to Mars are unknown. The following assumptions were derived for this work, or are similar to assumptions used for previous Altair (NASA's former Constellation Program lunar lander) analyses. They are adequate for conceptual design and analysis but should not be considered formal mission requirements.

The following assumptions apply to all lander vehicle components (MDM, MAV, and payload/cargo).

Table 4-18 General Configuration and Analysis Assumptions

System/Configuration	Assumption	Notes
Lander	73,000 kg	Lander (including cargo) + descent propellant
Aeroassist System*	10,000 kg	Does not include rigid aeroshell mass for horizontal lander
Launch	83,000 kg	Total mass supported by lander primary structure
Landed	53,000 kg	No LOX in MAV tanks (MAV LOX generated by ISRU)
Loads	**Assumption**	**Notes**
Axial Launch	5g	Similar to previous Ares V/Altair assumed axial load
Lateral Launch	2g	clocked 0, 45, and 90 degrees
Landing	1g	Sizing from tool developed for Altair lunar lander
Entry	3g	Not analyzed during this phase (future work)
Component	**Assumption**	**Notes**
Shell Structure (except tanks)	IM7 977-2 Gr-Ep	Sandwich, composite facesheets with AL honeycomb core
Propellant Tanks	Aluminum-Lithium	Composite tanks possible, but benefit unclear at this size
Struts	IM7 977-2 Gr-Ep	Similar to Altair composite strut assumptions
Fittings and Attachments	Aluminum-Lithium	Al, Al-Li, or Titanium possible for future trades

The design process for the Mars lander is constrained by assumptions about the mass of the lander, the forces to which it will be subjected, and the materials to be used in constructing its major components. These are thus made prominent in the design report. Courtesy NASA ("Human exploration of Mars" 2014).

Instead of generating a single design concept, the team's task in this case is to generate several different configurations. Before stating design assumptions, the need for this approach is explained.

The team here explains the difficulties involved in obtaining certainty on specifications—"Firm requirements for a human mission to Mars are unknown"—but explains how design specifications were generated.

The report then presents the chosen design—or, sometimes, several options—in sufficient detail for the audience to understand how it satisfies the need. Deeper discussion of this design might focus on possible future improvements, other potential applications, or an honest assessment of strengths and weaknesses.

Even in a design report for such a speculative scenario as a mission to Mars, selected designs must be tested to some degree—CAD models are presented to show packaging implications of both designs.

When discussing the selected design, the report's tone becomes unusually hesitant, with all claims being heavily qualified, emphasizing the very preliminary character of the work.

Figure 4-38 Final horizontal (H3) lander configuration (as-analyzed CAD model).

To analyze the H3 configuration, the lander was attached to a 10-m-diameter aeroshell structure assumed to also serve as the launch vehicle shroud as shown. This assumption results in a larger available volume for lander packaging, simplifies interfaces, and eliminates the need for two structural systems. It is admittedly a somewhat ambitious assumption, because the optimum shapes for the shroud and aeroshell will not be the same. Ultimately separate shroud and aeroshell structures may be necessary, resulting in larger system mass estimates than those obtained here.

As shown, the lander is attached to the aeroshell/shroud at six lateral attachment points along its length. The base of the aeroshell is attached to a cryogenic propulsion stage (CPS), or some other upper stage of slightly smaller diameter via a conical section as shown. For analysis purposes the end of the conical section is fixed in translation. The lander is attached to the aeroshell conical section via a partial-cone adapter and series of adapter struts. The partial cone and adapter struts carry axial launch loads, while the lateral attachment points increase lateral stiffness and frequency. Other attachment schemes were considered, including elimination of the lateral attachments resulting in a purely cantilevered configuration. This cantilevered configuration was analyzed but was rejected because of extremely low first-mode natural frequencies (see Section 4.6.5.4) that could not be raised significantly by adding mass to the adapter struts or cone. Alternate aeroshell configurations and integration scenarios are suggested topics

The presentation of the designers' decisions includes both the claims made for their major designs and some explanations of why other alternatives were rejected. Courtesy NASA ("Human exploration of Mars" 2014).

Work documentation for a design report includes details of research into the needs of users and stakeholders, the results of research on prior art and technical literature, and alternative concepts generated. Thorough documentation of performance modeling or prototype testing might also be presented here, along with detailed drawings and specifications of the chosen design.

Some design reports are produced for *internal recordkeeping*, with extensive work documentation so that the design choices can be understood well after the chosen design is complete. Other design reports inform the *sponsor* who asked for the design, focusing on the need statement and the chosen design details. Work documentation retains the detailed drawings and specifications but often includes less information about the design process.

12.5 Feasibility studies

A feasibility study analyzes whether a course of action is worth trying. All efforts require resources—not only money, but also people's time and attention—and a feasibility study is focused on determining whether the expected outcomes are worth the resources required. For example, a feasibility study would be useful when considering replacing a component in an assembly with a lower-cost version from a different supplier. Possible reduction in performance and ease of production should be weighed against the potential savings.

Asteroid Retrieval Feasibility Study

This feasibility study focuses on a way to identify, capture, and retrieve a small near-Earth asteroid (NEA).

2 April 2012

Prepared for the:

Keck Institute for Space Studies
California Institute of Technology
Jet Propulsion Laboratory
Pasadena, California

While the sponsors for this study are academic rather than industrial, this feasibility study uses an Answers First organization.

The context of the study is stated in the first section, with elaboration taking place in Section II.

The results are presented in Sections III through VI.

The discussion of expected challenges to implementing this plan—mainly technology challenges—occurs in Section VII.

Finally, discussion of next steps occurs in Section VIII.

Table of Contents

Feasibility cover page and table of contents.

How the sections are specialized

The *context* of a feasibility study arises from a need or opportunity being investigated—why the initiative's costs and benefits are being assessed and why the organization needs to understand that initiative's prospects for success.

The report's findings must enable a recommendation as to whether further effort on the initiative is warranted. Qualitative arguments will support the recommendation, weighing the outcomes and resources that are expected based on the quantitative results.

IX. CONCLUSIONS

The two major conclusions from the KISS study are: 1) that it appears feasible to identify, capture and return an entire ~7-m diameter, ~500,000-kg near-Earth asteroid to a high lunar orbit using technology that is or could be available in this decade, and 2) that such an endeavor may be essential technically and programmatically for the success of both near-term and long-term human exploration beyond low-Earth orbit. One of the key challenges—the discovery and characterization of a sufficiently large number of small asteroids of the right type, size, spin state and orbital characteristics—could be addressed by a low-cost, ground-based observation campaign identified in the study. To be an attractive target for return the asteroid must be a C-type approximately 7 m in diameter, have a synodic period of approximately 10 years, and require a ΔV for return of less than ~200 m/s. Implementation of the observation campaign could enable the discovery of a few thousand small asteroids per year and the characterization of a fraction of these resulting in a likelihood of finding about five good targets per year that meet the criteria for return.

Conclusions of a feasibility study. © 2012 Keck Institute for Space Studies, California Institute of Technology. Reprinted with permission.

The optimistic conclusions presented here summarize the overall initiative detailed earlier, with significant appeal to the potential impact on international cooperation, industrial development, and providing "a new rationale for global achievement and inspiration."

One variant of the genre is the *comparative feasibility study.* Rather than simply comparing a single proposed action to the status quo, comparative feasibility studies weigh multiple alternatives. A structured approach to comparing alternatives, like a decision matrix, would likely be prominently displayed in the analysis and discussion section of such a study, with the recommended "best" alternative discussed in more detail than the others. Such studies may be part of a larger design report or proposal.

Summary

Writing for decision-making audiences in industry

- Industry reports' readers need to make decisions quickly; to support such decisions, your writing must be credible and concise.
- Secondary audiences—such as senior executives, lawyers, regulators, and future project engineers—may need a more substantial understanding of how your findings or recommendations were reached.

Elements of the Answers First format

- The context of your project or problem must be shared to ensure the audience knows both what work was done and why it was needed.

- The results are given quickly, with an emphasis on the primary results and relevant discussion. Address the significance of the results within the overall context.
- The work documentation (content that supports the results and discussion) that the audience might be interested in viewing will appear at the end of a report in the Answers First format, or it may be stored separately in a project archive.

Progress and status reports

- Focus on the recent developments, since the audience already knows the context, prior progress, and goals.

Design reports

- Instead of chronicling the process used, emphasize the need statement and the criteria used to judge the success of the design.
- Present the chosen design in sufficient detail for the audience to understand how the criteria were met and the need was addressed.
- Details of the design might be placed in the work documentation section, along with alternative concepts, prior art findings, and performance analysis work.

Feasibility studies

- The feasibility of a course of action should be clearly stated so authorizers can decide whether further effort on the initiative is warranted.

Corresponding

13

Some engineers may see "written" correspondence as less frequent than it once was. It's true that letters and memos are likely to be dispatched and received electronically rather than in print—and more routine email is likely to become highly informal. Nevertheless, these documents' essential functions for engineers and their clients and employers are largely unchanged: They provide channels for the transfer of information and verifiable records that can be consulted later. (This function as a documentary record makes letters, memos, and email important to follow up conversations in person or on the phone.) As always, the appropriate type of correspondence depends on audience and purpose:

Letters	Communicate with an external audience such as a client or regulatory agency
Memos	Record a query, policy, decision, or finding within an organization
Email	Provide updates, seek information, and keep multiple parties informed both within and outside an organization
Phone calls	Communicate information that is time-sensitive or distressing or that might be misunderstood in an email
Social media	Collaborate and network with co-workers, communicate with clients, suppliers, and media during an emergency, and provide information during hiring

Objectives

- To select the appropriate correspondence—letter, memo, email, phone call—for the situation and to follow the conventions of the genre
- To maintain a professional tone in formal and informal correspondence, remaining mindful of audience concerns and priorities and likely reactions to one's own message
- To frame news so that its meaning is quickly understood and that the recipient is receptive to its message

13.1 Maintaining a professional tone in correspondence

Although individuals sign or initial their correspondence, all professional correspondence reflects not just on the individual author but the organization that he or she represents. Therefore, it is especially important to maintain a professional tone in this correspondence. While "professionalism" isn't always the easiest quality to define, a good rule of thumb is that your correspondence shouldn't reflect any individual attitude or disposition if generalized to your employer. (This doesn't mean that you can't ever question technical or business decisions—rather, that the tone of your writing should reflect that you're doing so from a position of thoughtful engineering judgment.)

In general, to maintain a professional tone, you should:

- Construct your message thoughtfully.
- Use complete sentences and appropriate punctuation.
- Avoid slang and derisive comments.
- Err on the side of formality: Use courteous greetings and closings, avoid contractions, etc.

Some companies may have other specific rules to which they expect their employees to adhere when communicating internally and externally.

Some variation in formality is to be expected: Daily email within your team won't perfectly match letters to a client in their tone. Remember, though, that all written correspondence carries the weight of potential liability. You're likely aware that written correspondence to a client or customer could be considered legally binding—but litigation might mean that internal email is searched and released as well.

13.2 Letters

Letters communicate with recipients outside of your own organization, whether clients or other external parties. Letters are formal documents that create an official record of exchanges. Many letters fall into one or more categories:

Cover letters	Accompany longer documents, such as proposals and reports, and provide major findings
Requests and inquiries	Ask for information, goods, or services
Acknowledgment letters	Confirm receipt of information or items from an external party, helping to maintain open communication between the parties
Adjustment letters	Respond to complaints from other parties and make amends
Justification letters	Provide an explanation in response to a question or complaint or in support of a request
Reference letters	Support an application by supplying a testimonial, detailing the strengths of an individual applying for a job or of an organization applying for a grant or contract

Conventions of professional letters

Generally, letters are only one or two pages in length, although they may include enclosures (i.e., other documents). You may encounter several other styles for professional letters, but it is always safest and easiest to use the most formal *full block* letter format:

- Left-align each line, with no indenting.
- Single-space within paragraphs; double-space between them.

- Before the salutation, include at least the date and the recipient's address. The sender's address may also be included (or may be printed on company letterhead).
- Precede your signature with a formal closing (such as "Sincerely") followed by a comma.

Full block letter format includes the date and the recipient's address in the upper-left-hand corner of the page. (The sender's address is typically included in the company's letterhead.)

Double-spacing separates paragraphs and other elements of the page (such as address blocks).

Letters typically address recipients with their formal titles, even if face-to-face conversation would be on a first-name basis.

Business correspondence should be concise, but making it too abrupt will compromise its professional tone. An opening should take particular care to express courtesy.

September 12, 2012

Mr. Humphrey Kyle
DuPage County Historical Preservation Agency
370 W. Frontier Rd.
Bolingbrook, IL 60440

Re: Draft First Amended Memorandum of Agreement
Washington St. Bridge No. 37
Des. No. 7389575, DHPA No. 9386

Dear Mr. Kyle:

Thank you for taking the time to review the draft of the First Amended Memorandum of Agreement for the Washington St. Bridge No. 37 Replacement Project previously sent to you on July 16, 2012. Comments from the Deputy Historic Preservation Officer were received and incorporated. An updated draft is enclosed for your review and comment. Please provide any remarks by October 15, 2012. If we do not receive a response by this date, it will be assumed that your agency feels the draft is adequate.

Once all draft comments have been incorporated and agreed to by the signatories, a final document will be distributed for signature.

If you have any questions, please call me at 812-555-8653 or contact me via e-mail at ardizzone@strandassociates.com.

Sincerely,

Marcela Ardizzone
STRAND ASSOCIATES, INC.®

This cover letter accompanying the draft document requests feedback and thanks the recipient for commenting on a previous draft. © 2015 Strand Associates, Inc. Reprinted with permission.

In general, letters and other forms of professional correspondence arrive quickly at a statement of the main point. Stating your message up front helps your reader know how to read the information that follows it. In particular, if you are addressing a technical problem and need to make sure that the recipient takes action, burying the problem several paragraphs into the document creates significant risks—most importantly, that a reader will never make it far enough in the document to identify the necessary action or even to recognize the need.

Sometimes, it is appropriate to preface the central message with some remarks to prepare your audience for what they'll be reading. For example, you are unable or unwilling to agree to a change in the conditions of a contract, you would likely want to introduce the factors that make those changes hard to accomplish. Regardless of the underlying issues, you risk a more negative reaction if you place the refusal first: The reader may become angry, seeing your firm as unreasonable and approaching the rest of the message with less willingness to negotiate and work out a mutually agreeable resolution.

Re: Influent Submersible Pumps

Dear Ms. Stenzel,

As you are aware, the large influent pumps installed at the Albany Wastewater Reclamation Facility (WRF) have been a source of concern. They are unable to reliably meet their specified capacity because the current exceeds the limits of the installed variable frequency drives.

The letter opens by stating that the pumps cannot meet their specified capacity. If informing the recipient of this fact for the first time, the writer would be slightly less direct, with a sentence or two to prepare him or her to receive bad news.

The specifications for the influent pumps state the pumps will be installed in the influent pump station wet well, will be used to lift raw unscreened wastewater in the influent pump station wet well to the screen influent channels, and shall be capable of handling solids and long stringy materials found in raw unscreened wastewater. The large pumps were specified to be supplied with 60 horsepower (hp) motors. The specifications also state that the specified motor hp is the minimum requirement and that the motor shall be large enough to not be overloaded at any point on the design curve. The specifications further state the equipment manufacturer shall coordinate with the variable speed drive supplier at the time of equipment start up to address minimum speed requirements to protect both motor and equipment and to meet specified design and performance requirements.

The first body paragraph states the specifications for the pumps and acknowledges that they do not meet minimum horsepower requirements.

Shop drawings were submitted (February 24, 2011) showing the large pumps were equipped with 60 hp motors that had a full load current

The second body paragraph details the shop drawings and the company's determination that the pumps should meet design performance requirements based on those drawings. However, in usage the pumps draw more amps than what the shop drawings show as the full load.

rating of 71.6 amps. The curves submitted with the shop drawings showed the pumps meeting the design capacity of 4,800 gallons per minute (gpm) at 37 feet of head. The curve also showed a shaft power (not motor power) of 55.7 hp at this design point. Test curves showing actual pump performance (submitted March 29, 2012, test dates July 25 and 26, 2011) showed the pumps meeting the design performance requirements. Each pump test power curve had points exceeding 60 hp. An additional factory test was performed on pump No. 5 at the WRF facility in Amarillo, Texas. This clean water test showed a power input of 48.8 kW (65.4 hp) and a current draw of 73 amps at 4,800 gpm and 39.3 feet total dynamic head. This test was performed at full speed using a full voltage, nonreversing starter. Neither the electrical feed used for this test, nor the fluid being pumped, are the same as those in practice at the Albany WRF.

After the pumps were put in service, it became apparent the current draw of the pumps was considerably more than the 71.6 amps shown as the full load current on the submitted shop drawings. Current as high as 84 amps has been observed during operation of the pumps. The capacity of the drives associated with the pumps is 77 amps, although they can run at a higher current for short periods.

Evidence shows that the motors do not match the shop drawings.

PWG has stated that "there is considerable reserve available in the motors" and that they could be renameplated to 65 hp. Correspondence from PWG has revealed that the motors are actually 80 hp motors. This information indicates the motors are not as indicated on the submitted shop drawings. Because of the electrical characteristics of AC induction motors, a derated motor, like the 80 hp motor provided for this application, will often operate at a reduced power factor at full load resulting in a higher current at full load than a motor that is not derated. This phenomenon could be contributing to the higher full load amps in this installation.

Prior to stating that the redesigned pumps should be supplied without cost, the letter provides an argument as to how they fell short.

To allow the pumps to operate without limiting their speed and capacity, the Township asks that the installed VFDs be replaced with larger drives having a current limit of 96 amps. The existing drives are Allen Bradley PowerFlex 700. To fit in the existing cabinets, the new Allen Bradley drives would have to be the PowerFlex 753 model. Because the motors provided with the pumps were not as specified, or as submitted, and because the motors cannot be operated in the system as designed, these changes should be made without any cost to the Village. As an alternative, the motors, or the pumps and motors, can be replaced with units that will adhere to the project specifications. Please provide a response within two weeks indicating the course of action to be taken.

Sincerely,

Marcela Ardizzone
STRAND ASSOCIATES, INC.®

In this letter, the company asserts that redesigned pumps should be provided without cost but first details the original pump's failure.

13.3 Memos

Within an organization, formal correspondence takes place not by letter but by memorandum—the Latin term indicating that something must be recalled or remembered—abbreviated as "memo." Memos often record policies, decisions, or findings as part of an organization's collective memory. Technical memos can be especially important to establish a record enabling future engineers to understand why decisions were made in a particular way.

The basic format and style of memos have been generalized to cover most formal correspondence, whether distributed via email or hard copy. Of course, email was designed to mimic the header format of the memo, starting the correspondence by naming the sender, recipient, date, and subject.

Achieving conciseness

Memo style privileges conciseness above all. Memos often concern a central policy, finding, or proposal that should be immediately apparent; moreover, memos are often written to busy executives who receive a large volume of correspondence, leading them to skim documents quickly for relevant information before moving on to the next item. For this reason, an organization or department will sometimes issue a guideline to limit memos to a page or two, although such rules are not universal.

"Concise" does not mean "short." Reducing word count is desirable if it can be done without removing important content—but it should never lead you to skimp on the substance and rigor of your technical analysis or policy thinking. If the memo cannot accommodate that thinking, follow these steps:

- Organize the memo itself around delivering and explaining your central message.
- Determine what data, findings, or calculations need to be included in attachments so that readers can understand and explore that message. Ideally, these can be managed in a small handful of pages.
- If you have much more data or technical material than that, you may inflict on your reader what engineering managers call "data dump," meaning that a memo may not suffice to explain it. Consider a slightly longer report format.

Organizing for clarity and ease of reading

Three techniques can help to draw a reader's attention quickly to the key points of a memo:

- *Forecasting* directs the memo's content and the reader's attention, stating the main argument up front, as clearly as possible. This is usually accomplished in a brief paragraph of only a few sentences, preceding the memo's first heading.
- *The "talking head" style of headings* states each section's main point in the heading—not necessarily in a complete sentence, but in enough detail to express a fairly complete idea. When this is done thoroughly, readers can grasp the memo's basic argument by reading the opening paragraph and headings alone.
- *Lists and tables* deliver some kinds of information or reasoning, especially quantitative data such as costs or technical specifications, more efficiently than paragraphs. (Including graphs or illustrations in a memo is unconventional but potentially effective.)

TO: Project File
FROM: Paige Hardy
DATE: September 21, 2012
RE: Centract AGT Mixer Evaluation

The memo heading provides key information for filing and retrieving a company's internal record.

This memo summarizes the results of a phone-solicited questionnaire to wastewater treatment plants (WWTP) currently using Centract AGM Mixers.

The opening sentence succinctly states the memo's purpose.

Evaluation

Reference WWTPs in the United States and Canada were provided by Centract. Operational staff from the WWTPs were contacted between September 6 and 14, 2012. Attached is a blank questionnaire that was used when contacting the respective WWTP staff. Table 1 summarizes the feedback from eight WWTPs.

The evaluation first provides the important context—the number of treatment plants surveyed, as well as the dates of the survey and the questions asked.

Five of the eight WWTPs indicated a favorable recommendation for the AGT mixer. Of the three that did not have overall favorable recommendations, one said it was too early to tell (Vancouver–installed in 2011), another recommended the performance but not mechanical aspects of the mixer (Tacoma–installed in 2011/2012), and one person said he would not recommend the mixer (Boise–installed in 2008 by Enersave). The negative responses were mainly based around mechanical issues, although most were covered under warranty. Three facilities indicated that Centract had a slow response time. The start-up dates for these mixers range from approximately 2006 to 2012.

The memo sums up the evaluation—identifying the number of favorable and unfavorable recommendations—before detailing problems.

Each paragraph then addresses a single problem.

Generally, there were no foaming issues, problems mixing during drawdown, or fluctuation in gas pressure associated with the use of the AGT mixer. Two WWTPs experienced foaming before installing the AGT mixer; one was believed to be caused by operating with a Stology mixer, and the other was believed to be caused by overfeeding one digester with high-strength (food) waste. One WWTP had a foaming issue after the AGT mixer was installed, but it was traced to the aeration basin and was fixed when the aeration basin was dosed with chlorine.

This memo is relatively dense—it's to be read closely rather than skimmed quickly—but relatively short paragraphs still help readers along.

Seven of the eight WWTPs were a retrofit, and modifications required for the cover and gas piping varied among installations. Only one

The conclusion of the memo's first section, not excerpted here, briefly summarizes the evaluation.

WWTP indicated there was an issue modifying its existing digester cover. Only three of the WWTPs had drained their tanks since the installation of the AGT mixer, and none indicated an excessive buildup of grit.

Four of the eight WWTPs have had some sort of mechanical issue with the mixer but most were still covered under the warranty. Several people mentioned the oil reservoir was initially too small (since they had to oil the mixer once a day) and had Centract provide a larger reservoir so that they only had to oil once or twice a week. . . .

This memo summarizes feedback from wastewater treatment plants that have used Centract mixers.

13.4 Email

Obviously, email has become our most frequent form of correspondence—absolutely pervasive in professional life—and a major workplace skill set. The volume of email alone presents challenges, and even managing a quick reply to the most urgent items can feel like a victory. Conducting your email correspondence, though, requires a keen awareness of its many different audiences, purposes, and contexts. Think of email as a *medium*, a channel of communication that can be used to deliver many different kinds of messages.

Choosing a level of formality

Only a few highly formal kinds of documents—such as formal contracts or letters offering employment—are routinely expected to be sent and received in print. Other kinds of letters and memos are increasingly handled electronically.

If you are using the body of an email message to send a letter to a client, or to deliver a memo within your company, treat the email as *highly formal*. You might initially regard such formatting details as a waste of time: In fact, though, the highly formal structure and design of such a document is one of the most important ways to make your message stand out as important in a crowded inbox.

Date: **September 2, 2014**
To: **All Faculty and Staff**
From: **Robert Coons, Senior Vice President & Chief Administrative Officer**
Subject: **Annual Certification: Computer Use and Conflict of Interest**

As approved by our Board of Trustees' Audit Committee, Rose-Hulman's Computer Use Policy and Conflict of Interest (COI) Policy require a mandatory annual acknowledgement and certification by Rose-Hulman faculty and staff. We have instituted an online disclosure process for both policies, available on your Banner Web account. Please read the attached Conflict of Interest policy, then access the online survey by doing the following:

1. Log on to Banner Web by clicking here
2. Choose the Agreements/Authorizations/Surveys tab
3. Select the Conflict of Interest Disclosure and the Annual Computer Use Policy-Acknowledgement Form
4. Be sure to click "Submit as Completed" when you are finished

Please read the attached policy and complete both online forms no later than **Wednesday, September 17, 2014.**

If you prefer to use paper disclosure forms rather than the online form, please contact Amy Timberman to receive printable disclosure documents. Printed forms should be submitted to Rob Coons by the above deadline.

Background: By their Charter, the Audit Committee of the RHIT Board of Trustees is responsible for assessment of the adequacy of internal accounting, compliance, and controls systems. This responsibility includes examination of policies/controls which govern the use of computers within each employee's control, protect the Institute from copyright infringement actions, and protect the Institute from conflict of interest issues. The computer use form and conflict of interest disclosure form streamline the audit process by sharing the compliance responsibility among all employees, rather than employing a team of auditors to engage in a costly and intrusive process of physically/digitally verifying the contents of each computer on campus, or interviewing employees regarding conflict of interest issues. [...]

Thank you in advance for your cooperation.

Rob

Robert A Coons
Senior Vice President and Chief Administrative Officer
PHONE: 812-555-8007
FAX: 812-555-3935
CELL: 812-555-3252

ROSE-HULMAN
INSTITUTE OF TECHNOLOGY
CM 20
5500 WABASH AVENUE
TERRE HAUTE, IN 47803-3920
www.rose-hulman.edu

In this case, the memo header information appears in the body of the email. This may seem redundant, but the formality tells recipients that senior administrators see the content as highly important.

Even though the memo is sent by a senior vice-president, its message opens by invoking an even higher authority—policies approved at the Board of Trustees level.

As in a print memo, document design features are used to make action items easy to identify and complete. In electronic form, the principle is extended with hyperlinks and colored text for emphasis.

The action items are presented above, recognizing that relatively few recipients will read this far. It's a courtesy, though, to explain in more detail why the action items are necessary.

Highly formal email—in which email is essentially the medium for a letter or memo as it would be sent in print—is often used for formal business procedures such as this annual audit for an organization's computer use and conflict-of-interest disclosure policies. Reprinted with permission.

At the other end of the spectrum, personal email is sometimes written as if the writer were sending text messages, with absent punctuation and even occasional "textspeak" abbreviations. There may be occasions where you're merely using email as a convenient substitute for face-to-face conversation or texting, and the content is trivial; in that case, it may be acceptable to compose email messages as *workplace chat*.

Many engineering professionals, however, err toward such excessively informal workplace email. As you write, consider:

- If the information in the email conversation had to be forwarded to someone, would my colleague or supervisor be able to do so without having to rewrite the content of the email?
- Could the email be made public or entered into a legal proceeding without embarrassment at its tone or style?

In most jobs, everyday workplace email seldom falls into the "highly formal" range of memos or letters; the trick to sending professionally credible email is to clearly differentiate it from mere chat, using relatively formal features even in brief notes:

- The message should be composed in complete sentences, supplemented by appropriate formats such as bulleted lists where needed.
- Include a greeting, a closing, and signature. (After two or three messages within an email thread, these are often dropped.)
- Subject lines should be specific so that they can be used for future reference and searching.

These guidelines are highly flexible: They allow your tone to remain casual and pleasant, where appropriate, while remaining professional.

From: Mullans, Alexander J
Sent: Wednesday, September 05, 2012 3:55 PM
To: JD Horowitz
Cc: Mohan, Sriram; Hopkins, Andrew J; Jacoby, Alexander M; Kennedy, Luke E
Subject: [SitAware] Access to source and documents

All members of the team and the professor are copied on the email.

A bracketed tag lets the recipient see which project an email addresses, but still allows for a more precise subject line.

Hi JD,

Since we spoke on the phone, we have determined which software we will be using to manage and document your project:

- Microsoft Project for scheduling
- GitHub private repository for source control and issue tracking
- Sky Drive for document storage and collaboration

To faciliate your access to the project, can you please send me the following:

- A GitHub username, so I can add you as a collaborator on the repository
- A Microsoft Account email address, so I can provide you acccess to the SkyDrive shared folder

Thanks,
Alex

Referring to a past conversation quickly places the message in context.

Bulleted lists make the email easy to use as a reference, perhaps during a team discussion.

The greeting and closing reflect the working relationship between the sender and recipient: Here, "Hi" and "Thanks!" are more appropriate than "Dear" and "Sincerely."

In this email to a client, a student team provides an update and requests information. © 2012 Alex Mullans. Reprinted with permission.

Recognizing when not to use email

Consider whether email is the most appropriate medium for your purpose. A conversation in person or on the phone can be faster than email, especially if an extended email thread is likely to ensue. You may also want to select a more appropriate medium when:

- Reporting an engineering problem that needs to be addressed. Most organizations have other channels for reporting problems to provide oversight/checks, whereas an email might be missed.
- The information is time-sensitive. Instead of depending solely on email, follow up by phone.
- The content or tone may be misunderstood. While its extensive use has led to a general education of users (e.g., that all caps is equivalent to shouting), tone is still frequently misunderstood. When you are concerned that might be the result, it is likely better to make a phone call or speak face to face, allowing more nuanced and accurate understanding of tone.

- The information is distressing. For example, it is insensitive for a company to notify employees of a layoff via email. Similarly, criticism of a colleague's work on a project might be received more gracefully if delivered in person.

13.5 Phone calls

The telephone was once the primary tool for requesting or providing information quickly, a role now assumed by email. However, situations occur in which a phone call is a superior or even a required method of correspondence. A telephone call may work best when:

- The information is time-sensitive
- The content or tone of an email might be misunderstood
- The information is distressing

In cases like these, voice communication gives better confirmation that your message has been understood and better ability to adjust your delivery quickly to meet audience needs.

Choosing a level of formality

Formality in telephone conversations is guided by the social conventions of face-to-face conversations rather than the conventions of written correspondence. For example, if you would use a formal title such as Dr. or Ms. in a face-to-face conversation, use it in a phone conversation. Similarly, use first names in a phone conversation only if you would use them in a face-to-face conversation.

The level of formality in conversation between engineers and their correspondents varies by firm and geographic location, so be alert to cues indicating you may have violated your listener's social conventions.

Keeping the conversation professional

Engineers should be mindful of their professional and ethical obligations to their employer and to their correspondent when conversing by phone (or in person).

First, be wary of delivering orders or reaching agreements verbally, since doing so leaves no written record. To ensure correct understanding and to protect your firm against liability:

- Avoid verbal orders.
- Document telephone conversations in written, follow-up memos or emails.

Second, consider the value of your recipient's time. Accepting your telephone call has an opportunity cost for the recipient—time he or she could spend on something else. Estimate the duration of the call first and ask if your call recipient can give you that time. If not, agree to a later callback time; if yes, stick to your estimate. Your correspondents will appreciate your thoughtfulness.

13.6 Social media

Business and government policies on the professional use of social media are still emerging, though some trends can be identified. A 2013–14 survey of social media in the workplace around the world (Proskauer, 2014) reports that:

- 90% use social media for business purposes.
- 80% have social media policies.
- 43% permit all employees to access social media.
- 36% actively block access to social media.
- 17% have provisions protecting them against misuse of social media by ex-employees.

While social media will obviously have their own characteristic formats, the underlying assumption for email holds true when using social media for business: Write as if representing your organization in public. Public embarrassment emerges rapidly from a tweet, Facebook post, or Instagram picture with a hint of impropriety or insensitivity, however unintentional.

Businesses also use social media during hiring (although it is illegal to use some information [a candidate's race, gender, national origin, religion, etc.] to make employment decisions). Engineers should be aware of their employer's social media policies and of their rights under equal employment opportunity laws. When using social media professionally, engineers are bound by their ethical obligations to act as faithful agents of their employers, to avoid deceptive acts, and to issue public statements only in an objective and truthful manner.

Summary

Maintaining a professional tone in correspondence

- In general, to maintain a professional tone, you should follow the expected format and conventions of the genre (letter, memo, email), construct your message thoughtfully, use full sentences and appropriate punctuation, and avoid slang and derisive comments.
- Written correspondence also carries the weight of a potential liability. Any written correspondence to a client or customer could be considered legally binding. In addition, all of a company's internal documents—memos and email—can be searched and released in the event of litigation.

Letters

- Letters communicate with recipients outside of your own organization, whether clients or other external parties.
- In general, stating your message upfront helps your reader know how to read the information that follows it.
- Sometimes, it is appropriate to prepare your audience in the opening of your correspondence for the message that follows.

Memos

- Memos often record policies, decisions, or findings as part of a "permanent record," serving as an organization's collective memory.
- Memo style privileges conciseness, but reducing a document's word count achieves conciseness only if it can be accomplished without removing important content.
- To draw a reader's attention quickly to the key points of a memo, forecast the main content of the memo, use "talking head" detailed headings, and include lists and tables when conveying quantitative data.

Email

- *Highly formal* email sends a fully formal letter or memo in electronic form, preserving all of its conventions and formatting.
- Everyday workplace email should include a greeting, a closing, and signature (at least on the first exchange of emails) and a specific subject line.
- Avoid the temptation to let your email lapse frequently into mere *workplace chat* more similar to text messaging.
- Professional email often needs to be forwarded. Provide enough context and the appropriate level of formality so that a colleague or supervisor can do so without rewriting.
- Avoid using email when reporting an engineering problem that needs to be addressed, when the information is time-sensitive or distressing, and when the content or tone of the email may be misunderstood.

Phone calls

- Phone calls are often used when an email would be inappropriate—when information is time-sensitive, when an email might be misunderstood, or when the information is distressing.

- The level of formality of a phone conversation is guided by the social conventions of face-to-face conversations.
- To meet your professional obligations, avoid verbal orders and be considerate of your call recipient's time.

Social media

- Write professionally on social media as if representing your organization in public.
- When using social media professionally, engineers are bound by their ethical obligations to act as faithful agents of their employers, to avoid deceptive acts, and to issue public statements only in an objective and truthful manner.

Proposing

14

Much of an engineer's professional life is dedicated to translating ideas into material reality. A design sketch or a circuit diagram, for instance, might yield first a prototype and then a manufactured system or device. Most of the time, we think of this as a technical process—but if we want to make things happen in the real world, we must first communicate technical concepts, often in the form of proposals.

Engineering proposals may target an external audience. If you are like many engineers in industry, you will write proposals to persuade a potential client or customer to hire your firm to perform work. Alternatively, you might write a proposal for readers inside of your own organization, arguing that a new practice should be adopted or that an existing process should be modified. (Such a proposal might concern a technical process or system but could just as easily focus on other institutional matters, from the organization of departments to employee benefits.)

Any person who has authority to make decisions for or against your proposal is considered an "authorizer." Whether internally or externally focused, authorizers are the primary audience for proposals. Thus the audience will need to be persuaded of the proposal's merits on several fronts:

- The plan is feasible. A proposed action, practice, or system can be implemented successfully.
- The plan solves a problem. It meets a need or advances a goal that the organization is facing.
- The cost is justified. The plan's costs have been fully considered and are reasonably consistent with the expected positive outcomes.
- The risks have been fully assessed. The proposal provides rationale for the risk assessment and a plan for mitigating the risks.

Objectives

- To write engineering proposals that include all common elements and follow the conventions of the genre
- To write effective external proposals that are guided by the content of the *request for proposals* even when adapting content from previous proposals
- To write effective internal proposals that persuasively argue for a course of action that aligns with the organization's priorities and values

14.1 Common elements of proposing

Sometimes the audience is expecting a proposal and knows what it's about, sometimes not. In either case, these three tasks define the writing of a proposal:

Context	Why the authorizer should consider the work important
Work	What you intend to do, when and where it will be done, who's affected, who's in charge, and what it costs
Evaluation	How the plan's success is measured

You may arrange these components differently for different audiences, but the goal is always the same: to persuade your audience to support, authorize, and fund your proposed work. In addition to these tasks, you must meet all requirements for content, form, time, and place of submission.

Addressing the problem or context for the project

Explain why the proposed work is desirable. Anticipate the authorizer's questions: What is the background story? What are the relevant social, economic, and environmental conditions? Why should the work be done now? Why by us? What is the total cost? What are the expected funding sources? What are the consequences of not doing the work? What are your proposal's advantages and disadvantages compared to alternatives? Who else has done similar work in the past or is doing so at present?

Describing the work to be performed

Describe the work to be done, including methods and materials, schedule, budget, and outcomes. Details should be explicit, unambiguous, and measurable. Explain relevant standards that the work must meet and establish that these standards will be satisfied. If the work includes systematic or procedural changes, describe in detail the new procedures and their expected effects on all stakeholders.

Planning for project evaluation

Include a plan for measuring the success of the project, including precise milestones and outcomes. The measurements should satisfy specific criteria based on research, prior experience, prototyping, or modeling. Make these measurements explicit, detailed, and easy to find—the authorizer will look for them. The evaluation plan must signal your credibility to all readers.

Meeting submission requirements

If a proposal fails to meet stated requirements or does not conform to a specified format, it is likely to be rejected without review. Meet *every requirement* for content, form, time, place, and receipt of your proposal. Don't guess! If a submission requirement is ambiguous, seek clarification from an appropriate authority well before the deadline.

14.2 External proposals and responding to requests for proposals

Requests for proposals

Many engineering projects are initiated and controlled by a client's or customer's *request for proposals* (RFP). An organization that wants to have work done issues an RFP to solicit proposals from engineering firms that want to compete for a contract to perform the work. RFPs are most widespread in civil engineering: Municipal governments issue RFPs to contract firms to design and build roads, wastewater systems, bridges, and other infrastructure. Private sector clients might similarly use RFPs to solicit the design and/or manufacture of any device, part, subsystem, or technical service. An automotive or aerospace company undertaking a large project might solicit proposals for mechanical parts, engine subsystems, circuits, microprocessors, databases, and web interfaces. Practically any engineering discipline could be involved.

Formats vary, but an RFP is likely to include:

- The work to be done
- Its background and context
- Specifications and criteria for successful solutions
- Criteria by which submitted proposals will be evaluated
- Deadlines for proposal submission

The phrase "Request for Proposals" will almost always be prominent.

The program title and description establish the scope of what the sponsors or authorizers expect to fund. Solicited projects may be somewhat open (as in the category of assistive technology); many other RFPs require the bidder to complete much more narrowly determined work.

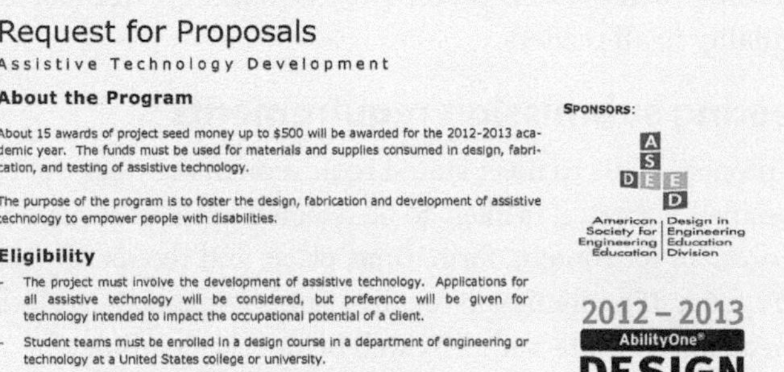

Request for Proposals

Assistive Technology Development

About the Program

About 15 awards of project seed money up to $500 will be awarded for the 2012-2013 academic year. The funds must be used for materials and supplies consumed in design, fabrication, and testing of assistive technology.

The purpose of the program is to foster the design, fabrication and development of assistive technology to empower people with disabilities.

Eligibility

- The project must involve the development of assistive technology. Applications for all assistive technology will be considered, but preference will be given for technology intended to impact the occupational potential of a client.
- Student teams must be enrolled in a design course in a department of engineering or technology at a United States college or university.
- Each institution may submit a maximum of three applications per year.
- Student teams must be supervised by a faculty member.

SPONSORS:

This RFP for student design teams is more modest in length and in project funding than those to which engineering firms in industry write responses. The principles, however, are identical: The RFP establishes the expectations of the sponsors. © 2012 Renee Rogge. Reprinted with permission.

RFPs can be lengthy documents. Some will be shorter up front and include appendices, while others will detail the requirements for a proposal in the main body of the RFP.

Application Requirements

A 3-page proposal must include the following:

1. ***Project objective***—a description of the functional need(s) for the device that is to be developed and for whom (and how) such a device or system will empower the intended person(s).
2. ***Work scope***—a narrative describing the work to be completed by the project team to design, fabricate and test the assistive device or system.
3. ***Schedule***—a description of the project time table.
4. ***Budget***—a description and justification for the requested funds.
5. ***Other funding***—identify other sources of funding, if any.
6. ***Project Team***—identify the students and faculty involved with the project.
7. ***External collaboration***—description of individuals or organizations external to the university, such as non-profit agencies, that will provide advice to the design team.

Submission information

Applications for financial assistance are accepted on a rolling deadline, i.e. applications will be considered for funding until the funds are depleted. Award details are available on the DEED website (http://coen.boisestate.edu/reggert/deed) under 'Assistance Program.'

Questions about the Financial Assistance Program for Assistive Technology Development should be directed to Renee Rogge (rogge@rose-hulman.edu).

Listed requirements will usually make the criteria clear. Here, the awards will go to technical projects judged as having the greatest potential to empower individuals with disabilities.

Authors of an RFP will usually not state explicitly that proposals' content should be organized in the same sequence with the same headings, but they will nevertheless expect submissions to adhere to that structure.

An RFP will normally specify in some detail the protocols that should be followed in writing a response. © 2012 Renee Rogge. Reprinted with permission.

Responding to an RFP

Engineering proposals are somewhat different from reports. Proposals have fewer definite rules governing the arrangement of the content: A potential client may organize its RFP requirements differently from others, depending on the nature of the work to be done, the problem to be solved, and its own organizational procedures and culture. The RFP may ask you for a definite sequence of information (specifying an exact sequence of sections and subsections) or it may define the arrangement of the content very broadly (specifying a few large sections such as a "project narrative").

Therefore, when responding to an RFP, keep in mind three related rules of thumb:

1. An RFP's issuers and reviewers—especially those from government agencies—may **prioritize exact compliance with detailed regulations and instructions,** and so should you. (This attention to precision and detail will typically be apparent in the RFP itself.)

2. When in doubt, **follow the exact sequence of content in the RFP itself.** Doing so will help reviewers locate specific information quickly and efficiently.
3. Much of a proposal's **content can be adapted from a firm's past proposals** or related work. As it appears in many documents, some proposal material will become *boilerplate text,* designed to be inserted into many documents.

Choose boilerplate text carefully and purposefully, especially in detailing the credentials of the firm and its engineers, and in showcasing past projects. For companies that often respond to RFPs, past proposals constitute a portfolio from which you can draw evidence. As with résumés and application letters, though, relying too heavily on unmodified boilerplate may backfire: Emphasize credentials and past project successes that best match the RFP, and tailor them as closely as possible to the client's needs and expectations.

Phase I results put the project in context and address the program criteria explicitly. "Funding is based on the results achieved in Phase I" as well as "the scientific and technical merit" of the Phase II proposal.

The program aims to "Increase private-sector commercialization of innovations derived from Federal research and development funding."

This section begins with the proposal's general technical area and then becomes progressively more specific and precise.

The first sentence of this last paragraph amounts to a kind of "thesis statement" for the proposal.

DOD Sample Phase II Proposal

ADVANCED DEVELOPMENT OF NEXT ACTUATORS FOR HUMAN SENSORY FEEDBACK

TOPIC NUMBER: A92-023A

PROPOSAL TITLE: ADVANCED DEVELOPMENT OF NEXT ACTUATORS FOR HUMAN SENSORY FEEDBACK

FIRM NAME: TECHNICAL RESEARCH ASSOCIATES, INC.

ABSTRACT

The development of actuators with enhanced capabilities is critical to the achievement of sensory feedback systems for intuitive, real time human operations of telerobotic systems. The objective of this research project is to continue development of new actuators using active materials which will enhance the capabilities of dexterous, exoskeletal feedback systems for telerobotic applications. In Phase I, feasibility of novel Terfenol-D driven actuators was demonstrated. The new actuators are efficient, responsive, small and exert relatively high forces. The actuators provide proportional forces and are easily interfaced with digital electronics because of low voltage requirements. Phase II will pursue advanced development of proportional force resist ive brake actuators and active linear actuators. The actuator designs will be optimized, fabricated and integrated into a digitally controllable exoskeleton demonstration test bed.

Commercial, small, light weight, responsive actuators could find application by military and commercial entities for telerobotic applications. The completed devices could also be used as actuators in a number of automated mechanical systems found in home , office and industrial environments.

KEYWORDS:

- Actuators
- Robotics
- Exoskeleton
- Active Materials
- Terfenol-D
- Force Feedback
- Magnetostrictive

(The sections above are on the Cover Sheet)

Identification and Significance of the Problem or Opportunity

The integration of human control and robots has demonstrated advantages effected by combining the cognitive capabilities of the human with the strength and durability of robots. The human is given the ability to operate real time in lethal and hazar dous conditions by remotely controlling the manipulation of the robot. Significant effort has been expended in developing robotic end effectors, sensors and exoskeletons to control the motions of the robotic arms. Human sized robotic hands have been dev eloped which can effect fine manipulations. For intuitive control and fine manipulation, small light weight exoskeletons with force feedback actuator mechanisms are needed to enhance operator awareness.

This Phase II project will adapt Terfenol-D magnetostrictive technologies to actuator mechanisms useful in sensory feed back systems. Magnetostrictive materials are materials which change shape when exposed to magnetic fields. Magnetostrictive mate rials have been available for over a century and came into common use in the 1950's with the use of nickel-based alloys. These alloys were limited to strains in the order of 50 ppm. The discovery of piezoceramics improved this with strains of approximat ely 250 ppm. Discovery of Terfenol-D a relatively new magnetostrictive material improved on the magnitude of the achievable strains as compared to previously available magnetostrictive materials by a factor of 40 times or about 10 times that of piezocera mics such as lead titanate zirconate (PZT). The Terfenol-D material was discovered during research into sonar by the U.S. Navy. Terfenol-D is an alloy of terbium, dysprosium, and iron. Actuators based on Terfenol-D have been shown to transmit large amo unts of energy in a small volume and are highly efficient. The Terfenol-D systems require relatively low intensity magnetic fields, and can operate at low voltages compared to piezoceramics. The new alloy is being applied to many applications which have been previously accomplished by piezoceramics. Piezoceramics are brittle while the Terfenol material demonstrates a high modulus. The material has a rapid response time (5KHz) unlike shape memory alloys such as Nitinol.

Some quantitative claims are made for Terfenol-D, but the major persuasive task is to show its advantages over piezoceramics.

This proposal begins by directly addressing the needs of the funding program. Courtesy US Department of Defense (SBIR/STTR, n.d.).

TASK 3. Linear Motor Development

The object of this task is to develop one or more specific linear motor actuators. The actuator(s) developed will be for the specific application and designed to the specifications developed in Task 1. Task 1 proved the feasibility of using Terfenol d. to construct a compact linear motor actuator. Because of the limited time and budget constraints of Phase I, the prototypes developed were built using available materials and constructed to prototype tolerances. In order to develop a practical device, every aspect of the design will need to be optimized. The linear motor concepts of Phase I will be improved and extended to include digital computer control.

The main purpose of this Task is clearly stated.

Reference to Tasks described earlier shows how the Tasks support each other and logically fit together.

TASK 3.1 ANALYTICAL MODELING

The analytical models for these devices will be improved to allow design alternatives to be simulated short of actual construction. The model will take into account brake surface characteristics, device leverage, expected force range, range of motion, shaft speed, bandwidth, power dissipation pre-stress conditions, tolerances, thermal expansion and structural stability of the device housing. Issues of noise reduction will also be analyzed. The models will be used to optimize the mass specific and volume specific performance of the final device. The project team has the expertise and computers available for this task.

Referencing prior successful work reinforces the firm's capabilities while also showing awareness of relevant work in the area—the context.

The writers explain the simulation's relevant parameters, showing their technical capability.

The writers explain what they're adding to prior work, establishing the project's intellectual merit.

Choosing what concepts to pursue is linked to design specifications developed earlier and the overall goal of the RFP—commercialization of technology that's developed here.

Design specifics will be developed with the desired performance in mind.

TASK 3.2 CONCEPT DEVELOPMENT

Will be devoted to the exploration of variations of the linear motor actuator concepts demonstrated in Phase I. Each promising concept will be modeled analytical and possibly prototyped. Improved designs will be sought. In Phase I we prototyped actuators with hard logic control that allowed for control of motor power and speed. A During Phase II we will extend our scope to look at general computer control of speed position and power. Because these devices can be turned on and off in a short time period (less than a thousandth of a second) it is possible to consider computer digital control where the actuator is turned on and off in computer generated pattern so that the device can made to exhibit a complex linear motion.

With the perfection of both an efficient programmable brake mechanism and an efficient programmable linear motor it is theoretically possible to incorporate both into a hybrid device that could simulate a generalized complex force. (Within its strength and range of motion limits.) Designs that could take advantage of this possibility will be considered during this concept development task.

The linear motor drive lends itself to application as either a direct actuator or as a remote actuator that applies a force through tendons or hydraulic fluids. These variations will also be considered during this task.

[. . .]

TASK 3.3 CONCEPT SELECTION

Will select those actuators that will be perfected and tested. The selection(s) will be made in conformation to the requirements of Task 1 with consideration for the products that have the best prospect for Phase II commercialization.

TASK 3.4 PROTOTYPE DESIGN

Will complete the design and engineering of the specific devices selected in Task 3.3. This will require mechanical design, magnetic circuit design, design of the drive electronics and control software. Each of these tasks will be directed to optimizing the specific performance of the device, i.e. reducing the size, weight, power requirement and increasing the band width and dynamic range of the device(s).

TASK 3.5 PROTOTYPE CONSTRUCTION

Prototype construction will be done in-house with the possible exception of a few high precision parts. The TRA machine shop is currently capable of performing most prototype work. If special machine work is required it will be purchased locally.

This task includes fabrication of the mechanical device, assembly of the drive electronics and development of control software. The robotics applications envisioned at this time will likely have a computer interface between the Master and Slave in order to track and translate commands and sensor feedback so it makes sense to put as much control logic as possible in the same computer.

Assessment of the completed designs is a necessary step and provides a final check against the results.

TASK 3.6 TEST/MODIFY

The actuators will be functionally tested to see if they meet the design specifications. We anticipate that there will be some iteration of tasks 3.6, 3.4 and 3.5 as the prototype is modified to improve its performance.

TASK 3.7 DOCUMENTATION

Test data will be analyzed and documented during this task. Device design documentation will be corrected to include any modifications made as a result the optimization process.

This proposal responds directly to an RFP, and thus has been tailored to directly address the expectations and fundamental concerns expressed therein. Courtesy US Department of Defense (SBIR/STTR, n.d.).

14.3 Internal proposals

When proposing a change or initiative in your own workplace, you may not have the same kind of detailed guidance that you have when responding to an RFP. You will thus likely face more strategic decisions about how to frame and organize your ideas. Internal proposals offer some definite advantages, though. You will more fully understand your own organization's priorities, values, and habits of mind, and those of the individuals and groups whose support is needed to win approval for your ideas.

Internal proposals vary widely in scope and formality. A proposal might well mean pitching a project to the vice-president of your division in a high-stakes meeting. However, you are also making a proposal when you email your project team with an idea about how to make the team's weekly meetings more efficient. Regardless of the level of formality, internal proposals have three key differences from those written externally in response to an RFP:

1. Most internal proposals must persuade your audience that **a problem or opportunity exists and merits action.** An RFP would more or less explicitly acknowledge such problems or opportunities—but if you are writing an internal proposal, you have probably perceived problems with the status quo that others haven't yet recognized.
2. Internal proposals may need to **anticipate more logistical requirements** than those solicited by a client or sponsor. Most RFPs will indicate the general requirements of the project and the resources that it is likely to demand. When you envision a new program or project for your own organization, you may have to start from scratch to determine the resources required—time, money, and personnel—and the likely returns from these investments.
3. A proposal will only succeed if it is logically and technically sound, but that soundness is not sufficient to see it through. **Your organization's business environment, history, and underlying values may be decisive** in determining whether your proposal is accepted and implemented. For example, if you are proposing robotic automation to cut labor costs, you will need to anticipate reactions to the loss of jobs and to a high initial investment. How will decision-makers weigh those factors relative to the eventual cost savings?

Many engineers (and other professionals) lament organizational "politics" that can stand in the way of good ideas. However, every workplace is defined to a large extent by institutional and interpersonal histories, relationships, and recurring conversations. If you want to improve assembly-line productivity, are you implicitly blaming workers? Their supervisors? The systems engineers who designed the manufacturing process? A general model for such constituencies and their objections, with suggested responses, is provided in *Buy-In: Saving Your Good Idea from Being Shot Down*, by John P. Kotter and Lorne A. Whitehead.

Summary

Common elements of proposing

- Before engineering projects are undertaken, they usually begin as a proposal—a plan of action that details the context (why the proposed work is important), the work plan (what exactly will be done), and the evaluation (how the success of the project will be measured).
- Detailed analysis may be needed to establish the existence of a problem or opportunity, as well as the plan's feasibility, cost, and risks.

External proposals and responding to requests for proposals

- An RFP will include general information about the work that needs to be done, the submission requirements, and the criteria by which submitted proposals will be evaluated.
- When a proposal is sent in response to an RFP, satisfying all of the submission requirements is important, even if the arrangement of content is flexible.
- A proposal includes information establishing the competency of the proposing firm—related prior work and biographical details of its personnel.

Internal proposals

- Internal proposals are less likely to be responding to the strict requirements of an RFP.
- Successful internal proposals must persuasively argue that a problem or opportunity exists and merits action.
- Successful internal proposals must anticipate logistical requirements and other resource needs.
- Beyond logical and technical soundness, successful internal proposals must consider the organization's business environment, history, and values.

15 Instructing

Instructions convey the *how* of a process, such as assembling or operating a piece of equipment. Giving instructions requires breaking down a complex task into small, discrete steps that can be easily followed. The process includes anticipating misunderstandings or common problems that the user of your instructions might encounter while performing the task or executing the procedure. To identify and help to solve such problems, instructions are often subjected to usability testing, in which the performance of test users determines their current effectiveness and where they need to be refined.

Objectives

- To break down a complex task into small discrete steps appropriate for the intended audience
- To anticipate possible misinterpretations and to explain these steps to the audience in a way that avoids them
- To design and administer usability tests, investigating whether users can correctly follow instructions to successfully perform a task
- To interpret the resulting user feedback and observations to improve the instructions

15.1 Principles of writing instructions

Common types of instructions that engineers will write are listed below. The principles of writing instructions apply to all of these types:

Assembly instructions	Explain how to connect component parts together in the correct orientation and order to make a new structure or device
Operation instructions	Describe how to perform a specific task using a piece of equipment
Experimental procedures	Communicate how to perform an experiment, similar to operation instructions
Method of analysis	Describe the analysis of data to reach a conclusion, often seen in research reports. By treating the analysis as a process, the audience better understands the validity of the results and the assumptions involved.
Safety information	Requires specific language, icons, and colors—mandated by technical standards—to communicate the specific danger and how to avoid it

Accommodating users with varying levels of knowledge and experience

Well-written instructions are crafted specifically to accommodate the intended audience. Instructions for operating a new piece of equipment that are intended for an experienced technician will look fundamentally different from instructions for the same equipment intended for new engineers just out of college.

If you have a single primary user, you can tailor the instructions to provide enough guidance to perform the task without explaining things he or she already knows or assuming knowledge he or she doesn't have. Multiple audiences make the job harder. Think about the range of possible audiences and their backgrounds. With careful thought and analysis, you can identify the full range of prior knowledge. Then, write the instructions with enough detail for the inexperienced but with the high-level steps easily accessible for the experienced. Careful planning about the structure of the instructions can serve both audiences at the same time—headings that communicate the high-level steps with detail presented in the following prose or in notes.

If the purpose of the instructions is to teach something, add explanations to the steps to help comprehension. While instructions are not usually intended to make the audience experts in a subject, providing reasons for the steps will help readers to adapt the process, for new situations. What are the objectives of the process, and how do you know they have been met? For example, instructions on how to set up and calibrate a particular strain gauge should help the audience perform the same calibration with strain gauges from a different supplier, applied to a different surface.

Defining steps and sub-steps

The key to writing effective instructions is to divide a complicated task into steps, each of which makes progress toward the goal. What constitutes an appropriately sized step depends on the audience. More experienced audiences will understand larger steps with more abstraction than inexperienced audiences. However, most audiences would be overwhelmed by a hundred-step procedure—if your instructions are this long, structure the steps into a hierarchy with no more than three levels (section, step, and sub-step).

The process step "measure the power output of the op-amp" might be sufficient for experienced audiences but would be too abstract for someone who has never used a multimeter before. Instead, the instructions might break down this step into the sub-steps of how to attach the leads to the correct locations on the breadboard (which might require explaining breadboard circuitry), how to put the multimeter on the correct setting, and how to read the units on the display.

Once you have divided the process into manageable steps, use document design principles (see Module 5: *Designing documents for users*) to make each step distinct and readily apparent to the user. For long instruction sets, the major steps might appear as headings with substeps presented as subsections. In particular, use:

- A larger size font for headings
- White space to separate the steps
- The same size font and style for all of the steps
- Bold or italics for emphasis.

The heading stands out more than the sub-steps. Experienced audiences would be guided by the headings without needing to read the details that follow.

The steps are numbered and reference the lettered parts that appear in the illustration.

The illustration shows the relative location of parts using standard engineering drawing conventions.

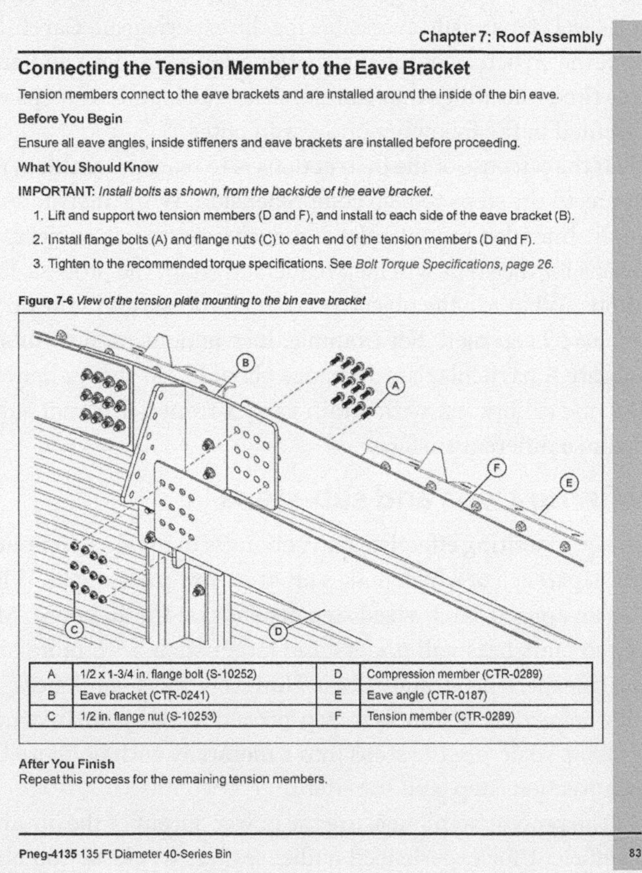

Chapter 7: Roof Assembly

Connecting the Tension Member to the Eave Bracket

Tension members connect to the eave brackets and are installed around the inside of the bin eave.

Before You Begin

Ensure all eave angles, inside stiffeners and eave brackets are installed before proceeding.

What You Should Know

IMPORTANT: *Install bolts as shown, from the backside of the eave bracket.*

1. Lift and support two tension members (D and F), and install to each side of the eave bracket (B).
2. Install flange bolts (A) and flange nuts (C) to each end of the tension members (D and F).
3. Tighten to the recommended torque specifications. See *Bolt Torque Specifications, page 26.*

Figure 7-6 *View of the tension plate mounting to the bin eave bracket*

A	1/2 x 1-3/4 in. flange bolt (S-10252)	D	Compression member (CTR-0289)
B	Eave bracket (CTR-0241)	E	Eave angle (CTR-0187)
C	1/2 in. flange nut (S-10253)	F	Tension member (CTR-0289)

After You Finish

Repeat this process for the remaining tension members.

Pneg-4135 135 Ft Diameter 40-Series Bin 83

This page from a grain storage bin construction manual illustrates one of many steps that must be followed. © 2015 The GSI Group, LLC. Reprinted with permission.

Using precise language

Unclear instructions are frustrating for all audiences. The lack of clarity might come from sloppy preparation where steps are missing, a misunderstanding of the audience's experience, or simply ambiguous writing.

Precision comes from using exactly the right word in exactly the right place. If your audience knows the difference, use the specific word instead of the general category—"lag bolt" instead of "fastener."

Begin each step with a precise active verb stating the action that the user is to take. Parallel structure will make your instructions direct and easy to read.

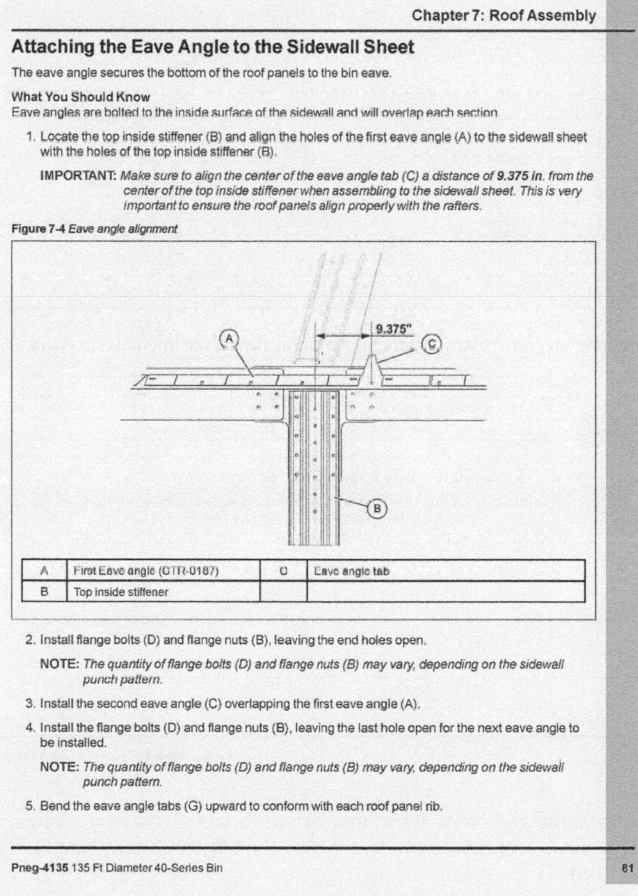

Chapter 7: Roof Assembly

Attaching the Eave Angle to the Sidewall Sheet

The eave angle secures the bottom of the roof panels to the bin eave.

What You Should Know

Eave angles are bolted to the inside surface of the sidewall and will overlap each section.

1. Locate the top inside stiffener (B) and align the holes of the first eave angle (A) to the sidewall sheet with the holes of the top inside stiffener (B).

   IMPORTANT: *Make sure to align the center of the eave angle tab (C) a distance of **9.375 in.** from the center of the top inside stiffener when assembling to the sidewall sheet. This is very important to ensure the roof panels align properly with the rafters.*

Figure 7-4 *Eave angle alignment*

A	First Eave angle (CTR-0187)	C	Eave angle tab
B	Top inside stiffener		

2. Install flange bolts (D) and flange nuts (B), leaving the end holes open.

   NOTE: *The quantity of flange bolts (D) and flange nuts (B) may vary, depending on the sidewall punch pattern.*

3. Install the second eave angle (C) overlapping the first eave angle (A).
4. Install the flange bolts (D) and flange nuts (B), leaving the last hole open for the next eave angle to be installed.

   NOTE: *The quantity of flange bolts (D) and flange nuts (B) may vary, depending on the sidewall punch pattern.*

5. Bend the eave angle tabs (G) upward to conform with each roof panel rib.

Pneg-4135 135 Ft Diameter 40-Series Bin 81

Each of the steps begins with an active verb—"locate," "install," "bend"—that indicates what action the user should take.

A critical measurement (9.375 inches) is discussed below the relevant instruction step.

To further ensure the correct alignment, the measurement is repeated in the illustration.

Precise instructions are needed in the construction of a grain storage bin because incorrect assembly will cause later components to not fit correctly. © 2015 The GSI Group, LLC. Reprinted with permission.

Incorporating illustrations

Illustrations are an important part of an instruction set. Use line drawings or computer models to illustrate spatial relationships that are hard to describe in words, particularly for assembly and operation instructions. The concepts presented in Module 29: *Illustrations* are useful for creating instructions with a visual component.

The illustration here shows a detailed view taken from a large structure.

The arrows, while not strictly following engineering drawing conventions, clearly indicate where the components in the detailed image are located in the large structure.

The illustration shows the parts of the structure as they come together, with the correct parts appearing in an "exploded" view.

Showing only the completed structure with a list of steps would be less helpful than the exploded view shown here.

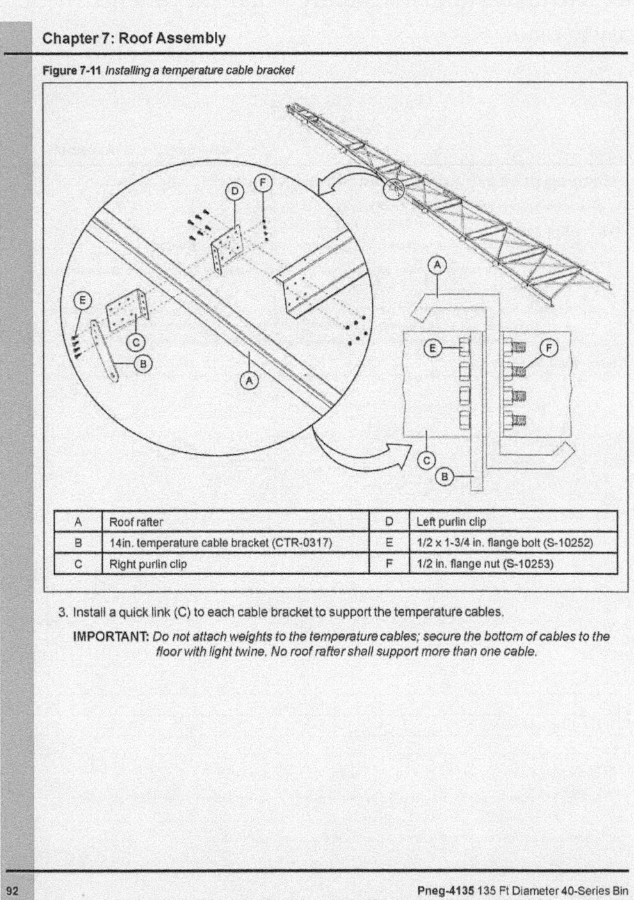

Chapter 7: Roof Assembly

Figure 7-11 *Installing a temperature cable bracket*

A	Roof rafter	D	Left purlin clip
B	14in. temperature cable bracket (CTR-0317)	E	1/2 x 1-3/4 in. flange bolt (S-10252)
C	Right purlin clip	F	1/2 in. flange nut (S-10253)

3. Install a quick link (C) to each cable bracket to support the temperature cables.

IMPORTANT: *Do not attach weights to the temperature cables; secure the bottom of cables to the floor with light twine. No roof rafter shall support more than one cable.*

92 **Pneg-4135** 135 Ft Diameter 40-Series Bin

Illustrations are effective at communicating correct spatial relationships that make instructions easier to follow.

Safety information and precautionary statements

Many technical tasks and procedures entail safety risks if performed incorrectly. One of the most important jobs in writing instructions is to notify readers of those risks in a way that captures their attention and guides them toward correct, safe execution of the process. In addition to protecting safety, precautionary icons and messages can help readers to anticipate easily made mistakes that degrade the end result of the procedure.

Safety information is governed by standards set by the American National Standards Institute (ANSI 535), the U.S. Occupational Safety and Health Administration (OSHA 1910), and the International Organization for Standardization (ISO 3864). The ANSI and OSHA standards specify the following hierarchy of safety information:

"Indicates a hazardous situation which, if not avoided, will result in death or serious injury" (ANSI)

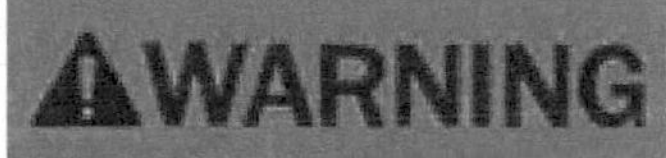

"Indicates a hazardous situation which, if not avoided, could result in death or serious injury" (ANSI)

"Indicates a hazardous situation which, if not avoided, could result in minor or moderate injury" (ANSI)

"Preferred to address practices not related to personal injury" (ANSI)

SAFETY INSTRUCTIONS

"Includes notices of general practice and rules relating to health, first aid, medical equipment, sanitation, housekeeping, and general safety" (OSHA)

The standards specify exactly how these notices are to be used. For instance, if the risk is only to property rather than to safety, a yellow CAUTION indicator may be used, but must not use the exclamation point icon, which indicates personal safety exclusively. Written instructions less frequently contain the lower-level green and blue signs, which appear more often on signs posted in a workplace.

Standards specify use of the red DANGER indicator only in "extreme" hazards that will cause serious injury if they occur.

The IMPORTANT notice is also a type of precaution, alerting readers to details that must be observed to obtain correct results.

Chapter 7: Roof Assembly

Assembling a Double A-Frame

To achieve the required spacing between each A-frame, a double A-frame will need to be assembled prior to installing and lift into place as one assembly.

Before You Begin

The double A-frame assembly will consist of three assembled roof rafters, purlin clips, and purlins. There are also two different purlin clips, a left purlin clip (B) and a right purlin clip (C). The left purlin clip has six holes, compared to the right, which has seven holes along the purlin clip mating surface.

NOTE: *The seventh hole located in the center of the right purlin clip (C) is for identification purposes only. No hardware is used in the location.*

What You Should Know

DANGER *The completed assembly is extremely heavy, and all lifting should be performed by a certified crane operator with prior bin construction experience. Never lift during windy conditions.*

When fully assembled, the double A-frame becomes very heavy.

1. Place a roof rafter (A) parallel to an assembled A-frame (I).
2. Install left and right purlin clips (B and C) at locations shown, and secure the left and right purlin clips to the roof rafter (A) using flange bolts (D) and nuts (E).Tighten to the recommended torque, see *Bolt Torque Specifications, page 26* specifications.
3. Install a purlin (H) at each location shown, and install flange bolts (F) and nuts (G).

IMPORTANT: *The center rafter in the double A-frame will need to be assembled on blocks raising it above the other rafters. This is necessary to adjust for the difference in rafter elevations due to the conical shape of the roof.*

For procedures entailing any degree of safety hazard, instructions include precautionary statements and symbols complying with ANSI and OSHA standards. © 2015 The GSI Group, LLC. Reprinted with permission.

15.2 Usability testing

No matter how careful you are when writing instructions, your first draft will likely need improvement and clarification. To minimize the amount of rework required, think of possible misinterpretations while you're writing. You might find it useful to compile a list of frequently asked questions (FAQs) or a troubleshooting guide.

Ask a colleague who is unfamiliar with the process being described but who matches your intended audience if he or she understands the instructions. Have this person repeat the steps in his or her own words.

When you hear your colleague's version of the instructions, misinterpretations might become obvious. If your colleagues differ greatly from your intended audience, they might try to approximate the audience's understanding rather than their own. However, such informal feedback is no substitute for a usability test that includes participants who closely approximate the intended audience.

Once you believe your instructions are complete and correct, conduct a usability test by asking representative users to follow the instructions without assistance. Observe their difficulties and misunderstandings to determine which deficiencies to address. A formal usability report might be needed to summarize the findings, particularly with many usability tests or when performing such a test for others.

iOS Usability Study Phase 2 for the *Henry* smart project management application

November 11, 2014

Andrew Yuk, Mason Stevens, Carter Gillespi, Kevin Trist, Tyler Davis, Samuel Beck

This is a usability study for the Henry iOS iPhone Application. Henry is a smart project management system that allows users to create and analyze projects in real time, directly from a version control system like Git. Projects, milestones, and tasks can be created and assigned to project collaborators using Henry. This study was intended to test user interactions with the Henry application and determine areas of strength and areas where improvement may be needed.

A usability study was conducted to evaluate the user interface of a software program—common in software engineering.

Specifically, Henry was evaluated based on how well users were able to navigate to specific pages like project detail views, a project's milestones list, and task detail views, and complete specific functions, such as adding tasks and modifying a task's status. As the user completed these tasks, the number of clicks and total time for task completion was recorded.

User interfaces are expected to be intuitive, so limited instructions are given, but the observations of user difficulty demonstrate the type of feedback that a usability test will provide.

NAVIGATION TO HOME PAGE

Users' difficulty navigating to home from specific pages, particularly the tasks display page was readily apparent. Problems with this feature were highlighted in the exit surveys and the total number of clicks data. The minimum number of clicks required to navigate home from

Observing user interactions illuminates which tasks were difficult and where the expected actions were not obvious.

the task detail page was 2, however, no users completed the task in 2 clicks. On average it took 6 clicks because users did not realize that swiping the top bar would open up a navigation screen that would allow them to immediately navigate home. Therefore, this feature of Henry is unintuitive. Subjectively, we determined that the feature was relatively simple to learn and use, and the main reason for its lack of use was because it was never explained to the users. As a result, we did not implement any changes for the next phase.

Some changes were made to the interface based on the observations made during the usability test, but design tradeoffs were also considered.

CHANGING TASKS STATUS

It was apparent to us during Phase 1 that after users selected an implementation status, they were expecting some sort of action to take place, instead of remaining on the same screen. As part of our improvements for Phase 2, we changed Henry so that once a user selected an implementation status, the app would return to the parent screen with the new implementation status displayed. This change did not significantly improve the total time to complete the task. In fact, the mean time for these tasks remained exactly the same across each phase, however, it did slightly reduce the total number of clicks and made the navigation of the app much smoother.

NAVIGATING TO DETAILS PAGE

Frequently throughout the study, users mistakenly navigated to the next sub-list rather than viewing the details. Users were supposed to click the information icon, but most of them did not. In between the phases, we changed the information icon to a "Details" button. This change slightly improved the users' performance in Phase 2, but it came at the cost of visual appeal and navigability.

This Usability Study reports on an iPhone application called Henry—a smart project management system. The study investigated the intuitiveness of the user interface.

Summary

Principles of writing instructions

- What qualifies as well-written instructions depends on the intended audience—the more experienced the audience, the more abstract the instructions can be.
- Breaking up a process into useful steps requires thinking about what steps an audience can take. Use document design principles to make the steps clear to the audience.
- Steps should begin with precise action verbs and use parallel structure.
- If spatial relationships are important for a set of instructions, use illustrations to clarify.
- Include relevant warnings and cautions where the audience will read them.

Usability testing

- Improvements in instructions are made by having someone else test them. Merely reading through the instructions can allow a potential user to supply some suggestions, but this cannot substitute for a formal usability test, in which the instructions are actually used to complete the procedure.
- Clarifications and elaborations can be added for steps where users make mistakes or appear unsure what to do.
- Usability tests generate the best feedback when they include participants who closely approximate the intended audience.

16 Applying for a job

Applying for a job or admission to graduate school can be a stressful experience: You're trying to convince complete strangers that you're the most qualified candidate while simultaneously trying to learn about the company's culture and opportunities. Add in the awareness of competing against hundreds of others just like you, and the task can seem overwhelming.

The best approach is to be prepared by making sure your written documents—résumés, application letters, curriculum vitae, and personal statements—stand out by making them easy to read and highlighting what makes you different from all the other applicants.

Objectives

- To craft a targeted résumé that appeals to an organization's values and priorities
- To apply design principles to a résumé that help the reader quickly locate, process, and understand important content
- To write application letters that confidently emphasize your best qualifications and professional persona
- To shape the tone of application letters deliberately to convey attractive professional qualities such as expertise and collegiality
- To craft a curriculum vitae that gives a comprehensive chronicle of academic and professional experiences and accomplishments
- To write a personal statement that provides a narrative, logical argument about the meaning that you place on your credentials

16.1 Targeting the audience

To attract serious attention to you as a potential employee, your application materials must target the particular organization to which you are applying. Many students apply for numerous positions with little or no revision to differentiate each résumé, curriculum vitae, or application letter. In doing so, though, they miss most of the situation's rhetorical opportunities:

- Appealing deliberately to an organization's values, priorities, and professional practices
- Showing their enthusiasm and diligence by having researched the organization
- Revising the résumé's all-purpose content to highlight credentials of particular interest

Effective application materials target not only the desired qualifications listed in a job ad but also the culture of the organization to which it is being submitted. While your credentials are obviously important in establishing that you are qualified for the position, your perceived fit within the organization depends more on the way that those credentials are presented. Do you emphasize your ability to reduce costs and maximize efficiency? To deliver excellent service and maintain relationships with customers? To innovate and develop exciting new designs?

Most organizations state their values on their website, both in their pages directed to potential employees and to the wider public. (A wider internet search can be useful in determining the extent to which an organization's stated values actually drive its operations. Enron listed "respect" and "integrity" at the beginning of its corporate motto, but journalists and employees alike report a ruthless, cutthroat environment.) Speaking with contacts at an organization, even at a career fair, can provide additional information about the nature of the workplace that will later be useful in targeting your application materials.

You will want to draft a general résumé and application letter (or curriculum vitae and personal statement) that becomes the platform for more targeted revisions:

- To impress hiring managers in industry, explain how your work on an internship created cost savings, improved quality control, or enhanced technical systems or processes.
- When targeting a university research community, use your experience conducting experiments to emphasize procedures and findings.
- For any technical or industrial field in which you lack paid employment experience, draw on your most relevant academic or extracurricular projects.

16.2 Résumés

The résumé is the central document detailing your credentials, experiences, and skills. In many employment searches—especially early in an engineering career—the résumé will guide the deliberations of prospective employers about your candidacy. For internships or entry-level positions, you will most likely have to submit a résumé electronically; an application letter may or may not be required.

Résumés face a uniquely demanding rhetorical situation: Recruiters and hiring managers report that they reach initial verdicts on a résumé—which can determine whether the writer remains in the pool of candidates for a position—in less than a minute. Document design features must therefore showcase important content with such rapid processing in mind. Page design principles, which may be of secondary importance when writing a report or memo, make a huge difference in the success of a résumé.

Removing obsolete and irrelevant credentials

Effective résumé writing begins with clear insight into which credentials and experiences will present you in the best light—as an *emerging engineering professional* rather than "just another student." A beginning college student's résumé will list credentials and experiences from high school—GPA and class rank, summer jobs, and local awards. All of these may well be impressive in a high school context—but to a recruiter or hiring manager, they paint a picture of someone young and inexperienced. Once you've enrolled in college and begun to pursue a definite professional path, you should cultivate, select, and showcase those experiences that are most relevant to that path:

High school	Retain only high school credentials that all readers can recognize as highly noteworthy—graduating as a valedictorian or salutatorian, winning a national competition—or are directly relevant to engineering.
Extracurricular activities	Many engineering students gain valuable experience by working on technical projects or competition teams, from aerial robotics to concrete canoes. Include these, and describe your contributions and achievements in some detail. Most other collegiate extracurricular activities merit a line or two in an "Activities" section. Greater detail can show leadership or teamwork skills if you can establish direct responsibility for substantial, concrete accomplishments, which impress readers more than simple titles (team captain, club treasurer).

Paid employment	As a rule, omit retail, food service, or other non-engineering jobs typically held by students. Make exceptions only for job descriptions that showcase important achievements—supervising other employees, being promoted, or completing special projects—and list these along with technical credentials.

Defining experiences and skills

Once you've held engineering jobs over a period of several years, the "Experience" section will list those positions and describe your accomplishments in each. Students, though, need not limit their "Experience" entries to paid employment: If your engineering skills are more evident in academic and extracurricular projects, give those projects priority over unrelated jobs.

Whether it covers work on campus or in industry, an effective "Experience" entry results from a series of purposeful decisions:

Employer (or sponsor) and title of your role	Decide which of these will be more impressive or recognizable to your intended audience, and use it to define the entry (usually in bold type). List the other below it (usually in smaller italics). Make this decision once for the "Experience" section: Don't vary it from one entry to the next.
Describing duties and achievements	Individual bullet points on a résumé can accomplish a surprising variety of goals in presenting you in the most favorable light. • Start with a meaningful context: What did your work achieve for your employer or sponsor? • Detail the most important tasks with an eye toward matching the job you want to acquire. For instance, describe what you've done

- with specific equipment, software, or methods of analysis mentioned in the job ad.
- Carefully select precise verbs to reflect both technical ("prototyped," "debugged") and interpersonal ("supervised," "corresponded") work. Verbs that first come to mind ("did," "helped," "worked") are often weak.
- Quantify your work where possible: "managed a $100,000 budget," "chaired a committee of ten members."

The "Skills" section can be more difficult to use effectively: Many applicants merely list abilities without sufficient precision to make an impression on readers. Formal, official credentials—as a Six Sigma Yellow Belt, or a Certified Software Process Engineer (CSPE)—will have a substantial impact without much effort; defining other skills requires a more deliberate process.

Software	Expertise in software and programming is not demonstrated by merely listing the programs in which you are proficient. For readers to take your abilities seriously, other means are needed. • Substantiate major experiences with that software in an "Experience" entry. • Subordinate the software tool (Xilinx Spartan) to the task it accomplishes (FPGA programming). • Address general-purpose tools only if you can identify specific high-level abilities: "programming macros in Excel VBA."
Laboratory and shop credentials	Listing competencies works better for these kinds of procedures and equipment. These credentials can range from specific items (lathes, oscilloscopes,

	centrifuges) to hands-on abilities that are often valuable in industry (welding, soldering).
Interpersonal skills	Many students list leadership, teamwork, and communication as skills. This can work to call attention to those topics, but the mere presence of these words on a list presents no evidence: These claims must be detailed and substantiated in the résumé's other sections.

The sequence of content

Résumé readers typically expect information to be organized in a traditional way. You need not use every one of the most frequent section headings—Objective, Education, Experience, Skills, Honors, Activities, and the like—but you might not want to deviate too far from that general sequence.

Only the quality of the credentials will convince the reader that it's worthwhile to hire you—but their sequence and placement on the page can maximize their impact. You don't want readers to have to search for a major credential that you want to showcase—or, worse, to miss it altogether.

Dakota Huckaby huckaby@rose-hulman.edu (812) 875-2538 5500 Wabash Avenue Terre Haute, IN 47803

Objective: To gain practical, first-hand production engineering experience through a summer internship in the chemical industry.

Education: **Bachelor of Science, Chemical Engineering** May 2016
Rose-Hulman Institute of Technology, Terre Haute, IN GPA: 3.96/4.0
Status: Junior
Related Courses: Applications of Heat and Mass Transfer, Kinetics and Reactor Design, Elements of Electrical Engineering
Anticipated Minors: Spanish Minor, Philosophy and Religion Minor

Skills:
Analysis and Modeling: Visual Basic, MATLAB, Minitab, Maple, Aspen, Spartan
Visualization: Canvas, Visio, Solid Edge ST5
Laboratory: Filtration, distillation, recrystallization; IR, UV-vis, and NMR spectroscopy; GC, HPLC, and column chromatography; titration and serial dilutions
Other: Communication of complex ideas, excellent leadership skills, project documentation, conversational Spanish

Experience: **Monument Chemical—Corporate Project Engineering Intern** 2014-Present
USF Warehouse Renovation:
— Managed all aspects of a capital project worth more than $360,000
— Chartered and led an eight-member, cross-functional team
— Strategically organized 54,000 ft^2 of warehousing space
— Designed interface and functionality of a real-time SAP barcoding solution

Specifying production engineering makes the objective statement specific; merely identifying an engineering discipline doesn't tell readers why a particular job speaks to your goals and interests.

In technical fields, numbers (54,000 ft^2, $360,000) can be an especially reliable way to indicate the significance of an accomplishment.

As an "Honors and Activities" section grows, it's often best to divide it. Credentials stand out in smaller "chunks" of text, and the broad themes—leadership, service, and academic achievement—become prominent.

Brandenburg Plant Water Balance Study:
- Conducted diagnostic experiments to explain a 30% discrepancy in the previous water balance
- Isolated a reporting error and implemented an adjustment equation, now called the Hooper Correlation
- Prepared a box flow diagram detailing flow to and from unit operations in the new, accurate water balance

Summary: Successfully executed seven additional projects, all pertaining to the Brandenburg site

Activities:	**Alpha Chi Sigma Professional Chemistry Fraternity**	2013-Present
	Vice Master Alchemist (Vice President)	
	– Second in command of chapter of more than 100 active members	
	– Planned and led all pledge education meetings	
	– Served as advisor to Master Alchemist as member of Executive Committee	
	Omega Chi Epsilon Honorary Chemical Engineering Fraternity	2014-Present
	Active Member	
Service:	**Chemistry on Wheels Program**, Volunteer Instructor	2013-Present
	Terre Haute Children's Museum, Volunteer Staff Member	2010-2012
Awards:	Heminway Bronze Medal—Highest Freshman Grade Point Average	May 2014
	AXΣ Wyvern Award, Level Three—Chemistry-Related Community Service	Oct. 2014
	Mr. Rose Pageant Winner—Campus-Wide Charity Competition	Jan. 2014
	Lilly Endowment Community Scholarship—Merit-Based, Full-Tuition Scholarship	May 2012
	2012 American Chemical Society Chemistry Olympiad National Finalist	Mar. 2012

Only a few highly distinctive experiences and awards have been retained from high school.

Students applying for summer internships usually focus their résumés on detailed accounts of their one or two most relevant jobs (or academic engineering projects). Lists of academic and extracurricular credentials round out the content. © 2015 Dakota Huckaby. Reprinted with permission.

Layout and white space

Quick reading by a prospective employer can be accommodated by *chunking* information into small segments that are easy to comprehend. Inserting more space between units of text signals that audiences should read them separately; progressively smaller space allowances signify higher degrees of logical continuity or connection.

As a general rule, then, line spacing should reflect logical hierarchy, with ample space between main sections and very little within a single list item. You might start with something like this:

- 24 points of space between main sections
- 14 points between subsections (e.g., for different groups of skills)
- 10 points between individual entries, such as jobs
- 6 points between bullets in the job descriptions

Caroline Winters

2001 Evans Road
Cary, NC 27513

(256) 555-7359
winters@gmail.com

Objective To acquire an entry-level position in Advanced Aerospace Systems

Education **Bachelor of Science** *Mechanical Engineering,* November 2011
Thermal and Fluid Sciences Concentration
Rose-Hulman Institute of Technology, Terre Haute, IN **3.60 GPA**

Skills Component modeling; SolidWorks; Pro Engineer Wildfire 4.0
Controls: LabView10; MPLAB IDE
Analyzing and modeling data: MATLAB; Microsoft Excel
Simulations: MCNP5 (Monte Carlo n-particle)

Relevant Coursework: Advanced Thermodynamics, Introduction to Electromagnetism, Propulsion Systems, Foundations of Fluid Mechanics

Experience **NIF Summer Scholar** June–August 2011
Laurence Livermore National Laboratory

- Designed a two collimator system to create a narrowband, directional gamma ray beam for Nuclear Resonance Fluorescence
- Simulated complete geometry in MCNP5 to prove design feasibility
- Designed a simulation of the full spectrum gamma ray beam, ~~and facility is~~ being used by the MEGa-Ray Project to predict experimental behavior
- Presented results at the Lawrence Livermore National Laboratory Poster Symposium

Department of Energy Scholar June–August 2010
Laurence Livermore National Laboratory
Summer Undergraduate Laboratory Internship (SULI)

- Designed neutron radiation shield doors to attenuate a highly directional gamma ray source
- Led a team to refurbish existing shield doors including: performing an ultrasound to test structural integrity
- Published an abstract and a poster of the project

National Science Foundation Scholar June–August 2009
College of Optical Sciences, University of Arizona
Integrated Optics for Undergraduates REU

- Researched non-linear dispersion effects in fiber optic cable
- Designed and executed an experiment in 10 weeks having no previous experience in optical engineering

AIAA Design/Build/Fly Competition Team
Project Leader: Flight Report

- Designed a collapsible UA V and completed a full design process, including fabrication, testing, analysis and revision 2010–2011
- Outlined a method for testing propulsion configurations in a low speed wind tunnel for future DBF teams
- Researched non-linearity in battery behavior

When pursuing an entry-level position in industry, most successful applicants emphasize credentials and experiences outside of their schools. Caroline's emphasis on research, though, reflects a more "academic" identity than that projected by most jobseekers. © 2015 Caroline Winters. Reprinted with permission.

Simply listing the names of software programs is the equivalent of listing tools—hammer, saw, wrench. Caroline makes clear what she can do with these software tools. Programs will likely vary between companies, but the acquired skills will transfer.

All bulleted job descriptions should be in parallel structure, with the same part of speech beginning each point.

As an outline of qualifications using sentence fragments, résumés omit end punctuation (periods); commas and semicolons can be used when needed.

Descriptions of professional tasks target the audience (the prospective employer) by being situated in the "big picture" of the value created for each employer.

Typography and visual hierarchy

White space is most important in separating your résumé's sections and subsections from one another. Within a section, the best tool for clear hierarchical distinctions is typography. The overall principle is for the weight of the type to reflect the size of the concept, so that the reader's eye moves from overall categories ("Education," "Skills") to the main entries ("Bachelor of Science in Computer Engineering") to the explanatory points that support them ("Planned and created a 16-bit multicycle processor on an FPGA").

Most résumés create these levels of hierarchy with distinct styles of a font at a single size, in some combination of ALL CAPS, SMALL CAPS, **bold,** *italic,* and roman (or "normal") type. The size of the font can be varied just as effectively, though—from as little as 10 points for individual items in a bulleted list up to 28 points or so for your name at the top of the page.

A résumé may easily accommodate two different fonts. As in other documents, the most common approach is to use type with serifs for the main body text, and to complement them with sans serif type for high-level headings. Because the résumé functions as an outline, though, most highly readable fonts will work, whether or not they have serifs.

16.3 Application letters

While a résumé provides an outline of your qualifications, an application letter adds your personal voice, giving a clearer sense of what kind of employee and colleague you would be. A well-written letter can also provide evidence of effective communication skills.

The application letter should thus maintain a professional voice and tone throughout—cordial and personable but not too informal—especially in the opening and closing paragraphs. You should demonstrate your enthusiasm for the position for which you are applying, as well as expressing confidence in your abilities without sounding overconfident or pompous. This apparent attitude, combined with credentials that are carefully chosen to be distinctive, can set you apart from other qualified candidates with similar education and work experience.

Feature an especially important credential or experience in your opening paragraph. Katelyn highlights her leadership of Sustainability Club, showing that her interests and experiences match Pall's stated values while also demonstrating her initiative.

September 18, 2013

Pall Corporation

Dear Hiring Manager,

Because of Pall Corporation's commitment to sustainable design and innovative techniques, I am interested in applying for the mechanical engineering internship for the 2014 summer. I am a mechanical engineering student at Rose-Hulman Institute of Technology with a focus on sustainable design. Sustainable development is at the core of my future career. At Rose-Hulman, I initiated Sustainability Club to advocate a "triple bottom line" approach to studying engineering.

At ARPAC, an engineering company designing and manufacturing packaging equipment, I enhanced my knowledge of manufacturing processes and operational standards. To save money and enhance performance for ARPAC, I standardized parts for machines and composed bill of materials. During my experience, I worked closely with another company and became more effective in professional communications in person and email.

While working at Rose-Hulman Ventures, I further developed my skills in team cooperation, ingenuity for design, organization, and keeping projects fun and interesting within the team. As an engineering intern for NICO, I was able to prototype neurosurgical devices that removed brain tumors and decreased surgery time. In addition, I also advanced my skills in SolidWorks, machining, and 3D printing. My work at NICO was especially rewarding because the results of my designs went to saving human lives.

My interpersonal skills, technical training and desire to learn will be an asset for Pall Corporation. I look forward to meeting with you to discuss the summer internship further. I may be reached at 859-555-3935 (cell) or by email at stenger@rose-hulman.edu. Thank you for your time.

All the best,

Katelyn Stenger

Katelyn Stenger

In her descriptions of her internships, Katelyn incorporates specific skills and traits that were listed in the job ad—creativity, organization, and documentation. If invited for an interview, she would likely be asked to elaborate upon these experiences further.

Include your contact information in the closing paragraph: your letter may be separated from your other application materials.

Katelyn's letter includes descriptions of her skills and experiences but emphasizes her personal priorities in her work. She states not only what she's done, but also the reasons that she has found value and meaning in her professional experiences.

Emphasize the benefits that you will *provide* to the organization—not the ones that you hope to *receive*. It's fine to explain how you believe your capabilities will grow in the job that you hope to land. Just be sure that those explanations support the larger argument that you will be a valuable asset for the company or institution to which you're applying.

Organizing your argument

An application letter makes the argument that you are a qualified candidate for the position and that you are a good match for the organization. Each section serves a particular purpose.

The *opening paragraph* declares your candidacy for a specific position and commonly notes how you found out about the opening, whether through a specific job posting or a professional contact. If the job announcement notes a specific location or includes a requisition number, include this information.

This opening is also a good place for demonstrating your knowledge of the organization, indicating how its values and priorities align with your own professional goals and background. Most importantly, emphasize what makes you unique as a job candidate. Anticipate that other applicants will merely state their majors and graduation dates; instead, describe strengths that couldn't be understood through such basic credentials. For example, readers will take notice of an engineering or computer science degree combined with a background in entrepreneurship or intercultural communication. Expertise or special interest in a particular technical field or industry activity—heat transfer, semiconductor fabrication, or life cycle analysis—has similar potential.

In the *body paragraphs*, each topic sentence should make an overall claim about the applicant, which is then supported with specific, memorable details. Take particular care to avoid reliance on generic skill sets that anyone can claim ("I am a team player" or "I am an effective problem solver"). Instead, make more specific statements: Indicate that an internship "enhanced my knowledge of manufacturing processes and operational standards." Then dedicate several sentences to the relevant details, explaining what you took from those aspects of the experience. Rather than merely repeating the résumé's overview of an experience, reflect on certain parts of the experience to explain how you developed an important professional ability or achieved a particular goal. (In doing so, you are arguing about what you'll be able to do in the job you're pursuing. Make a strategic decision about whether to state such claims explicitly.)

The *closing paragraph* provides the applicant's contact information, thanks the hiring managers for their consideration, and ends with a formal closing. In addition, it may be an appropriate place to provide an additional comment about your suitability for the position or organization.

Companies want to know if you're replying to a specific job ad, ensuring that each hiring manager receives the correct materials. (A requisition number, if the ad provides one, belongs in the opening paragraph.)

Greetings:

I would like to apply for the Space-X Internship/Co-Op Program that is advertised on your website, specifically to work within the structures group. I am currently a Junior at Rose-Hulman Institute of Technology studying Civil Engineering and Computational Science with a specialization in Structural Engineering. As part of my specialization, I am currently engaged in structures coursework, and I plan to move into introductory finite element analysis in the winter. I know that your company is primarily interested in Mechanical and Aerospace Engineers, but I feel that my major is a declaration of my competencies rather than my intended industry, and that my familiarity with structural engineering would make me a productive member of your team.

Louis reframes what might be perceived as a weakness and reframes it as a strength.

I have gained significant hands-on structural and technical experience from my work with the Rose-Hulman Human Powered Vehicle Team. In my Freshman Year, I was involved in manufacturing custom parts and performing carbon fiber layups. In my Sophomore Year, I took a more active role on the team, leading drivetrain and ergonomics design, testing, and manufacturing. This year, I have been elected treasurer, and I am learning to maintain a budget and manage material stock. In addition, I have taken on prototyping and testing custom-fit composite bicycle cranks and designing the vehicle's structural composite fairing.

The job ad list requests applicants with "hands-on design experience." Louis's composite/carbon fiber work fits the job description for the structures group, which includes working with composites.

For the past two summers, I have worked as both an Assistant and Journeyman Water Well Repair Technician for EGIS PA. I was primarily in charge of design and construction of a dual-use geophysical and video logging system. The bulk of my work was reverse-engineering an outdated unit and locating critical components, while working with professional fabricators to recreate custom mechanical systems. Through this work, I gained experience with component design, tooling, and sourcing.

Enclosed with this letter is a copy of my resume; I will follow up within the next few days to answer any questions or concerns you might have. In the meantime, if you are interested, call me at (919) 555-0563, or e-mail me at vaught@rose-hulman.

Sincerely,

Louis Vaught

Because his design experience is more pertinent to the position to which he's applying, Louis includes it above the paragraph covering his work as a technician.

Louis's letter includes more detail about each experience, part of a sustained argument that he is qualified for a position that isn't explicitly advertised to civil engineers. © 2014 Louis Vaught. Reprinted with permission.

16.4 Academia: The curriculum vitae and statement of purpose

The process of applying to graduate or professional schools calls for a slightly different set of application documents than a job in industry or government. Instead of a résumé, you will submit a *curriculum vitae*, and instead of an application letter, you will create a *personal statement* or *statement of purpose*. The techniques and principles that govern other employment documents still pertain, but application materials in higher education have their own conventions.

Applications for graduate admission (or for fellowships supporting graduate students) are read and evaluated by a faculty committee in the department to which you're applying. This audience profile can inform your writing in useful ways. You know that you're targeting readers with extensive expertise in highly specialized technical fields and that they will respect such expertise in others.

To present yourself as an ideal candidate for graduate education, you will want to establish a path to such expertise. If you have worked in research labs in industry or at a university, or presented at an undergraduate research conference, these kinds of credentials will argue strongly for you. If these activities are still in the future for you, you can show that your internships, courses, and projects have provided you with the necessary skills to undertake them as a new graduate student.

The curriculum vitae (CV)

A curriculum vitae (Latin for "life's study," often abbreviated as a "CV" or "vita") is broadly comparable to a résumé but covers the applicant's experiences and accomplishments comprehensively, rather than the selected career highlights that define a résumé. A CV, rather than a résumé, will be expected of students continuing on to a terminal degree (Ph.D. or M.D.) and a career in an academic, research, or medical environment. Over the course of such a career, the CV may accumulate several pages as you gain research publications, continuing education, grants, fellowships, and the like. While a résumé would typically shed older or less impressive credentials to make room for newer ones, the CV will largely just grow to accommodate them.

For an undergraduate applying to graduate programs, the primary difference from a résumé may be the headings used to organize the content. As with a résumé, there is some freedom to organize and order your credentials in the way that best showcases your own unique combination of skills and experiences.

EDUCATION
SKILLS (ideally subdivided to cover software, laboratory equipment and procedures, etc.)
TEACHING EXPERIENCE
RESEARCH EXPERIENCE
PUBLICATIONS
CONFERENCE PRESENTATIONS
HONORS AND AWARDS (including grants and fellowships)
PROFESSIONAL AFFILIATIONS or MEMBERSHIPS

Almost every CV will include these sections, which can be arranged for prominent placement of the most important and most impressive credentials—usually on the first page.

The personal statement or statement of purpose

As the CV approximates the résumé, the personal statement broadly resembles an application letter: It accompanies the list of credentials and provides a more narrative, logical argument about the meaning of those credentials for your personal development (and, implicitly, the intellectual value that you will bring to the academic department or organization that sponsors you).

As you apply to graduate programs, you may notice differences in their requirements. Some applications may supply specific questions that your statement is to address and may require statements of different lengths. It is important to meet these specific requirements, but in many cases you will be able to adapt your responses from a baseline or default statement of purpose. At a minimum, you should understand exactly what you want to convey in statements for each school:

- Your own central ideas about your technical interests, background, and experiences
- The most important and distinctive accomplishments that you want to discuss
- How your technical interests and skills relate to the specific offerings of the school—its research emphases, laboratories, etc.

The two typical names for the genre—"statement of purpose" and "personal statement"—reflect the two goals that define it.

Statement of purpose	Conveying *purpose* means addressing the goals that motivate your pursuit of an advanced degree. You may tell readers about the career that you envision: Do you imagine teaching and performing research at a university, or working in industry or government? Readers may be most interested, though, in the specific kinds of technical work that define you as an emerging scholar and researcher. Naturally, these plans and interests may change through your graduate career, but you can already establish scholarly promise. Describe technical projects clearly and in context, but use some specialized jargon in a way that communicates the expertise that you have acquired.
Personal statement	The statement is *personal* in distinguishing its writer from other applicants to the program—telling a story about how the writer developed defining interests, character qualities, and habits of mind.

Many writers approach this side of the statement by producing a life story, often describing technical interests or personal qualities emerging in early childhood. You will need to shape these carefully: Executed properly, they can be engaging and memorable, but they can also be trite.

Emma here identifies with her discipline, explaining her allegiance to the values of engineering—practicality and problem solving for human benefit.

Her original aspirations to art school are memorable and differentiate her from other engineering graduates.

Engineering faculty at a research university—the decision-makers targeted by the document—are technical specialists. Even if an applicant's technical focus may change, specifying a precise, current technical interest is beneficial.

As in industry, an applicant immediately stands out by explicitly showing knowledge of the targeted organization.

Before my senior year of high school, I volunteered for a summer camp for adults with physical disabilities. I was paired with a woman with Cerebral Palsy and was responsible for her complete care throughout the week. At seventeen years old, the experience of tending to another individual and putting all of her needs before my own was completely exhausting and overwhelming but also intensely rewarding. Essentially, my world was rocked. So many things that I had once taken for granted, I could now see through the eyes of my camper, and everywhere around me I noticed not symptoms that needed to be treated, but problems in the world that needed to be fixed. I quickly learned that engineering was the vehicle through which the problems that I had identified could be solved. I abandoned my applications to art school and with the idealism that only a seventeen year old can have, decided that I would become an engineer.

Course work in materials science channeled my interest to the fields of biomaterials and tissue engineering and led me to join Dr. Nolte's lab to work with polymer thin films. The independent study, exploration of problems, and equipment troubleshooting typical in laboratory research are all reasons that I have enjoyed the research experience. My academic and research interests correspond with the research activities of the IIT biomedical engineering faculty, specifically in the areas of biomaterials, drug delivery, and tissue engineering. This year, I am one of two inaugural candidates electing to complete a biomedical engineering undergraduate research thesis. My research in polymer thin films is a cross disciplinary project that focuses on how the mechanical properties of thin films change as they are exposed to mildly agitated environments. This approach serves as an *in vivo* surrogate and is largely applicable to drug delivery. I will present my findings at the Rocky Mountain Bioengineering Symposium this spring.

In graduate school, I would like to continue to study biomaterials and methods for drug delivery and I am particularly interested in Dr. Perez-Luna's work with hydrogels for biomedical applications. I am also very interested in cellular and tissue engineering and in the research being conducted in Professor Karuri's lab. These professors' areas of investigation represent a combination of engineering and science, which inherently promotes an interdisciplinary approach to research, which I value. I look forward to a school like Illinois Institute of Technology because of its multidisciplinary approach and emphasis not only on academic rigor but on service and hands-on education as well.

While avoiding prolonged technical details about her research, Emma successfully conveys scholarly expertise relevant to the work done in the department to which she's applying.

The graduate committee may admit a student to work in a particular laboratory. Declaring these interests makes Emma an attractive candidate and may influence her eventual placement.

This statement balances the personal and the professional: Readers develop a strong sense of Emma's research interests, but only after a memorable anecdote that helps to define her as a person. © 2011 Emma Dosmar. Reprinted with permission.

Summary

Targeting the audience

- Your research into the hiring organization should be evident in your application materials. Such research lets you know which credentials to emphasize and typically leads readers to notice your diligence.
- Applicants can show a good match with the hiring organization by appealing to the values and priorities evident on its website and in other materials.

Résumés

- Until you have extensive industry experience, your résumé should not exceed one single-sided page. Use that page to display your most important credentials and experiences for a particular job: Don't try to be comprehensive in covering your entire life.
- Cut high school content and other obsolete or irrelevant information from your résumé, freeing up space for more technical credentials.
- Instead of listing all paid jobs that you've held, build the "Experience" section around your work that applies most directly to engineering.
- Define and support items in the "Skills" section carefully; listed skills are claims about yourself that require support.

- Résumé page design must make it easy for a prospective employer quickly to locate, process, and understand important content.
- Résumés have more levels of *hierarchy* than most other documents. Their content is divided into sections with subheadings followed by supporting information.
- White space separates sections and subsections from one another; within a section, change the size and style of type to emphasize headings or main entries over supporting points.

Application letters

- The application letter allows you to establish your professional identity and to give a clearer sense of what kind of employee and colleague you would be. In addition, a well-written letter can provide evidence of effective communication skills.
- The letter provides greater detail about your qualifications, making the argument that you are a qualified candidate for the position and that you are a good match for the organization to which you are applying. Specific examples and details make your qualifications memorable.

Academia: The curriculum vitae and statement of purpose

- A curriculum vitae ("CV" or "vita") is broadly comparable to a résumé but aims for a much more comprehensive chronicle of the applicant's experiences and accomplishments and grows lengthier to accommodate them.
- The personal statement is a more narrative, logical argument about the meaning of the credentials on the CV. It should convey your own central ideas about your technical interests, background, and experiences, your most important and distinctive accomplishments, and how your technical interests and skills relate to the specific offerings of the school.

Processes

. . . of researching, drafting, revising, editing, and collaborating with others can help you to produce excellent documents even when you aren't sure where to start. Some writers often begin with uncertainty, trying to visualize the end product at which they hope to arrive. Others become stuck after producing a first draft, able to correct minor stylistic errors but finding it difficult to revise central ideas or reorganize sections and paragraphs to create substantial improvements on a large scale. Almost all professionals experience similar challenges when meeting or writing with teammates. In each case, breaking through such an impasse requires knowledge of proven techniques—as with a differential equation or a free body diagram, a solution requires patient, methodical application of proven steps in an appropriate order.

Processes

17 Researching

In all professional disciplines, current work is informed by the careful study and meticulous citation of scholarly documents. In engineering, grounding your work in prior art will enhance your credibility and may stimulate new ideas, especially in the design process. As a designer, effective research will enable you to use your time more efficiently—avoiding time spent "reinventing the wheel"—and minimizes the risk of infringing on someone's patent (and being sued). Moreover, within the organization in which you work, accurate and up-to-date knowledge is always needed to inform decisions about how to allocate money and other resources.

Objectives

- To gather preliminary information about a research topic by consulting with colleagues, experts, reference librarians, and consultants
- To generate a list of potential scholarly sources using internet search engines
- To search databases to find more rigorous peer-reviewed sources from journals, conference publications, books, and patents
- To integrate sources using paraphrases and direct quotations and citing sources using an appropriate citation style

17.1 Consulting with experts

Talking to experienced colleagues is the easiest and quickest way to gather preliminary information about a project, task, or technology that is new to you. Seek out colleagues who have extensive experience in the subject and who are respected by their co-workers. An initial email or phone call is appropriate to ask whether they are willing to help, but face-to-face meetings are more effective for transferring background knowledge quickly. Explain what you're researching; then ask what they think about the problem you're facing or what they know about a particular subject.

In addition to experienced colleagues, consult with your reference librarian. Explain the context and purpose of your research task, the background information that you already have, and the information you are seeking. Librarians can help you find an appropriate reference book, help you to find technical journal articles using the most relevant databases, or introduce you to other online resources. As a student, you have the advantage of easy access to an academic reference librarian. As a practicing professional, your company may employ such a librarian in-house; if not, consider visiting the technical library at the nearest university.

If consulting with experienced colleagues and reference librarians does not produce the desired result, you may choose to hire some outside experts as consultants. In fields that are very new or rapidly evolving, current knowledge may not yet be recorded in journals or books, so up-to-date knowledge may have to come from consultants. Professors at research institutions may be willing to share some of their technical knowledge. Both professional and academic consultation may be expensive, so use their expertise strategically.

17.2 Finding scholarly sources

In daily life, we often turn first to internet search engines when seeking information, and technical professionals might still use these tools as a first step when learning the basics in a new area of study: User-generated content found on internet forums and Wikipedia can provide enough basic content to develop your initial knowledge base. However, a document that references only user-generated webpages and Wikipedia does not demonstrate thorough research of a topic. In fact, in academic and industry settings alike, citing such sources at all may be damaging to your credibility as a technical professional. Your readers will know that Google and other search tools cannot filter out out-of-date, incomplete, incorrect, or misleading information. (Moreover, the search engines' proprietary ranking algorithms leave some mysteries: Are your first results prominent because these pages have high-quality information, or merely because someone has successfully optimized them for search engines?)

The searches below all use the same search phrase, "drinking water purification by ultraviolet light," showing the results yielded by different search tools:

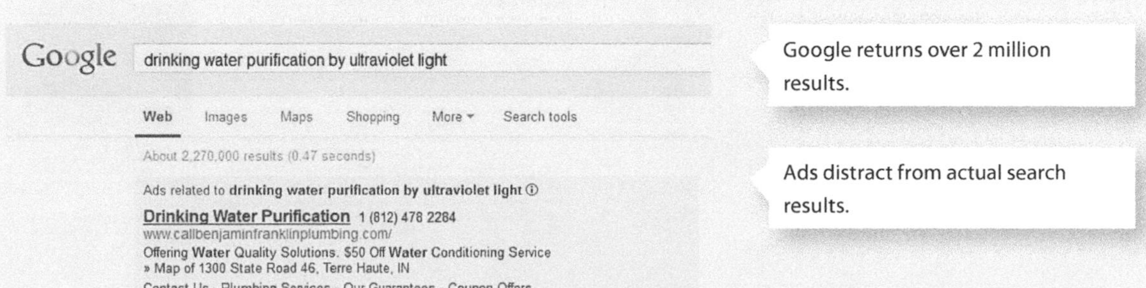

Three of the first five results are commercial products. These pages might be useful as an introduction to existing solutions, but skepticism is necessary when writers are trying to sell their products.

UV Water Disinfection - 150+ Units From $155
www.todayshealthyhome.com/
Chemical Free Safe Easy **Water Treatment** Free Ship
View this ad's deal - valid as of Jun 12, 2013

Tap water filter - Brita.com
www.brita.com/
Get Great-Tasting **Water** Without the Cost or Waste of Bottled **Water**.
Buy Products Online - Get a Colored Pitcher - Get a New Brita Bottle

SteriPEN: **Water Purifier** Portable **Water Purification** System
www.steripen.com/
SteriPEN **water purification** technology harnesses the brilliant power of **ultraviolet light** to make water safe to **drink**. It's the same technology used by leading ...

Ultraviolet germicidal irradiation - Wikipedia, the free encyclopedia
en.wikipedia.org/wiki/**Ultraviolet**_germicidal_irradiation
It is used in a variety of applications, such as food, air and **water purification**. ... Using ultraviolet **(UV) light** for **drinking** water disinfection dates back to 1916 in the ...

Ultra violet light for reverse osmosis **water filter** systems
www.free**drinkingwater**.com/ww-fi-**uv-light**.htm
The primary usage for a **UV light** is to disinfect filtered **water** at a certain flow rate. Bacteria, viruses, and other microorganisms are destroyed by the UV ...

A PDF from an .edu domain is likely to be authored by an expert and thus might be a useful reference.

[PDF] **Ultraviolet** Disinfection - National Environmental Services Center
www.nesc.wvu.edu/ndwc/pdf/ot/tb/ot_tb_f00.pdf
Using ultraviolet **(UV) light** for **drinking water** disinfection dates back to 1916 in ... **water** to radiation from **UV light**. ... ventional UV **treatment** as "emerging" tech-.

Ultraviolet Water Purification Systems - ESP Water Products
espwaterproducts.com/**uv-water-purification**.htm

Google and other general search engines will often link most prominently to commercial products rather than analysis by credible experts. Google and the Google logo are registered trademarks of Google Inc., used with permission.

The relevant article has an unexpected title, making it harder to find.

WIKIPEDIA
The Free Encyclopedia

Create account Log in

Article Talk Read Edit View history Search

Main page
Contents
Featured content
Current events
Random article
Donate to Wikipedia
Wikimedia Shop
Interaction
Help
About Wikipedia
Community portal
Recent changes
Contact page
Tools
Print/export
Languages
العربية

Ultraviolet germicidal irradiation

From Wikipedia, the free encyclopedia

This article **may contain too much repetition or redundant language**. Please help improve it by merging similar text or removing repeated statements. *(August 2011)*

Ultraviolet germicidal irradiation (UVGI) is a disinfection method that uses ultraviolet (UV) light at sufficiently short wavelength to kill microorganisms.[1] It is used in a variety of applications, such as food, air and water purification. UVGI utilises short-wavelength ultraviolet radiation (UV-C) that is harmful to microorganisms. It is effective in destroying the nucleic acids in these organisms so that their DNA is disrupted by the UV radiation, leaving them unable to perform vital cellular functions.

The wavelength of UV that causes this effect is rare on Earth as the atmosphere blocks it.[2] Using a UVGI device in certain environments like circulating air or water systems creates a deadly effect on micro-organisms such as pathogens, viruses and molds that are in these environments. Coupled with a filtration system, UVGI can remove harmful micro-organisms from these environments.

A low pressure mercury vapor discharge tube floods the inside of a biosafety cabinet with shortwave UV light when not in use, sterilizing microbiological contaminants from irradiated surfaces.

The application of UVGI to disinfection has been an accepted practice since the mid-20th century. It has been used primarily in medical sanitation and sterile work facilities. Increasingly it was employed to sterilize drinking and wastewater, as the holding facilities were enclosed and could be circulated to ensure a higher exposure to the UV. In recent years UVGI has found renewed application in air sanitization.

Deutsch
Español
Français
Bahasa Indonesia
Italiano
Bahasa Melayu
Edit links

Contents [hide]
1 History
2 Method of operation
3 Effectiveness
3.1 Inactivation of microorganisms
4 Weaknesses and strengths
4.1 Advantages
4.2 Drawbacks
5 Safety
6 Uses
6.1 Air disinfection
6.2 Water sterilization
6.3 UV tube project
6.3.1 Technology
6.4 Wastewater treatment
6.5 Aquarium and pond

The section on "water sterilization" would likely be most useful.

The appropriate Wikipedia article exists under the title "Ultraviolet germicidal irradiation." It's the 11th suggested article when searching on the original topic, below things like "Glossary of fuel cell terms." Four references are provided in the "Water sterilization" section. Reprinted under CC license 3.0 ("Ultraviolet," n.d.).

To maintain professional credibility, you should move beyond searching the internet to *scholarly* research. In engineering, as in other intellectual disciplines, the most advanced work in any particular field is often first published in a technical journal, whose authors, editors, and primary audience are expert researchers. Articles that appear in reputable journals have been *peer-reviewed*—vetted and validated by experts in the field, undergoing rounds of comment, critique, and revision. Peer review acts as a quality check on information, filtering out content that fails to meet standards in the field. Journal articles are the classic example of peer-reviewed sources, and the professional societies in each engineering discipline (ASCE in civil engineering, ASME in mechanical engineering, and so on) publish a variety of peer-reviewed journals in various technical fields. University presses and other scholarly publishers also employ experts in the field to review the completed text before publication. (Even reputable general-audience publications employ fact-checkers, but the term "peer review" is usually applied only when expert specialists have formally approved publication.)

Using databases to find peer-reviewed sources

When searching peer-reviewed journals, you can still begin within a search engine, using Google's Scholar tool to survey available materials. Google Scholar searches journals, books, theses, and abstracts of all types at the same time; on subjects about which you're not very knowledgeable, such a general search is appropriate.

Google Scholar returns almost 30,000 results, but the search can be refined to show more recent results or specific types of documents.

[PDF] links on the right sometimes provide free access to the article.

Quotes from linked article can help identify most relevant results.

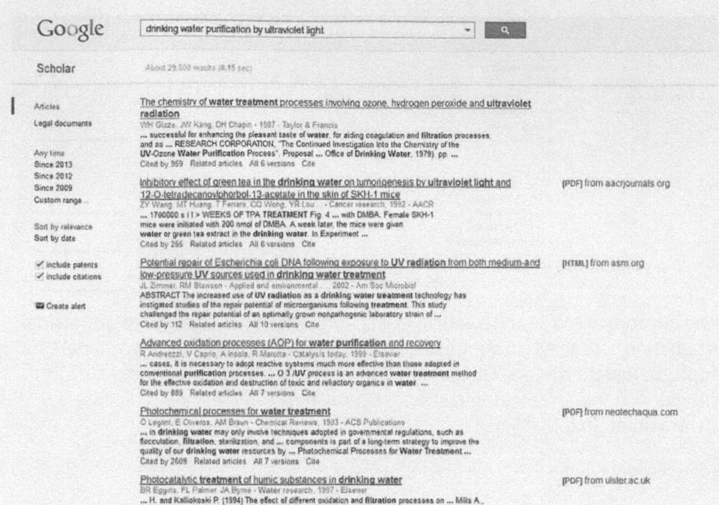

Entering the same search phrase into Google Scholar returns ten pages of journal articles and patents. The most recent article on the first page appeared in 2008 in *Nature*, one of the world's premier scientific journals; the oldest was published in 1987. Retrieved January 7, 2014. Google and the Google logo are registered trademarks of Google Inc., used with permission.

As you become more familiar with a subject (and know some of the commonly used jargon, terms of art, and acronyms), you can move to scholarly databases accessible only from subscribing libraries like those at universities: ScienceDirect, SciFinder, or a database specific to one engineering field, such as IEEE Xplore. If you're looking for only the most recent work on a topic, library databases have a significant advantage over Google Scholar: They can show articles "in press"—accepted but not yet published. Your reference librarian can help you to pinpoint the most relevant database for your particular research topic.

Over 2,000 results are returned, but Science Direct offers powerful tools to refine the results.

The first source in the list is in the journal *Water Research*—likely a great resource on state-of-the-art techniques for UV water purification.

The "show preview" button displays the article's abstract, letting you quickly check an article's relevance.

ScienceDirect (a library database) search results can help you narrow your search. Retrieved January 7, 2014. © 2015 Elsevier B.V. ScienceDirect® is a registered trademark of Elsevier B.V.

Getting the most out of any search tool requires strategic use of its search fields and refinement options. Begin with keyword searches, targeting the words or phrases that would necessarily appear in an article that is relevant to your topic. (You may be able to identify the most relevant items by looking for such keywords in the article title and abstract, rather than just in the full article text.) Limiting the search to results published in the last few years can help show only the most current results—at the risk of excluding a seminal paper that is still important in the field. If you find a particular journal that is especially relevant to the topic of your search, try browsing the contents of its recent issues: This will often turn up useful results that would elude other search strategies.

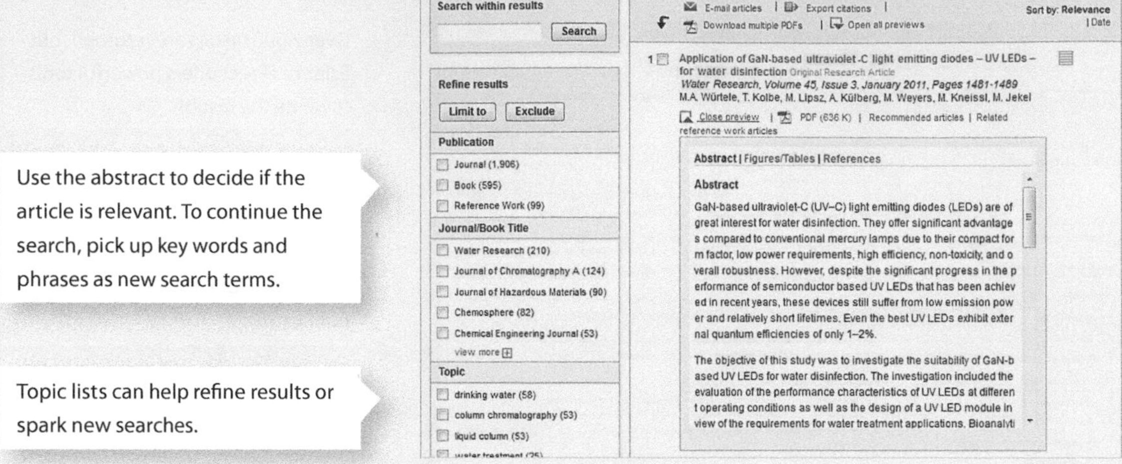

The ScienceDirect database offers many different ways to continue the search, based on the relevance of the current results. © 2015 Elsevier B.V. ScienceDirect® is a registered trademark of Elsevier B.V.

Refining results using Boolean logic

Database searches often yield very large sets of results that need to be narrowed down. You may be able to extract the best resources from the first page or two of search results simply by studying titles and skimming abstracts. Beyond this, you will want to refine the search itself. The best way to do so is to use *Boolean* searches, using basic logical terms called *operators* ("and," "not," "or"):

- If your initial search results contain an identifiable group of irrelevant documents addressing a different topic, use the "not" operator with a keyword to filter them out.
- Combining two important keywords with the "and" operator will produce a smaller, more manageable, and more relevant list of search results than either keyword alone.

Expanding results with keywords and bibliographies

The keywords themselves may also need to be refined or replaced. As you research the topic, adjust the scope of your search depending on the precise information that you need and the number of documents that you are able to locate. You may want either to narrow or to expand the conceptual category, casting a wider net ("composite materials") or being more selective ("fiber-reinforced epoxy"). You can also identify keywords used by the most relevant and useful articles you've found, and then use those keywords to initiate new searches.

Additionally, some search tools allow you to explore citations in the literature on your topic, identifying which articles are most often cited in other articles, or finding all newer articles that have cited the present article with which you are working. Working backward from any article is even easier: You can simply look at the references section for other potentially relevant articles. (This technique can even take you quickly from a Wikipedia article to a peer-reviewed reference suitable for persuading expert audiences.)

17.3 Using patents to review prior art

Research in the field of engineering also commonly includes searching for patents, which is called a search of *prior art*—meaning that you're finding out what other designers have already done. This is important for several reasons:

- To check the "landscape" (since patents disclose the operational details of inventions)
- To check the "freedom of operation"—to discover areas where you are prevented from manufacture, sale, or use of a particular invention
- To begin investigating whether an existing patent is valid

Any plan to manufacture a product should include a patent search. Inventors who patent an invention disclose that invention's details in exchange for a government-granted exclusive right to the invention for a period of time. A patent is granted only to a *novel*, *useful*, and *non-obvious* invention and is valid for approximately 20 years. When the patent expires, then the invention becomes open for others to produce, practice, or adapt.

Patent terms can vary significantly from the commonly considered 20-year default, depending on adjustments, extensions, and the payment of associated fees.

The patent may be issued to the inventors or to their employer.

"CL" refers to the classifications assigned by the Patent Office, categorizing the invention.

The Field of Search shows how the applicants themselves searched patents for prior art.

Related patents are cited by this patent in the left column; non-patent references appear in the right column.

A brief abstract describes the invention, and its parts are numbered in a diagram.

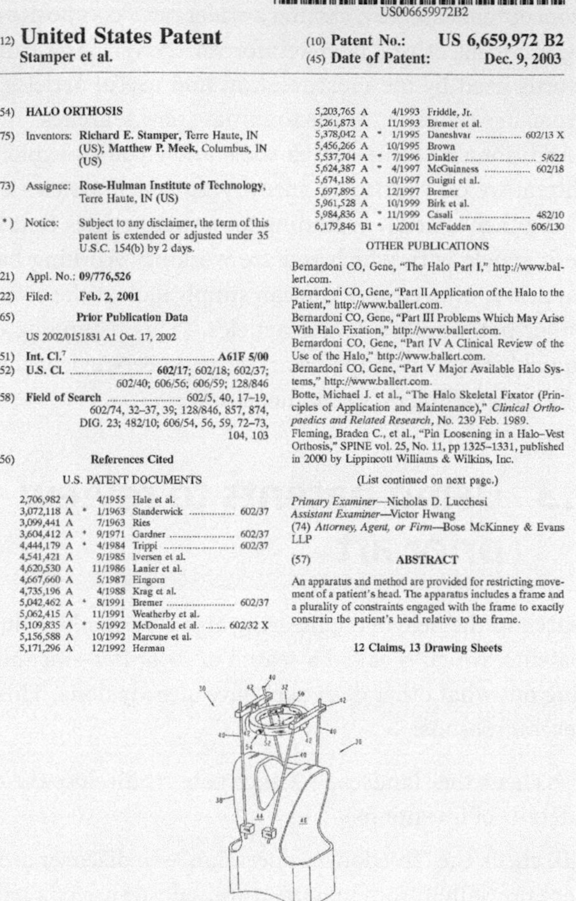

US006659972B2

(12) **United States Patent**
Stamper et al.

(10) **Patent No.: US 6,659,972 B2**
(45) **Date of Patent: Dec. 9, 2003**

(54) **HALO ORTHOSIS**

(75) Inventors: **Richard E. Stamper**, Terre Haute, IN (US); **Matthew P. Meek**, Columbus, IN (US)

(73) Assignee: **Rose-Hulman Institute of Technology**, Terre Haute, IN (US)

(*) Notice: Subject to any disclaimer, the term of this patent is extended or adjusted under 35 U.S.C. 154(b) by 2 days.

(21) Appl. No.: **09/776,526**

(22) Filed: **Feb. 2, 2001**

(65) **Prior Publication Data**

US 2002/0151831 A1 Oct. 17, 2002

(51) **Int. Cl.**[7] **A61F 5/00**
(52) **U.S. Cl.** **602/17**; 602/18; 602/37; 602/40; 606/56; 606/59; 128/846
(58) **Field of Search** .... 602/5, 40, 17–19, 602/74, 32–37, 39; 128/846, 857, 874, DIG. 23; 482/10; 606/54, 56, 59, 72–73, 104, 103

(56) **References Cited**

U.S. PATENT DOCUMENTS

2,706,982	A		4/1955	Hale et al.	
3,072,118	A	*	1/1963	Standerwick	602/37
3,099,441	A		7/1963	Ries	
3,604,412	A	*	9/1971	Gardner	602/37
4,444,179	A	*	4/1984	Trippi	602/37
4,541,421	A		9/1985	Iversen et al.	
4,620,530	A		11/1986	Lanier et al.	
4,667,660	A		5/1987	Eingorn	
4,735,196	A		4/1988	Krag et al.	
5,042,462	A	*	8/1991	Bremer	602/37
5,062,415	A		11/1991	Weatherby et al.	
5,109,835	A	*	5/1992	McDonald et al.	602/32 X
5,156,588	A		10/1992	Marcune et al.	
5,171,296	A		12/1992	Herman	
5,203,765	A		4/1993	Friddle, Jr.	
5,261,873	A		11/1993	Bremer et al.	
5,378,042	A	*	1/1995	Daneshvar	602/13 X
5,456,266	A		10/1995	Brown	
5,537,704	A	*	7/1996	Dinkler	5/622
5,624,387	A	*	4/1997	McGuinness	602/18
5,674,186	A		10/1997	Guigui et al.	
5,697,895	A		12/1997	Bremer	
5,961,528	A		10/1999	Birk et al.	
5,984,836	A	*	11/1999	Casali	482/10
6,179,846	B1	*	1/2001	McFadden	606/130

OTHER PUBLICATIONS

Bernardoni CO, Gene, "The Halo Part I," http://www.ballert.com.
Bernardoni CO, Gene, "Part II Application of the Halo to the Patient," http://www.ballert.com.
Bernardoni CO, Gene, "Part III Problems Which May Arise With Halo Fixation," http://www.ballert.com.
Bernardoni CO, Gene, "Part IV A Clinical Review of the Use of the Halo," http://www.ballert.com.
Bernardoni CO, Gene, "Part V Major Available Halo Systems," http://www.ballert.com.
Botte, Michael J. et al., "The Halo Skeletal Fixator (Principles of Application and Maintenance)," *Clinical Orthopaedics and Related Research*, No. 239 Feb. 1989.
Fleming, Braden C., et al., "Pin Loosening in a Halo–Vest Orthosis," SPINE vol. 25, No. 11, pp 1325–1331, published in 2000 by Lippincott Williams & Wilkins, Inc.

(List continued on next page.)

Primary Examiner—Nicholas D. Lucchesi
Assistant Examiner—Victor Hwang
(74) *Attorney, Agent, or Firm*—Bose McKinney & Evans LLP

(57) **ABSTRACT**

An apparatus and method are provided for restricting movement of a patient's head. The apparatus includes a frame and a plurality of constraints engaged with the frame to exactly constrain the patient's head relative to the frame.

12 Claims, 13 Drawing Sheets

The cover page for a U.S. Patent lists basic information about an invention in a standardized format with which engineers should be familiar. Courtesy U.S. Patent Office (Stamper & Meek, 2003).

You will likely want to search patents granted by the U.S. Patent and Trademark Office (USPTO), the European Patent Office (EPO), and the Japanese Patent Office (JPO). The Chinese Patent Office (CPO) is growing rapidly. Each of these organizations maintains its own up-to-date, complete search engine. However, other commonly used search engines—Free Patents Online, Google Patents, and World Intellectual

Property Association—offer greater ease of use, more customizable searches, and coverage of international patents. Search for both issued patents, which are enforceable, and patent applications, which may become enforceable in the near future.

17.4 Integrating sources: Paraphrase and direct quotation

As you incorporate your sources into a document, you will need to record your use of that source in a bibliography and in in-text citations. In addition to satisfying the ethical obligation to indicate the sources that you have used, your citations enhance your credibility, showing that you are informed by and connected to prior work. Furthermore, they allow readers to independently evaluate your sources and your use of them, and they can also provide a useful list of references for future researchers who may want to mine your bibliography just as you have mined others'. Technical journal articles typically include many references, especially in the introduction, where the authors ground their research in previous work.

Compiling an extensive bibliography of high-quality sources is only the first step, however, in using those sources effectively—even if that list correctly follows the specified citation style. To show readers that you are considering your sources critically and credibly, avoid the pitfalls sometimes experienced by novice researchers.

Avoid insufficient paraphrase

Whenever you are paraphrasing from a source—that is, describing its content in your own way—you must change both the author's wording and the sentence structure. A paraphrase that too closely resembles the original—for example, substituting synonyms for major words but otherwise keeping the original author's sentence structure—plagiarizes the original. Likewise, reordering a sentence's clauses is insufficient to avoid plagiarism if the original language is largely retained. Although professional technical writing rarely includes direct quotations, it is safer to quote directly than to risk inadvertent plagiarism.

The following sentences demonstrate insufficient and sufficient paraphrases of this original sentence, from the discussion section of

"Effects of Robotic Knee Exoskeleton on Human Energy Expenditure" in the *IEEE Transactions on Biomedical Engineering*:

> Our results show that using the robotic knee exoskeleton significantly reduces physiological responses during squatting if an appropriate control approach is applied.

- *Substitution of synonyms but keeping original word order is plagiarism*: "Gams et al. (2013) showed that an assistive exoskeleton device definitely lowers physiological demands while squatting, provided a reasonable control scheme is used."
- *Modifying the sentence structure but keeping original language is plagiarism*: "Gams et al. (2013) showed that if an appropriate control approach is applied, physiological responses during squatting are significantly reduced when using a robotic knee exoskeleton."
- *Changing the sentence structure and wording is necessary to avoid plagiarism*: "Gams et al. (2013) performed experiments with users performing a squatting motion while using a knee-based assistive exoskeleton device with different control schemes. A reduction in physiological responses was achieved for some, but not all, control schemes."

Avoid misuse of direct quotations

As long as your paraphrase is accurate and avoids inadvertent plagiarism, paraphrase is preferable to direct quotation. Technical documents rarely quote their sources directly: Not only will an accurate paraphrase suffice, but it makes the reader's job easier. Unnecessary direct quotation in a technical document often even creates doubts as to whether the author fully understands the material being cited. The journal articles excerpted in this module do not use any direct quotations.

Avoid combining multiple citations for one reference

Technical documents include a citation for *every* reference to a source. Because it is critical to distinguish which technical content comes from the author and which comes from a source, it is necessary to include a citation for every reference—even if it means citing multiple sources in a single sentence or the same source several times within one paragraph. (Again, though, citation styles are designed for different intellectual purposes. If the source is introduced carefully, some citation styles may enable you to cite it only once, even if it is referenced multiple times throughout a paragraph or even an entire page. When in doubt, though, repeat the reference for every sentence containing information from the source.)

Triantafyllidou et al. (2007) showed that particulate lead can be captured in aerators and serve as a potential source for lead release. This was based on the realization from previous work (Edwards and Dudi 2004) that Pb(IV) was being released from the pipe following a disinfectant change from free chlorine to chloramines. In this study, the lead solder was exposed to only free chlorine or chloramines. Thus, Pb(IV) would likely be present in the free chlorine copper pipe but less likely in the chloramine copper pipe. Wang et al. (2012) reported that 55–93% of the overall lead leached in recirculating lead pipes with brass coupled systems, was particulate lead. Wang et al. (2012) observed lead carbonates in the scales of the lead pipe used in their experimental study, whereas in the present, the feed water had a very low DIC and was buffered with blended ortho-polyphosphate. Thus, it is anticipated that the Pb(IV) was being formed in the chlorinated copper system, whereas the chloraminated system would be likely to have lead associated with deposited phosphate or ammonia complexed with Pb.

Findings of other researchers' work are paraphrased, not quoted directly.

It is not unusual to see two or more consecutive sentences cite the same source.

A citation must be used for every reference to a source, even if it requires repeating the reference. © 2013 American Society of Civil Engineers, reprinted with permission (Woszczynski, 2013).

Integrating direct quotes into the text

When direct quotes are included in a technical document, they need to be integrated into the text just as they would in any other document, into a grammatically correct sentence and a meaningful context.

Quotes that have not been introduced are abrupt and confuse the reader:

> Water utilities companies need to evaluate the safety of secondary disinfectants prior to introducing them into the public water supply. "Results show that lead concentrations were approximately of an order of magnitude lower in the free chlorine system than the chloramine systems, even though a corrosion inhibitor was used" [1].

Introducing a source and integrating it into a sentence of your own makes it easier for the reader to follow the logical progression of your ideas:

> Water utilities companies need to evaluate the safety of secondary disinfectants prior to introducing them into the public water supply. Woszczynski et al. found that "lead concentrations were approximately an order of magnitude lower in the free chlorine system than the chloramine systems, even though a corrosion inhibitor was used" [1].

The third sentence of the paragraph is the original text that is quoted below.

Conclusions

This study provides an analysis of lead release in a pipe rig system that was disinfected with either chloramines or free chlorine utilizing phosphate for corrosion control. The study used pipe loops and copper pipe rigs that had lead solder. The lead solder was the only source of lead. Results show that lead concentrations were approximately an order of magnitude lower in the free chlorine system than the chloramine systems, even though a corrosion inhibitor was used. In addition, the role of stagnation time was a factor for lead

The full original text from which a quotation is taken. © 2013 American Society of Civil Engineers, reprinted with permission (Woszczynski, 2013).

Software resources are available to automate parts of the process of preparing citations. Some library databases will provide a properly formatted citation, and even Microsoft Word has a reference feature that will format citations (if the identifying information is entered correctly). Nevertheless, you will need to know the conventions of the citation style to confirm that the output of these tools is correct.

17.5 Citing sources

Although they all include the same basic publication information that readers expect, citation styles vary according to the needs of different intellectual disciplines. Use whichever citation style is expected by your audience. Technical journals require authors to use a certain citation style, or they provide a style sheet that resembles one of the dominant styles (although it may differ on some of the details). Similar style sheets govern technical reports, white papers, and other documents produced within engineering companies and organizations. The major citation styles are updated continually, and many reference manuals can be found online. Accordingly, this book lists only a few sample entries in each citation style.

IEEE

IEEE style is most specific to the fields of engineering and computer science. (IEEE style may be most widely used in the electronic and computing technology fields but is not limited to them.) In this format, bracketed citations assign each source a number in the first sentence within the text in which it is referenced: one refers to source [1], then [2], and so on. These numbers are then maintained throughout the entire document—each number appearing in the text wherever the authors are referring to that source—and the bibliography is then arranged in this order, listing sources in order of their first appearance in the text.

1636 IEEE TRANSACTIONS ON BIOMEDICAL ENGINEERING, VOL. 60, NO. 6, JUNE 2013

Effects of Robotic Knee Exoskeleton on Human Energy Expenditure

Andrej Gams*, Tadej Petrič, Tadej Debevec, and Jan Babič

Abstract—**A number of studies discuss the design and control of various exoskeleton mechanisms, yet relatively few address the effect on the energy expenditure of the user. In this paper, we discuss the effect of a performance augmenting exoskeleton on the metabolic cost of an able-bodied user/pilot during periodic squatting. We investigated whether an exoskeleton device will significantly reduce the metabolic cost and what is the influence of the chosen device control strategy. By measuring oxygen consumption, minute ventilation, heart rate, blood oxygenation, and muscleEMG during 5-min squatting series, at one squat every 2 s, we show the effects of using a prototype robotic knee exoskeleton under three different noninvasive control approaches: gravity compensation approach, position-based approach, and a novel oscillator-based approach. The latter proposes a novel control that ensures synchronization of the device and the user. Statistically significant decrease in physiological responses can be observed when using the robotic knee exoskeleton under gravity compensation and oscillator-based control. On the other hand, the effects of position-based control were not significant in all parameters although all approaches significantly reduced the energy expenditure during squatting.**

Index Terms—**Adaptive control, exoskeletons, metabolic cost, oscillators, squatting.**

I. INTRODUCTION

WEARABLE robots, also known as the exoskeleton devices, are being developed to either augment the abilities of able-bodied humans or to substitute/improve condition of people with impaired physical abilities [1]. Such devices have been the subject of research for many years and various mechanism design and control issues have been discussed in the literature, see, for example, a comprehensive overview by Dollar and Herr [2]. Whether augmenting abilities or improving disabilities, the purpose of assistive exoskeleton devices is to provide useful mechanical power by means of synchronous operation with the person that is wearing it [3].

The first source referenced is assigned the number one, and any time that source is referenced again throughout the document it is cited with [1].

Each new source is assigned the next sequential number, so the second source referenced is assigned a [2].

In-text citations can occur in the middle of a sentence when only part of the sentence references a source, or where the second part of the sentence references a different source—as is the case for [7].

IEEE style does not require referencing each source by that author's name, although it is acceptable to reference the authors for significant sources (such as [12]). For a team of more than two authors, follow the first author's name with "et al."

Dollar and Herr, referenced early in the first paragraph, are again referenced after [12]—so the originally assigned [2] is used again to indicate that it is the same source.

The numbering order then resumes. Following source [2], a new source is referenced, and so it is assigned [13].

The human–robot interface is crucial for a safe and comfortable interaction with a wearable robotic device [4]. Various strategies for detecting user intentions were proposed. Highly invasive strategies, that detect user intentions through the central or peripheral nervous system were demonstrated on primates to control a robotic manipulator [5]. Slightly less invasive are approaches based on surface electromyography (EMG) [6]. The user intentions can be correlated with his/her EMG data using either model-based approach [7] or model-free approach [8]. EMG was also used for the control of the HAL-5 exoskeleton device in prediction of the motion [9] and for adjusting the impedance around the knee joint [10]. An effective fuzzy-neuro controller to automatically control an exoskeletal shoulder robot with EMG signals of the human shoulder muscles was proposed in [11]. Petrič et al. [12] show how EMG measurements can be applied for the control of periodic tasks.

Noninvasive methods, on the other hand, use different force and contact sensors to control the device [2]. Several control methods have been applied using noninvasive methods, for example, [13] discusses a controller which uses inverse dynamics of the exoskeleton as a positive feedback controller so that the loop gain approaches unity, but requires a very accurate model of the system. Banala et al. [14], on the other hand, propose just gravity compensation. Ronsse et al. [1], [15] propose a method based on adaptive frequency oscillators. The method is only applicable for periodic tasks. In this paper, we propose and evaluate a similar yet considerably modified method of controlling a robotic knee exoskeleton.

Simultaneous frequency extraction andwaveform learning allow filtering of ameasured signal with predicting and estimating the output in real time, without delays between the input and the output. Few systems allow simultaneous frequency extraction and waveform learning [16]. For example, Righetti et al. [17] propose a system with pool of adaptive frequency oscillators in a feedback loop. A similar approach was used for the first layer of a two-layered movement imitation system [16]. A novel design of the first layer introduced in [18] uses a single oscillator . . .

In IEEE citation style, references are assigned consecutive numbers placed in brackets. © IEEE, reprinted with permission, from (Gams, et al., 2013).

[1] is a journal article with multiple authors.

[3] appears in a conference proceedings—such papers are written by experts but are generally less authoritative than journal articles.

[27] is a book; books are cited less often than articles in engineering research.

1644 IEEE TRANSACTIONS ON BIOMEDICAL ENGINEERING, VOL. 60, NO. 6, JUNE 2013

REFERENCES

[1] R. Ronsse, T. Lenzi, N. Vitiello, B. Koopman, E. van Asseldonk, S. De Rossi, J. van den Kieboom, H. van der Kooij, M. Carrozza, and A. J. Ijspeert, "Oscillator-based assistance of cyclical movements: Model-based and model-free approaches," *Med. Biol. Eng. Comput.*, vol. 49, pp. 1173–1185, 2011.

[2] A. Dollar and H. Herr, "Lower extremity exoskeletons and active orthoses: Challenges and state-of-the-art," *IEEE Trans. Robot.*, vol. 24, no. 1, pp. 144–158, Feb. 2008.

[3] J. Babič, T. Petrič, T. Debevec, and A. Gams, "Kinematic adaptations during repetitive squatting motions using robotic knee exoskeleton," in *Proc. 21th Int. Workshop Robot. Alpe-Adria-Danube Reg.*, 2012, pp. 313–317.

[4] S. De Rossi, N. Vitiello, T. Lenzi, R. Ronsse, B. Koopman, A. Persichetti, F. Vecchi, A. Ijspeert, H. Van der Kooij, and M. Carrozza, "Sensing pressure distribution on a lower-limb exoskeleton physical human-machine interface," *Sensors*, vol. 11, no. 1, pp. 207–227, 2011.

[5] J. M. Carmena, M. A. Lebedev, R. E. Crist, J. E. O'Doherty, D. M. Santucci, D. F. Dimitrov, P. G. Patil, C. S. Henriquez, and M. A. L. Nicolelis, "Learning to control a brainmachine interface for reaching and grasping by primates," *PLoS Biol.*, vol. 1, no. 2, p. e42, 2003

[6] J. Rosen, M. Brand, M. Fuchs, and M. Arcan, "A myosignal-based powered exoskeleton system," *IEEE Trans. Syst. Man Cybern. Part A: Syst. Humans*, vol. 31, no. 3, pp. 210–222, May 2001.

[7] E. Cavallaro, J. Rosen, J. Perry, and S. Burns, "Real-time myoprocessors for a neural controlled powered exoskeleton arm," *IEEE Trans. Biomed. Eng.*, vol. 53, no. 11, pp. 2387–2396, Nov. 2006.

[8] C. Kinnaird and D. Ferris, "Medial gastrocnemius myoelectric control of a robotic ankle exoskeleton," *IEEE Trans. Neural Syst. Rehabil. Eng.*, vol. 17, no. 1, pp. 31–37, Feb. 2009.

[9] C. Fleischer, C. Reinicke, and G. Hommel, "Predicting the intended motion with emg signals for an exoskeleton orthosis controller," in *Proc. IEEE/RSJ Int. Conf. Intell. Robot. Syst.*, Aug. 2005, pp. 2029–2034.

[10] S. Lee and Y. Sankai. "Power assist control for walking aid with hal-3 based on emg and impedance adjustment around knee joint," in *Proc. IEEE/RSJ Int. Conf. Intell. Robot. Syst.*, 2002, vol. 2, pp. 1499–1504.

[11] K. Kiguchi, K. Iwami, M. Yasuda, K. Watanabe, and T. Fukuda, "An exoskeletal robot for human shoulder joint motion assist," *IEEE/ASME*

[19] R. Ronsse, N. Vitiello, T. Lenzi, J. van den Kieboom, M. Carrozza, and A. Ijspeert, "Adaptive oscillators with human-in-the-loop: Proof of concept for assistance and rehabilitation," in *Proc. IEEE 3rd RAS EMBS Int. Conf. Biomed. Robot. Biomechatron.*, Sep. 2010, pp. 668–674.

[20] Y. Muramatsu, H. Kobayashi, Y. Sato, H. Jiaou, T. Hashimoto, and H. Kobayashi, "Quantitative performance analysis of muscle suit—estimation by oxyhemoglobin and deoxyhemoglobin," in *Proc. IEEE Int. Conf. Robot. Biomimet.*, Dec. 2011, pp. 293–298.

[21] C. J. Walsh, K. Endo, and H. Herr, "A quasi-passive leg exoskeleton for load-carrying augmentation," *Int. J. Humanoid Robot.*, vol. 4, no. 3, pp. 487–506, 2007.

[22] M. Bernhardt, M. Frey, G. Colombo, and R. Riener, "Hybrid force-position control yields cooperative behaviour of the rehabilitation robot lokomat," in *Proc. 9th Int. Conf. Rehabil. Robot.*, Jun./Jul. 2005, pp. 536–539.

[23] J. Veneman, R. Kruidhof, E. Hekman, R. Ekkelenkamp, E. Van Asseldonk, and H. van der Kooij, "Design and evaluation of the lopes exoskeleton robot for interactive gait rehabilitation," *IEEE Trans. Neural Syst. Rehabil. Eng.*, vol. 15, no. 3, pp. 379–386, Sep. 2007.

[24] C. Krewer, F. Müller, B. Husemann, S. Heller, J. Quintern, and E. Koenig, "The influence of different lokomat walking conditions on the energy expenditure of hemiparetic patients and healthy subjects," *Gait Posture*, vol. 26, no. 3, pp. 372–377, 2007.

[25] G. Sawicki and D. Ferris, "Powered ankle exoskeletons reveal the metabolic cost of plantar flexor mechanical work during walking with longer steps at constant step frequency," *J. Exper. Biol.*, vol. 212, no. 1, pp. 21–31, 2009.

[26] T. Lenzi, D. Zanotto, P. Stegall, M. Carrozza, and S. Agrawal, "Reducing muscle effort in walking through powered exoskeletons," in *Proc. IEEE Annu. Int. Conf. Eng. Med. Biol. Soc.*, Sep. 2012, pp. 3926–3929.

[27] D. Winter, *Biomechanics and Motor Control of Human Movement*, 4th ed. New York, USA: Wiley, 2009.

[28] P. Costigan, K. Deluzio, and U. Wyss, "Knee and hip kinetics during normal stair climbing," *Gait Posture*, vol. 16, no. 1, pp. 31–37, 2002.

[29] P. Wretenberg and U. Arborelius, "Power and work produced in different leg muscle groups when rising from a chair," *Eur. J. Appl. Physiol. Occupat. Physiol.*, vol. 68, pp. 413–417, 1994.

[30] J. Pratt, B. Krupp, C. Morse, and S. Collins, "The roboknee: An exoskeleton for enhancing strength and endurance during walking," in *Proc. IEEE Int. Conf. Robot. Autom.*, Apr./May 2004, vol. 3, pp. 2430–2435.

Trans. Mechatronics, vol. 8, no. 1, pp. 125–135, Mar. 2003.

[12] T. Petrič, A. Gams, M. Tomšič, and L. Žlajpah, "Control of rhythmic robotic movements through synchronization with human muscle activity," in *Proc. IEEE Int. Conf. Robot. Autom.*, May 2011, pp. 2172–2177.

[13] H. Kazerooni, J.-L. Racine, L. Huang, and R. Steger, "On the control of the berkeley lower extremity exoskeleton (bleex)," in *Proc. IEEE Int. Conf. Robot. Autom.*, Apr. 2005, pp. 4353–4360.

[14] S. Banala, S. Agrawal, A. Fattah, V. Krishnamoorthy, W.-L. Hsu, J. Scholz, and K. Rudolph, "Gravity-balancing leg orthosis and its performance evaluation," *IEEE Trans. Robot.*, vol. 22, no. 6, pp. 1228–1239, Dec. 2006.

[15] R. Ronsse, N. Vitiello, T. Lenzi, J. van den Kieboom, M. Carrozza, and A. Ijspeert, "Human-robot synchrony: Flexible assistance using adaptive oscillators," *IEEE Trans. Biomed. Eng.*, vol. 58, no. 4, pp. 1001–1012, Apr. 2011.

[16] A. Gams, A. J. Ijspeert, S. Schaal, and J. Lenarčič, "On-line learning and modulation of periodic movements with nonlinear dynamical systems," *Autonom. Robot.*, vol. 27, no. 1, pp. 3–23, 2009.

[17] L. Righetti, J. Buchli, and A. J. Ijspeert, "Dynamic hebbian learning in adaptive frequency oscillators," *Phys. D*, vol. 216, no. 2, pp. 269–281, 2006.

[18] T. Petrič, A. Gams, A. J. Ijspeert, and L. Žlajpah, "On-line frequency adaptation and movement imitation for rhythmic robotic tasks," *Int. J. Robot. Res.*, vol. 30, no. 14, pp. 1775–1788, 2011.

[31] R. Horst, "A bio-robotic leg orthosis for rehabilitation and mobility enhancement," in *Proc. IEEE Annu. Int. Conf. Eng. Med. Biol. Soc.*, Sep. 2009, pp. 5030–5033.

[32] J. J. Jeka and J. R. Lackner, "Fingertip contact influences human postural control," *Exper. Brain Res.*, vol. 79, no. 2, pp. 495–502, 1994.

[33] B. Ainsworth, W. Haskell, S. Herrmann, N. Meckes, D. Bassett Jr., C. Tudor-Locke, J. Greer, J. Vezina, M. Whitt-Glover, and A. Leon, "2011 compendium of physical activities: A second update of codes and met values," *Med. Sci. Sports Exercise*, vol. 43, no. 8, pp. 1575–1581, 2011.

[34] S. Almosnino, D. Kingston, and R. B. Graham, "Three-dimensional knee joint moments during performance of the bodyweight squat: Effects of stance width and foot rotation," *J. Appl. Biomech.*, 2012, To be published

[35] P. Reberšek, D. Novak, J. Podobnik, and M. Munih, "Intention detection during gait initiation using supervised learning," in *Proc. IEEE-RAS 11th Int. Conf. Humanoid Robot.*, Oct. 2011, pp. 34–39.

[36] K. N. Gregorczyk, L. Hasselquist, J. M. Schiffman, C. K. Bensel, J. P. Obusek, and D. J. Gutekunst, "Effects of a lower-body exoskeleton device on metabolic cost and gait biomechanics during load carriage," *Ergonomics*, vol. 53, no. 10, pp. 1263–1275, 2010.

Authors' photographs and biographies not available at the time of publication.

[34] is a journal article not yet published.

A typical references page in IEEE citation style lists sources according to their order of appearance in the body of the text. © IEEE, reprinted with permission, from (Gams, et al., 2013).

IEEE style is recognizable by:

- Bracketed numbers designating each source
- Authors' initials with periods
- Use of quotation marks for the titles of short works (articles, chapters, individual webpages) and italics for long works (journals, books)
- Sentence case (with only the first word and proper nouns capitalized) for short works, and title case (with all major words capitalized) for long works

Article	Author names, "Title of article," *Title of Journal*, vol. [volume number], pp. page numbers, year (and month, if present) of publication [1] S. R. Small, M. E. Berend, M. A. Ritter, C. A. Buckley, and R. D. Rogge, "Metal backing significantly decreases tibial strains in a medial unicompartmental knee arthroplasty model," *J. Arthroplasty*, vol. 26, pp. 777–782, 2011.
Book	Author names, *Title of Book*. City of publication: publisher name, year of publication. [2] S. A. Broughton and K. M. Bryan, *Discrete Fourier Analysis and Wavelets: Applications to Signal and Image Processing*. Hoboken, NJ: Wiley, 2008.

Chapter or part of a book	Author names, "Title of chapter," in *Title of Book*, editor names, Eds. City of publication: publisher name, year, pp. page numbers. [3] C. A. Berry and D. J. Walter, "Application of operational amplifiers," in *Fundamentals of Industrial Electronics*, B. M. Wilamowski and J. D. Irwin, Eds. Boca Raton, FL: CRC Press, 2011, pp. 1–30.
Website content	Website author or sponsor. (Date of access). "Page title." Available: URL. [4] Firebase. (2015, March 6). *Objective-C and Swift iOS Guide: Understanding Security* [Online]. Available: http://www.firebase.com/docs/ios/guide/understanding-security.html.
Technical report	Author names, "Title of report," Organization name, City, State/Nation, Report number, year. [5] S. Mohan, A. Sengupta, and Y. Wu, "Access control for XML: a dynamic query rewriting approach," Indiana University Dept. of Computer Science, Bloomington, IN, Technical Report No. 609, 2005.
Patent	Author names, "Title of patent," U.S. Patent x xxx xxx, Month, day, year. [6] R. E. Stamper and M. P. Meek, "Halo orthosis," U.S. Patent 6 659 972, Dec. 9, 2003.

Author–date

Other widely used citation styles in technical fields prioritize the author and the date, including this information within in-text citations comprising either footnotes or parenthetical references. The bibliography is then usually alphabetized. Such author–date citation styles are widely used across many disciplines, including engineering. Author–date styles used in technical documents are likely to be based on the specific formats defined by the Chicago Manual of Style, the American Psychological Association (APA), or the Council of Science Editors (CSE).

Case Study

Comparison of Chlorine and Chloramines on Lead Release from Copper Pipe Rigs

Meghan Woszczynski[1]; John Bergese[2]; and Graham A. Gagnon[3]

Abstract: The objective of this study was to assess lead release in a pipe rig system that was disinfected with either chloramines or free chlorine. The study was carried out using pipe loops and copper pipe rigs that had lead solder, which provided the only source of lead. The water quality of the treated water had a low alkalinity (<5 mg/L as $CaCO_3$), neutral pH, and low hardness (<5 mg/L as $CaCO_3$). However, the study used a corrosion control program that consisted of dosing with 0.8-mg PO_4/L of zinc orthophosphate and controlling the pH to 7.3, that was consistent with the corrosion-control program operated in the host water treatment plant. Key findings from the study confirmed that chloramines would result in lead release under the current corrosion-control program, whereas free chlorine was not as compromising. Lead concentrations were approximately an order of magnitude lower in the free chlorine system than the chloramine systems. In this study, chloramines with a target residual concentration of 5 mg/L released 382 μg/L and 49 μg/L following stagnation times of 24 h and 30 min, respectively. Furthermore, chloramines with a target residual concentration of 1 mg/L resulted in a lead release of 73 μg/L and 14 μg/L following a stagnation time of 24 h and 30 min, respectively. By comparison, the pipe rigs that were dosed with a free chlorine target residual concentration of 1 mg/L had lead concentrations of 12 μg/L and 2 μg/L for the 24-h and 30-min stagnation times. This project demonstrates that care needs to be taken when evaluating secondary disinfectants, particularly for those water systems having a low dissolved inorganic carbon concentration. **DOI: 10.1061/(ASCE)EE.1943-7870.0000712.**

CE Database subject headings: Lead; Copper; Corrosion; Chlorine; Drinking water; Water pollution; Pipes.

Author keywords: Lead; Copper; Corrosion control; Chloramines; Free chlorine; Orthophosphate.

Introduction

Lead and lead-tin solder were widely used in premise plumbing in North America until the 1980s. This soldering practice has declined significantly over the years, but lead release from older premise plumbing remains a concern. To address concerns associated with lead release in drinking water, corrosion control strategies such as pH adjustment (Knox et al. 2005; Tam and Elefsiniotis 2009), ensuring adequate buffering capacity of dissolved inorganic carbon (DIC) and pH (Korshin et al. 2000; Dryer and Korshin 2007), and the use of chemical inhibitors, such as phosphate (Maddison et al. 2001; Nguyen et al. 2011), stannous chloride (Hozalski et al. 2005), and sodium silicate (Lintereur et al. 2010) have been reported in the literature. For systems having a very low DIC, chemical inhibitors such as phosphate have been reported as means to mitigate corrosion and metal release (Maddison et al. 2001). For systems utilizing phosphate for corrosion control, the operating pH range to ensure its effectiveness is narrow and is typically in the range 7.0–7.5 (AWWARF 1996).

To remain in compliance with current and future regulations concerning disinfection by-products (DBPs) [i.e., trihalomethanes (THMs) and/or haloacetic acids (HAAs)] many water providers have switched, or are considering switching, to chloramines as an alternative to chlorine for disinfection. Disinfection with chloramines produces fewer DBPs (i.e., THMs and HAAs), is an effective disinfectant and provides a stable and persistent residual along the distribution system. Within Nova Scotia, there is evidence that HAA formation poses significant challenges to utilities (McDonald et al. 2006), which is largely related to the high content of hydropillic organic matter in the raw water (Lamsal et al. 2012). As a treatment solution, there is evidence that chloramination can have negative impacts on the delivered drinking water quality (Renner 2004). It has been shown that chloramines have a dramatic effect on lead services even after 1 min flushing (Edwards and Dudi 2004). Decomposition of chloramines has been found to be proportional lead [as Pb(II)] release through a reduction of lead oxide (Lin and Valentine 2008). Lead oxide, PbO_2, reduces to Pb(II). Pb(II) is the lead ion that is soluble and contributes to the amount of lead. While Edwards and Dudi (2004) found that chlorine reacts with soluble Pb(II) to rapidly precipitate to insoluble lead at the bench scale level.

To address lead release, phosphate has been offered as a corrosion inhibitor to prevent lead release under specific pH conditions (Nadagouda et al. 2009). However, there is limited information for drinking water containing low levels of DIC and the associated ability of phosphate inhibitors to prevent lead release in the presence of chloramines.

Thus, the objective of this paper was to compare lead release in premise plumbing that was disinfected with either free chlorine or chloramines. The only source of lead for this study was from lead/tin solder, which was used to connect copper pipe and simulate premise plumbing. The study was conducted at the pilot-scale level with a finished drinking water that has an alkalinity of less than 2 mg/L as $CaCO_3$ and effectively utilizes phosphate in their distribution system for lead control. The role of stagnation time as a

[1]Dept. of Civil and Resource Engineering, Dalhousie Univ., Halifax, NS, Canada B3H 4R2. E-mail: meghan.wos@gmail.com

[2]Dept. of Civil and Resource Engineering, Dalhousie Univ., Halifax, NS, Canada B3H 4R2. E-mail: jbergese@dal.ca

[3]Dept. of Civil and Resource Engineering, Dalhousie Univ., Halifax, NS, Canada B3H 4R2 (corresponding author). E-mail: graham.gagnon@dal.ca

Note. This manuscript was submitted on September 18, 2012; approved on March 11, 2013; published online on March 13, 2013. Discussion period open until January 1, 2014; separate discussions must be submitted for individual papers. This paper is part of the ***Journal of Environmental Engineering***, Vol. 139, No. 8, August 1, 2013. © ASCE, ISSN 0733-9372/2013/8-1099-1107/$25.00.

J. Environ. Eng. 2013.139:1099-1107.

If multiple sources are referenced in a single sentence or part of a sentence, they can be referenced within the same parenthetical citation. For a team of more than two authors, follow the first author's name with "et al."

The third sentence in the first paragraph references eight sources. Each parenthetical citation applies to the part of the sentence that directly precedes it.

These references summarize the primary findings of the authors cited, as the introduction surveys what has already been accomplished in the research field. Such citations do not quote articles directly.

A typical article using an author–date citation style, where the author's last name and date of the reference are parenthetically included in the text. © 2013 American Society of Civil Engineers, reprinted with permission (Woszczynski, 2013).

applied, dissolved copper was the predominant form present, which is consistent with results presented by Wang et al. (2012).

Conclusions

This study provides an analysis of lead release in a pipe rig system that was disinfected with either chloramines or free chlorine utilizing phosphate for corrosion control. The study used pipe loops and copper pipe rigs that had lead solder. The lead solder was the only source of lead. Results show that lead concentrations were approximately an order of magnitude lower in the free chlorine system than the chloramine systems, even though a corrosion inhibitor was used. In addition, the role of stagnation time was a factor for lead and copper release in premise plumbing: a higher lead and copper concentration was found in a 24-h stagnation time (long) than a 30-min stagnation time (short). For the long stagnation time (24 h), the highest total lead concentration was 382 μg/L when the system was dosed with chloramines at a high target dose of 5 mg/L. In comparison, the system dosed with free chlorine at a target value of 1 mg/L released a total lead concentration of 12 μg/L for the same stagnation time (24 h). The particulate fraction of lead was higher than the dissolved fraction in systems dosed with chlorine and chloramines. For copper, the most predominant form was dissolved copper for both disinfectants. These results indicate that water utilities need to evaluate secondary disinfectants' impact on lead corrosion in premise plumbing. This is particularly important for water systems having a low dissolved inorganic carbon concentration.

Acknowledgments

The authors thank Halifax Water and the Natural Sciences and Engineering Resource Council of Canada (NSERC) for financial support to the NSERC/Halifax Water Industrial Research Chair. Additional funding support for this project was obtained from the Canadian Water Network (CWN). Additionally, the authors also acknowledge and thank research support staff at Dalhousie University (Halifax, Nova Scotia, Canada), specifically Heather Daurie, and at the J.D. Kline Water Supply Plant (Halifax, Nova Scotia, Canada).

References

Arnold, R. B., and Edwards, M. (2012). "Potential Reversal and the Effects of Flow Pattern on Galvanic Corrosion of Lead." *Environ. Sci. Technol.*, 46(20), 10941–10947.

AWWARF. (1996). *Internal corrosion of water distribution systems*, 2nd Ed., American Water Works Association Research Foundation, Denver.

Berthouex, P. M., and Brown, L. C. (2002). *Statistics for environmental engineers*, 2nd Ed., Lewis Publishers, Boca Raton, FL.

Boulay, N., and Edwards, M. (2001). "Role of temperature, chlorine, and organic matter in copper corrosion by-product release in soft water." *Water Res.*, 35(3), 683–690.

Boyd, G. R., Shetty, P., Sandvig, A. M., and Pierson, G. L. (2004). "Lead in tap water following simulated partial lead pipe replacements." *J. Environ. Eng.*, 130(10), 1188–1197.

Cartier, C., Arnold, R. B., Triantafyllidou, S., Prevost, M., and Edwards, M. (2012). "Effect of flow rate and lead/copper pipe sequence on lead release from service lines." *Water Res.*, 46(13), 4142–4152.

Churchill, D. M., Mavinic, D. S., Neden, D. G., and MacQuarrie, D. M. (2000). "The effect of zinc orthophosphate and pH–alkalinity adjustment on metal levels leached into drinking water." *Can. J. Civ. Eng.*, 27(1), 33–43.

Deshommes, E., Laroche, L., Nour, S., Cartier, C., and Prévost, M. (2010). "Source and occurrence of particulate lead in tap water." *Water Res.*, 44(12), 3734–3744.

Dryer, D. J., and Korshin, G. V. (2007). "Investigation of the reduction of lead dioxide by natural organic matter." *Environ. Sci. Technol.*, 41(15), 5510–5514.

Edwards, M., and Dudi, A. (2004). "Role of chlorine and chloramine in corrosion of lead-bearing plumbing materials." *J. Am. Water Works Assoc.*, 96(10), 69–81.

Eisnor, J. D., and Gagnon, G. A. (2003). "A framework for the implementation and design of piot-scale distribution systems." *J. Water Supply: Res. Technol.–AQUA*, 52(7), 501–519.

Eisnor, J. D., and Gagnon, G. A. (2004). "Impact of secondary disinfection on corrosion in a model distribution system." *J. Water Supply: Res. Technol.–AQUA*, 53(7), 441–452.

Gagnon, G. A., and Doubrough, J. D. (2011). "Lead release from premise plumbing: a profile of sample collection and pilot studies from a small system." *Can. J. Civ. Eng.*, 38(7), 741–750.

Gagnon, G. A., et al. (2008). "Disinfectant efficacy in distribution systems: a pilot-scale assessment." *J. Water Supply: Res. Technol.–AQUA*, 57(7), 507–518.

Halifax Water. (2010). *Halifax Water, 14th Annual Rep.*, Halifax Water, Halifax, NS.

Hozalski, R. M., Esbri-Amador, E., and Chen, C. R. (2005). "Comparison of stannous chloride and phosphate for lead corrosion control." *J. Am. Water Works Assoc.*, 97(3), 89–103.

Kim, E. J., Herrera, J. E., Huggins, D., Braam, J., and Koshowski, S. (2011). "Effect of pH on the concentrations of lead and trace contaminants in drinking water: A combined batch, pipe loop and sentinel home study." *Water Res.*, 45(9), 2763–2774.

Knowles, A. D., MacKay, J., and Gagnon, G. A. (2012). "Pairing a pilot plant to a direct filtration water treatment plant." *Can. J. Civ. Eng.*, 39(6), 698–700.

Knox, G., Mavinic, D., Atwater, J., and MacQuarrie, D. (2005). "Assessing the impact of corrosion control measures on tap drinking of the Greater Vancouver Regional District." *Can. J. Civ. Eng.*, 32(5), 948–956.

Korshin, G. V., Ferguson, J. F., and Lancaster, A. N. (2000). "Influence of natural organic matter on the corrosion of leaded brass in potable water." *Corrosion Sci.*, 42(1), 53–66.

Lamsal, R., Montreuil, K. R., Kent, F. C., Walsh, M. E., and Gagnon, G. A. (2012). "Characterization and removal of natural organic matter by an integrated membrane system." *Desalination*, 303, 12–16.

Lin, Y. P., and Valentine, R. L. (2008). "Release of Pb(II) from monochloramine—mediated reduction of lead oxide (PbO_2)." *Environ. Sci. Technol.*, 42(24), 9137–9143.

Lintereur, P. A., Duranceau, S. J., Taylor, J. S., and Stone, E. D. (2010). "Sodium silicate impacts on lead release in a blended potable water distribution system." *Desalination and Water Treatment*, 16(1–3), 427–438.

Maddison, L. A., Gagnon, G. A., and Eisnor, J. D. (2001). "Corrosion control strategies for the Halifax regional distribution system." *Can. J. Civ. Eng.*, 28(2), 305–313.

McDonald, B., MacDonald, J., and Gagnon, G. A. (2006). "Haloacetic acid occurrence in drinking water in Nova Scotia and the implications for small treatment systems." *Proc., 12th Nation. Drinking Water Conf.*, Canadian Water and Wastewater Association (CWWA), Ottawa, ON.

McNeill, L. S., and Edwards, M. (2004). "Importance of Pb and Cu particulate species for corrosion control." *J. Environ. Eng.*, 130(2), 136–144.

Nadagouda, M. N., Schock, M. R., Metz, D. H., DeSantis, M. K., Lytle, D., and Welch, M. (2009). "Effect of phosphate inhibitors on the formation of lead phosphate/carbonate nanorods, microrods, and dendritic structures." *Cryst. Growth Des.*, 9(4), 1798–1805.

Nguyen, C. K., Clark, B. N., Stone, K. R., and Edwards, M. (2011). "Acceleration of galvanic lead solder corrosion due to phosphate." *Corrosion Sci.*, 53(4), 1515–1521.

Renner, R. (2004). "Plumbing the depths of D.C.'s drinking water crisis." *Environ. Sci. Technol.*, 38(12), 224A–227A.

1106 / JOURNAL OF ENVIRONMENTAL ENGINEERING © ASCE / AUGUST 2013

J. Environ. Eng. 2013.139:1099-1107.

The first reference listed is a journal article with two authors.

The second reference listed is a book published under the name of an organization rather than an individual.

A report or white paper issued by an organization or company, such as Halifax Water, is listed under the organization's name and has its title italicized.

The abbreviations "Proc." and "Conf." typically indicate a paper published in the proceedings of a technical conference.

In author–date citation style, references pages list sources alphabetically by the last name of each item's first listed author. © 2013 American Society of Civil Engineers, reprinted with permission (Woszczynski, 2013).

Many publications use subtle variants of the major author–date styles. However, it usually isn't too difficult to tell which style is being used or adapted.

Chicago author–date style is recognizable by:

- Authors' full names
- Use of quotation marks for the titles of short works (articles, chapters, individual webpages) and italics for long works (journals, books)

- Title case (with all major words capitalized) for all titles

Article	Author names. Year of publication. "Title of Article." *Title of Journal* volume number: page numbers. Small, Scott R., Michael E. Berend, Merrill A. Ritter, Christine A. Buckley, and Renee D. Rogge. 2011. "Metal Backing Significantly Decreases Tibial Strains in a Medial Unicompartmental Knee Arthroplasty Model." *The Journal of Arthroplasty* 26: 777–782.
Book	Author names. Year. *Title of Book*. City of publication: publisher name. Broughton, S. Allen and Kurt M. Bryan. 2008. *Discrete Fourier Analysis and Wavelets: Applications to Signal and Image Processing*. Hoboken, NJ: Wiley.
Chapter or part of a book	Author names. Year. "Title of Chapter." In *Title of Book*, edited by editor names. City of publication: publisher name. Berry, Carlotta A. and Deborah J. Walter. 2011. "Application of Operational Amplifiers." In *Fundamentals of Industrial Electronics*, edited by Bogdan M. Wilamowski and J. David Irwin. Boca Raton, FL: CRC Press.
Website content	Website author or sponsor. Year. "Page title." Date of access or last modification. URL. Firebase. 2015. "Objective-C and Swift iOS Guide: Understanding Security." Accessed March 3, 2015. http://www.firebase.com/docs/ios/guide/understanding-security.html.
Technical report	Author names. Year. "Title of Report." Report number, City of publication: organization name. Mohan, Sriram, Arijit Sengupta, and Yuqing Wu. 2005. "Access Control for XML: A Dynamic Query Rewriting Approach." Technical Report No. 609. Bloomington, IN: Indiana University Dept. of Computer Science.
Patent	Author names. Year. Title of patent. U.S. Patent x,xxx,xxx, filed [date], and issued [date]. Stamper, Richard E. and Matthew P. Meek. 2005. Halo orthosis. U.S. Patent 6,659,972, filed February 2, 2001, and issued December 9, 2003.

APA style is recognizable by:

- Authors' initials followed by periods
- Parentheses around years
- Absence of quotation marks for short works but use of italics for long works
- Sentence case (with only the first word and proper nouns capitalized) for article and chapter titles

Article	Author names. (Year of publication). Title of article. *Title of Journal,* volume number, page numbers. Small, S. R., Berend, M. E., Ritter, M. A., Buckley, C. A., & Rogge, R. D. (2011). Metal backing significantly decreases tibial strains in a medial unicompartmental knee arthroplasty model. *The Journal of Arthroplasty,* 26, 777–782.
Book	Author names. (Year). *Title of Book.* City of publication: publisher name. Broughton, S. A. and Bryan, K. M. (2008). *Discrete Fourier Analysis and Wavelets: Applications to Signal and Image Processing.* Hoboken, NJ: Wiley.
Chapter or part of a book	Author names. (Year). Title of chapter. In editor names (Eds.), *Title of Book* (pp. page numbers). City of publication: publisher name. Berry, C. A. and Walter, D. J. (2011). Application of operational amplifiers. In B. M. Wilamowski and J. D. Irwin (Eds.), *Fundamentals of Industrial Electronics* (pp. 1-30). Boca Raton, FL: CRC Press.
Website content	Website author or sponsor. (Year). *Page title.* Retrieved [date of access] from URL. Firebase. (2015). *Objective-C and Swift iOS Guide: Understanding Security.* Retrieved March 3, 2015, from http://www.firebase.com/docs/ios/guide/understanding-security.html.
Technical report	Author names. (Year). *Title of Report* (Report Number). City of publication: organization name. Mohan, S., Sengupta, A., and Wu, Y. (2005). *Access Control for XML: A Dynamic Query Rewriting Approach* (Technical Report No. 609). Bloomington, IN: Indiana University Dept. of Computer Science.

Patent	Author names. (Year). *U.S. Patent x,xxx,xxx*. City of issue: nation of issue. Stamper, R. E. and Meek, M. P. (2005). *U.S. Patent 6,659,972*. Washington, DC: U.S.

CSE style is recognizable by:

- Authors' initials without periods
- Titles using neither italics nor quotation marks
- Sentence case for all titles, including books

Article	Author names. Year of publication. Title of article. Abbreviated Title of Journal volume number: page numbers. Small SR, Berend ME, Ritter MA, Buckley CA, Rogge RD. 2011. Metal backing significantly decreases tibial strains in a medial unicompartmental knee arthroplasty model. J Arthroplasty 26: 777–782.
Book	Author names. Year of publication. Title of book. City of publication (state/nation): publisher name. Broughton SA, Bryan KM. 2008. Discrete Fourier analysis and wavelets: applications to signal and image processing. Hoboken (NJ): Wiley.
Chapter or part of a book	Author names. Year of publication. Title of chapter. In: [editor names], editors. Title of book. City of publication (state/nation): publisher name. p. page numbers. Berry CA, Walter DJ. 2011. Application of operational amplifiers. In Wilamowski BM, Irwin JD, editors. Fundamentals of industrial electronics. Boca Raton (FL): CRC Press. p. 1–30.
Website content	Page title. [Updated [date of last update]]. Website author or sponsor; [cited [date of access]]. Available from: URL. Objective-C and Swift iOS guide: understanding security [Internet]. Firebase; [cited 2015 March 3]. Available from http://www.firebase.com/docs/ios/guide/understanding-security.html

Technical report	Author names. Year. Title of report. City of publication: organization name. Report number. Mohan S, Sengupta A, Wu Y. 2005. Access control for XML: a dynamic query rewriting approach. Bloomington, IN: Indiana University Dept. of Computer Science. Technical Report No. 609.
Patent	Inventor names, inventors; [patent holder], assignee. Date issued. Title of patent. United States patent US x,xxx,xxx. Stamper RE, Meek MP, inventors; Rose-Hulman Institute of Technology, assignee. 2003 Dec. 9. Halo orthosis. United States patent US 6,659,972.

Summary

Consulting with experts

- Researching prior and current work in a technical field will stimulate new ideas, improve processes, and help you make better decisions.
- Experienced colleagues and outside consultants can often supply expertise from their own experience and prior work.
- When you need to locate published work on a technical topic, reference librarians can help you to identify appropriate professional resources, such as journal articles and patents.

Finding scholarly sources

- Standard search engines and online resources, such as Google and Wikipedia, can provide very basic introductions to technical content, However, credible engineering research effort should identify up-to-date, expert-level sources such as peer-reviewed articles from technical journals.
- As you search, refine your results. As you learn about a topic, you will become familiar with commonly used keywords, phrases, and ideas; all can be used to create new searches that meet your research needs more precisely.
- One of the easiest ways to find relevant resources is to locate the sources cited in the bibliographies of the articles that you have already found.

Using patents to review prior art

- Patent searches are useful to identify inventions that have addressed technical needs similar to the ones on which you're working. They can also determine whether an existing patent might prevent the manufacture, sale, or use of a particular concept or invention.

- Patents follow a highly structured, information-dense format—but once you can follow that format, you can understand the history of related inventions.

Integrating sources: Paraphrase and direct quotation

- Correctly and completely citing your references is an important step in satisfying the ethical obligations of research.
- References connect your work to past work, showing that you are aware of what has been done before, enhancing your credibility.
- When citing prior work, care must be taken to avoid insufficient paraphrase, misusing direct quotations, combining citations for multiple sources, and orphan quotes.

Citing sources

- In the document text, IEEE style cites each source with a bracketed number that indicates the order in which sources are first cited.
- Author–date styles—including Chicago author–date, APA, and CSE—cite sources with the author's last name and year of publication in parentheses.
- Each style's bibliography entries have their own characteristic formats for the order of information, capitalization, and punctuation.

Drafting

18

Polished, professional documents require several drafts. The need for multiple drafts is most obvious for major engineering reports and proposals that might represent weeks or months of work. The most successful technical professionals, though, often use a multi-draft process even for a one-page progress report, or an important email—any message that merits a bit of thought about how best to help audiences to understand content, to evaluate its significance, and to take necessary actions.

The exact nature of your drafting process matters less than the fact that you use a deliberate process. Effective writers recognize from the outset that their final drafts most often look quite different from what they first produce, and they have developed routines and habits of mind that lead them purposely through the intervening stages.

As a rule, the early stages of drafting involve gathering ideas and information and trying out different ways of organizing this preliminary content. As long as it's subject to a rigorous process of revision, the quality of an early draft's prose matters little: More than anything else, such drafts represent the writer's process of working through the various parameters shaping the writing task, from the substance of the main ideas to different audiences' likely reactions.

Objectives

- To use outlining, free-writing, and idea-mapping techniques for generating, organizing, and prioritizing ideas at the early stages of drafting documents
- To generate prose for early drafts while retaining an open mind about whether and how it might appear in later drafts
- To apply criteria for evaluating scope, argument, and evidence to decide when a draft is sufficiently complete to shift the decision-making process from major content decisions to clarity and style

18.1 Planning the argument

Many engineers bring two misconceptions to the writing component of a project:

- Writing is a secondary task that follows real technical work in the lab or at the plant; we arrive at answers first, then write them up.
- Technical writing is about transmitting accurate information, not making arguments.

We may develop these beliefs from the familiar routine of writing lab reports in high school science classes: One performs an experiment, obtains results, and then writes them up. Reflections on the content are largely limited to questions about whether the report matches what the instructor has in mind. Moreover, because the whole activity has been designed to illustrate well-understood principles and techniques, there isn't much to argue about. The writing process thus appears to involve little more than "filling in the blanks" in a template—perhaps explaining students' general lack of enthusiasm for the activity of writing lab reports.

Authentic engineering design and research, however, don't lend themselves to such a straightforward, sequential process. Even if you *are* writing up an experiment and expect to use the well-known IMRaD format, the whole activity involves much more uncertainty; you must continually ask yourself the same kinds of critical questions that you expect from others:

- Have the results been tainted by flaws in the design of the experiment?
- Have I taken the correct measurements using the correct techniques?
- Do my findings contradict prior work by other researchers?
- Are my recommendations fully warranted by the evidence?

A complex project will require competent engineers to reflect on these and many other questions. It is, at best, highly unlikely that we will have answered all such questions before writing the experimental report. If we postpone our writing to the last step before a deadline and such questions *do* arise, of course, there's little that we can do: There's no time to revise the design of the experiment, make new measurements, or review the relevant published research in the area.

The intellectual demands of industry and of university research therefore urge a process in which writing proceeds in parallel with the other tasks that define the project. This may happen in a long series of project documents (proposals for the activities that you want to conduct, instructions for procedures that technicians will need to follow, correspondence about early experimental results, etc.), or you may just draft the project report or documentation in a continuous process as work proceeds.

Techniques for generating ideas

Whether starting a new project or just a new section of an ongoing report or proposal, it can be difficult to sit down to a blank screen and to begin composing. (Even professional writers find this daunting.) Techniques are available, though, for initiating and advancing the writer's process of problem solving and deliberation—often called "pre-writing," acknowledging that we are not yet attempting to craft the prose that audiences will eventually read.

Student writers often attempt to produce such prose right away, attempting to comfort themselves with progress toward a required page or word count. Measuring progress in this way, though, creates incentives to be wordy, inefficient, and aimless. Pre-writing counters this mindset, categorizing whatever words we produce as raw material for further processing. Pre-writing strategies also benefit your thinking by freeing you from the constraints of a formal document, allowing for broader creativity while feeding your productivity with the constraints of a clear, concrete task. (To advance that creativity further, it often helps to engage in these processes on paper or a whiteboard rather than the computer.)

Lists	Write down as many topics as you can think of that you need to address in the document. You might have a main ideas list and a supporting ideas list.
Idea maps	Write down key terms and group them together. The purpose of this exercise is to allow you to represent your ideas visually and to draw relationships between ideas. You might arrange your ideas hierarchically, sequentially, or in a web to illustrate the multiple connections between ideas.
Timed writing	Select a relatively short, specific amount of time to write—such as 10 or 15 minutes—and during that time write continuously. It doesn't matter what you write as long as you write. Don't worry about topic sentences, sentence structure, or word choice. The purpose of this exercise is to free you from the constraints of trying to write a polished document and to get your ideas on paper. At the end of the allotted time, review what you've written and identify the most useful ideas that emerged.

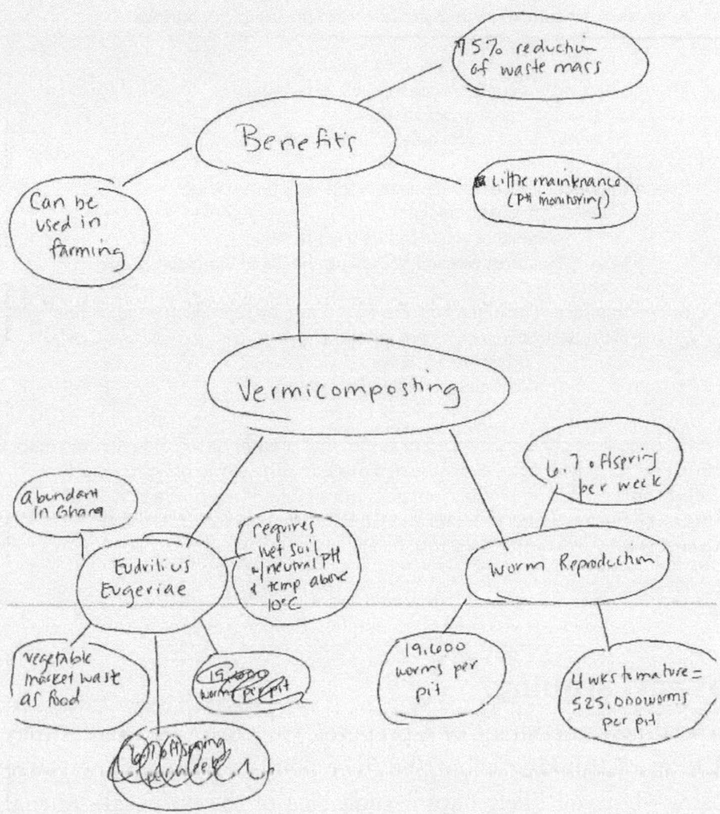

For an Engineers without Borders project in Ghana, this idea map begins to organize a proposal for a vermicomposting system—one that produces compost as earthworms digest waste and excrete castings that can be used as organic fertilizer. © 2014 Rebecca Hecht, Noura Sleiman, and Giuliana Watson. Reprinted with permission.

Changing one's medium—hand-writing ideas rather than entering them within software—can help to generate new ideas and perspectives.

In idea maps, you place the broadest main topic in the center of a page, then define major content areas that radiate outward. Each of these then likewise forms the center of a cluster of subtopics.

For documents of large scale, it might work best to use several idea maps, one for each major content area.

Mapping ideas can be a visibly messy process. For instance, adding detail may lead to the realization that an idea formerly treated as a secondary subtopic actually needs to be more prominent.

Many engineering students use these techniques for other forms of writing, but they seldom think of them as applicable to technical documents. In fact, these strategies work well even for documents that follow heavily formulaic structures: Even if the sections for your proposal are dictated by a sponsor's RFP, an idea map or list of ideas may be the best way to start within any section.

Formulating main ideas in extended phrases (or even complete sentences) can help to generate a structure of well-defined headings.

The outline form arranges topics in a more linear sequence than an idea map. Nevertheless, the same structure of hierarchy subordinates main ideas, supporting ideas, and details.

Ideas' logical relationships and relative importance are more important than any particular system of numbering items at different levels.

Main idea: Vermicomposting would extend the life of pit latrines

Supporting idea: Benefits of vermicomposting
Details: Earthworm compost can be used in farming
75% reduction of waste mass
Monitoring pH levels only maintenance required

Supporting idea: Eudrilius Eugeniae earthworm best choice
Details: Abundant in Ghana
Vegetable waste can be used for food
Requires wet soil with neutral pH and temp above 10C

Supporting idea: The initial cost can be reduced by reproducing the worms
Details: 19,600 worms per pit initially
6.7 offspring per week
4 wks to mature = 525,000 per pit

A list or outline can accommodate somewhat greater detail than an idea map. Such detail may only be available at a more mature stage of a project, though, after some time for development and refinement of major ideas. © 2014 Rebecca Hecht, Noura Sleiman, and Giuliana Watson. Reprinted with permission.

What am I arguing?

When writing something like a proposal, your drafting process may entail a lot of thinking about the finer points of the actions you're proposing—but you likely have a good idea of the proposal's crucial main ideas. This means that you can begin with the key reasons that you think audiences will need to acknowledge and accept in order to assent to your call for action.

Other genres may require more upfront consideration of the persuasive aim of your writing. If you're reporting the results of an experiment, for instance, you may spend a lot of time describing the experimental procedure before you ever worry about interpreting its results. Eventually, however, your work's direction can really only be determined and refined once you've settled on its central purpose. Technical documents don't normally define a single-sentence "thesis statement" like analytical papers in other academic disciplines, but drafting a document becomes easier and more efficient as its key assertions come into focus for you.

Building a logical structure

Outlining will enable you to further organize your ideas. Identify the needed sections for the document as well as the subsections for each section. For each section and subsection, identify the needed evidence and, if relevant, sources that will need to be cited. If you can then transform your list of section topics into assertions, you will advance both your thinking about the content and the speed with which you can draft sections of your first draft.

18.2 The sequence of drafts

Some time spent listing, diagramming, outlining, and free-writing can capture important thinking that we don't want to lose—but that thinking isn't yet in a coherent form ready to show to others. To move your content into that stage, it helps to have a sense of the rough stages through which your series of drafts will pass.

Producing the original prose: Early "prototype" drafts

Building a prototype is a crucial milestone in engineering design: However many hours are spent working it out on paper, the design must become a real, tangible object before its viability can be judged. The first versions of a document are prototypes in the same way: You're advancing ideas into the new stage of sentences and paragraphs and establishing the logical connections between your ideas.

As you write, you may be most comfortable drafting your document in a linear fashion from introduction to conclusion, but that sequence is not necessary. You might try starting with the sections dealing with the content in which you have the most confidence or the most expertise. Alternatively, you may decide to begin with the sections that state the most important findings, crafting your central message even if that part of the document is more challenging. Doing so may help you to make decisions about the evidence and reasoning that you need to present.

> Sentences in this draft lack polish, but they successfully capture information from the authors' sources and the specifications of their design.

Earthworm selection

A survey of earthworms in the Ghana was done and many different types of earthworm species were found that were capable of vermicomposting. Of the types of worms species found on location the "Eudrilius Eugeniae", also known as "the African Night crawler", is the best choice for the community since it is abundant in different areas in Ghana especially in Accra. [3] The earthworms require wet soil that has a neutral pH and a temperature above 10 degrees Celsius to survive. Ten degrees Celsius was lower than the average soil temperature on site. All of these conditions are met in the Gomoa Gyaman which makes this project feasible for the community.

> The authors' early draft follows the sequence in which the design would be implemented—choosing earthworms, building the latrine to the design specifications, and then introducing the worms into the pit.

Latrine Design

The community presented EWB with four possible sites to implement latrines. At any given site, we have designed a twin-pit system - two pits per site - for optimal availability to the community. Each pit will support ten stalls. Each of these latrines will be properly ventilated with space above and below each door, as well as PVC pipe leading from the pit to the roof. The roof will be slanted away from the front of the latrine to encourage runoff towards the back of the pit. There will be a series of cement slabs at the front and the back of each structure in order to access the pit.

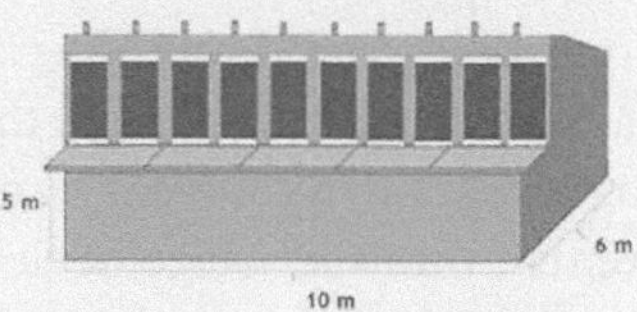

We have designed each pit to be 10 m long, 6 m wide, and 3.5 m deep, for a total size of 210 m^3. Each stall will be 1 m long, 2 m wide, and 2 m tall. These dimensions require 2.5 m of PVC pipe per stall. These dimensions were calculated to sustain the local population at each location of 625 people. Using the average waste production of 0.1 m^3 per person per day, and the worm conversion rate, each pit will fill after 2.4 years. After this time, the each pit will need to be closed for 0.46 years. Our suggestion is to have these on a rotation in such a way that the closed months do not overlap. If this schedule is chosen, then both pits will be in use 75% of the 3 year cycle. Since this is the case, the latrines will be able to sustain for a longer period of time, and has the potential to extend the 3 year cycle to perhaps a 5 year cycle.[1]

> The authors have entered rough calculations into early draft paragraphs but haven't yet adapted them to make them meaningful to readers ("0.46 years" is an unwieldy measurement of time).

Optimization of Earthworms

Since the African Nightcrawler earthworms are readily available in Ghana we would not need to introduce an invasive species. For the design of our latrines we would like to optimize the earthworms breeding and decomposition of waste. The more earthworms in the pit the larger the amount of waste that can be decomposed. If there are too many earthworms in the pit becomes harder for the earthworms to move around and they become less likely to reproduce. In order to balance the two, 150 adult earthworms per cubic feet are required in a pit [4]. At the optimum production volume the earthworms with reach sexual maturity thirty-five days after hatching.

> The problems with the chronological order emerge as the first draft proceeds. For instance, this structure steers the reader away from the vermicomposting worms only to return to them later.

The "prototype" draft of a design proposal puts the design team's decisions into formal written form so that the team can evaluate the major content and organization.

It is appropriate to do some rewriting of your prose while you work with early drafts, as long as your focus is on the larger picture. (Some level of consciousness regarding the precision of your language can help with refining your ideas and is important to the document as a whole.) Remember, though, that some sentences and perhaps even entire sections may need to be cut when the document is revised, so stressing over minor details is unwarranted at this stage. Instead, make a note to return to sentences and paragraphs with which you are not satisfied.

Like a design prototype, an early draft has the crucial purpose of detecting and analyzing unforeseen problems. Most likely, critical examination will reveal that some paragraphs or information will prove to be unnecessary or ineffective. As you would with a design prototype, be willing to take parts of a draft "back to the drawing board": It may be frustrating to abandon material that contains some of your hard work, but you'll eventually be more satisfied with the final product for having done so.

Identifying a logical organization

The most effective use of a prototype draft is often to study and revise the organization of ideas. Often the initial outlining of topics is not the most logical or efficient and needs to be reworked. "Reverse outline" your document by extracting the main idea for each paragraph, ideally the topic sentence. In evaluating your outline, consider where a reader might have trouble following your ideas. Does a reader need to be familiar with one concept to understand another? Is there a logical progression between topics? Are there topics introduced that are not developed or that seem tangential? In addition, watch out for any repetition of topics so that you can eliminate any redundancy.

Successive stages of revision: Later "pre-production" drafts

As the problems in early prototypes are studied and solved, a design team moves carefully out of prototyping and into work with more advanced models. The design may not yet be ready for production, but the team likely knows what works well and what needs incremental improvements. Likewise, you'll reach a complete draft in which you have defined the scope of your topic and the major sections of content that you are certain will remain in its final version. Early "prototype" drafts will give way to later "pre-production" drafts once you've made a few decisions about them:

- You've thought critically about the necessity of all of the information present in the draft.
- You've made all necessary adjustments to the scope of the document.
- You've identified sections or content areas that need the most additional development.
- You've settled major content decisions—how to frame problems, for instance—and can dedicate your attention more completely to the craft of writing at the paragraph and sentence levels.

The specific techniques of revision at this stage are discussed in Module 19: *Revising*.

Formerly separate sections have been combined; the organization is now topical, making it easier for audiences to follow.

Some rearrangement has occurred at the sentence level: For instance, the point about avoiding invasive worm species has moved into a new paragraph.

Plenty of room for revision remains at this stage: The sequence of ideas within sections and paragraphs can still change. Such considerations are more immediate than grammar and punctuation.

Earthworm Selection and Optimization

After researching different types of earthworms capable of vermicomposting we recommend the "Eudrilius Eugeniae" which is also known as "the African Night crawler". This earthworm is the best choice for the community since it is abundant in different areas in Ghana especially in Accra. [3] Since the African Night crawler earthworms are readily available in Ghana, we would not need to introduce an invasive species. The earthworms require wet soil that has a neutral pH and a temperature above ten degrees Celsius to survive. Ten degrees Celsius is lower than the average soil temperature on site. All of these conditions are met in Gomoa Gyaman, which makes this project feasible for the community.

For the design of our latrines we would like to optimize the earthworms breeding and decomposition of waste. The size of the pit corresponds directly to the amount of earthworms and waste that can be decomposed. If there are too many earthworms in the pit it becomes harder for the earthworms to move around and they become less likely to reproduce. In order to balance the two, 150 adult earthworms per cubic feet are required for every pit [4]. At the optimum production volume, the earthworms will reach sexual maturity thirty-five days after hatching.

Benefits of Using Vermicomposting

Using a dual pit latrine can reduce the time people need to wait for the worms to finish composting. With each individual community member producing 0.5 kg per day of waste the pits can remain open for two and a half years without maintenance. At this rate, the pits will require half a year of closure to allow the waste to completely decompose. Most of the maintenance on the pits occurs at the beginning building stage of the latrines and after the latrines are filled. Between these times there is little maintenance requirements for the system. After the pits are filled and the waste has finished decomposing, the compost must be removed from the pits and then the latrines will be available for use again. Besides emptying the pits, the only maintenance needed for the system is to monitor the pH levels.

The draft has advanced into a "pre-production stage" once fundamental revisions have adjusted the sequence of content. © 2014 Rebecca Hecht, Noura Sleiman, and Giuliana Watson. Reprinted with permission.

Summary

Planning the argument

- The intellectual demands of industry and of university research urge a process in which writing proceeds in parallel with the other tasks that define the project. The exact nature of your drafting process matters less than the fact that you use a deliberate process.
- Techniques are available for initiating and advancing the writer's process of problem solving and deliberation—often called "pre-writing." Lists, timed writing, and idea maps can help to generate ideas that will make the process of drafting easier.
- Outlining will enable you to further organize your ideas. If you can then transform your list of section topics into assertions, you

will advance both your thinking about the content and the speed with which you can draft sections of your first draft.

The sequence of drafts

- Like a design prototype, an early draft has the crucial purpose of detecting and analyzing unforeseen problems. Most likely, critical examination will reveal that some paragraphs or information will prove to be unnecessary or ineffective. As you would with a design prototype, be willing to take parts of a draft "back to the drawing board."
- The most effective use of a prototype draft is often to study and revise the organization of ideas. "Reverse outline" your document by extracting the main idea for each paragraph, ideally the topic sentence. In evaluating your outline, consider where a reader might have trouble following your ideas.
- Early "prototype" drafts will give way to later "pre-production" drafts after you have defined the scope of your topic and the major sections of content that you are certain will remain in its final version.

19 Revising

As your draft's major ideas and sections fall into place, the nature of your writing work will gradually shift: You'll spend less time crafting wholly new paragraphs and more time shaping and reworking existing material. Eventually, your draft will be ready for the proverbial "fine-toothed comb": you'll be able to read through the material with an eye toward catching stylistic mistakes in such areas as punctuation and capitalization. It takes practice, though, to manage this progress in an efficient and purposeful way.

Engineering problem solving encourages a focus on the correctness of details: You're well trained to insist on precision and accuracy in handling units of measurement and significant figures. These analytical habits help many engineers to realize the same virtues of precision and accuracy in their writing. Other engineers, though, become preoccupied too early with the details of writing mechanics, resulting in wasted effort. It isn't productive to correct the commas or dashes in a paragraph that is later rewritten from scratch—or that is cut from your document altogether. You can avoid this pitfall by working methodically down from the argument, evidence, and organization, through the clarity and coherence of paragraphs and sentences, and into the elimination of small stylistic errors.

Objectives

- To distinguish among the individual tasks within the revision process—revising (improving substance and clarity), editing (polishing the details), and proofreading (correcting stylistic errors)
- To use feedback from experts and peers to revise content and argument
- To identify structure and organization revisions to headings and paragraphs that strengthen the document
- To improve clarity by strengthening transitions, refining word choices, and proofing for errors

19.1 From revising to editing to proofreading

Common usage often treats "revising" and "editing" as loose synonyms, but expert writers understand a crucial distinction. The questions below illustrate the types of thinking that define the stages of your process, but they present only a few of the many inquiries that you're sure to make at each stage.

Revising *Re-vision* asks you to "see anew" your material—adding to it, giving it a new shape, or otherwise transforming it. You may be working within an individual paragraph or even crafting a single sentence, but your priorities continue to reside in the "big picture":

- Does the draft capture accurate information and sound ideas?
- Can readers distinguish main points from supporting details?
- Is evidence supplied for all assertions that need it?

Editing	Editing optimizes your material to create prose that is more efficient, more economical, and easier to use: • Where are readers likely to misunderstand a concept or an inference? • Which paragraphs' topic sentences don't yet introduce and explain a main idea? • Which sentences need to be streamlined because they have grown unwieldy and hard to follow? • Do all of the subject–verb–object sentence cores reflect the most important content?
Proofreading	Proofreading identifies and corrects mistakes: • Do all verbs and pronouns agree in number with their subjects and antecedents? • Are independent clauses within sentences separated with colons and semicolons, avoiding comma splices? • Are proper nouns correctly and consistently capitalized throughout the document?

It is possible for experienced writers to move back and forth among these mindsets. Adjustments at even the smallest scale—adding a punctuation mark or changing a preposition—can lead you to read a few sentences, redirecting your attention to larger-scale issues of clarity, coherence, or persuasiveness. Such small-scale adjustments also require some double-checking to confirm that they work within the larger document. On the other end of the spectrum, focusing on higher-order substance shouldn't prevent you from correcting misspellings or punctuation errors when you spot them.

Almost all writers, though, benefit from working through the stages in a disciplined way. It might seem easier to jump from a first draft to the proofreading stage, and you might be more confident in your ability to "fix mistakes" than in your ability to revise more substantively. Giving in to that temptation, though, means that you're missing out on potential improvements that can make a much bigger difference.

19.2 Revising content and argument based on feedback from experts and peers

Plenty of productive revisions can be made without others' input: As you revise, you're developing and refining an internal sense for the characteristics of clear, well-crafted prose. (It's especially effective to return to a draft after some time away: A fresh point of view aids meaningful re-vision.) Still, even the most insightful writers test their material by recruiting readers to deliver feedback.

In academic settings and in the world of published research, *peer review* is an intrinsic part of technical writing: Technical journals publish contributors' work only after it has passed stringent review by experts in the field (see Module 17: *Researching*). In industry, such formal review processes are less common—but feedback from colleagues is highly valued and is a standard part of most engineering work.

Whether in writing or engineering courses, student engineers can develop revision and editing skills through peer review activities. In this case, peers serve a different function from expert reviewers: They're not verifying your content, but letting you test your material on real readers. A few basic habits of mind will help you to get the most out of peer review:

- As a reviewer, focus on substance—not on grammatical mistakes, spelling errors, and awkward sentence mechanics. As a writer, ask your reviewers to do the same.
- The time that you spend reviewing others' work isn't just a charitable service for them: Your thinking about their work will benefit your own writing at least as much.
- Reviewers often enable us to understand and develop more fully our initial ideas and can suggest new directions that we had not considered. You don't have to implement every specific suggestion, but you should always consider the possibility that the suggestion identifies a real shortcoming.
- Don't dismiss peers' feedback if they misunderstand your content: Their inability to grasp your point may indicate that your intended primary audience will experience similar misunderstandings. (Non-expert peers' advice is thus especially valuable on documents intended for a non-expert audience.)

Disagreement or counterargument

If a reviewer disagrees with your argument or offers a counterargument, try to identify the source of the underlying differences that separate your position from the reviewer's. Try to understand the nature of the differences: Do you begin with different assumptions? Interpret the evidence differently? Emphasize different implications?

You may not need to completely abandon your argument, but you will at least need to refine it to address this disagreement. Generally, arguments are stronger after being tested—especially if your revisions can respond to reviewers' concerns or ideas in a way that recognizes their validity or importance, avoiding any hint of defensiveness or dismissiveness.

Adding evidence and clarifying reasoning

One of a critical reviewer's most important services is to indicate where evidence is insufficient or unpersuasive. Start responding by supplying evidence where it is absent: Intelligent readers will be deeply skeptical of unsupported assertions. If you believe that decisive evidence is already present, you may need to add emphasis or explicitness to your reasoning about it, or to explain key concepts or conclusions more clearly.

If possible, ask the reviewer directly what would make your argument more persuasive. Interactive, two-way conversations about a document are almost always more productive than written feedback alone: Author and reviewer alike can explore ideas and work past any misconceptions or faulty theories about what might not be working.

Earthworm Selection and Optimization

After researching different types of earthworms capable of vermicomposting we recommend the "Eudrilius Eugeniae" which is also known as "the African Night crawler". This earthworm is the best choice for the community since it is abundant in different areas in Ghana especially in Accra. [3] Since the African Night crawler earthworms are readily available in Ghana, we would not need to introduce an invasive species. The earthworms require wet soil that has a neutral pH and a temperature above ten degrees Celsius to survive. Ten degrees Celsius is lower than the average soil temperature on site. All of these conditions are met in Gomoa Gyaman, which makes this project feasible for the community.

For the design of our latrines we would like to optimize the earthworms breeding and decomposition of waste. The size of the pit corresponds directly to the amount of earthworms and waste that can be decomposed. If there are too many earthworms in the pit it becomes harder for the earthworms to move around and they become less likely to reproduce. In order to balance the two, 150 adult earthworms per cubic feet are required for every pit [4]. At the optimum production volume, the earthworms will reach sexual maturity thirty-five days after hatching.

Commented [A1]: Include why you recommend this earthworm in your topic. Your rationale and bibliography will demonstrate that you did your research

Commented [A2]: This phrasing is confusing because one sentence uses "above" and the next uses "lower" in comparison to 10 C. Either state what the average soil temperature is or at least use the same kind of comparison

Merely stating that research was done is much less effective than demonstrating the results of that research.

In this case, the reviewer's advice about word choice appears sound. Whenever a reviewer reports confusion, explore possible revisions for clarity.

A reader's fresh viewpoint can be especially helpful in assessing organization and logical priority. This is often evident when a draft creates confusion about which points are most central to the argument.

The reviewer notices an attempt to minimize a significant problem—revision here will significantly improve the reader's trust in the solution.

Benefits of using Vermicomposting

Using a dual pit latrine can reduce the time people need to wait for the worms to finish composting. With each individual community member producing 0.5 kg per day of waste the pits can remain open for two and a half years without maintenance. At this rate, the pits will require half a year of closure to allow the waste to completely decompose. Most of the maintenance on the pits occurs at the beginning building stage of the latrines and after the latrines are filled. Between these times there is little maintenance requirements for the system. After the pits are filled and the waste has finished decomposing, the compost must be removed from the pits and then the latrines will be available for use again. Besides emptying the pits, the only maintenance needed for the system is to monitor the pH levels.

Commented [A3]: Is this the primary benefit of vermicomposting though? Begin with the most compelling argument for vermicomposting before supplementary arguments.

Commented [A4]: In the last sentence you state that the community will need to monitor pH levels. In addition, 2 paragraphs above you state that the project is feasible because "all of these conditions are met" but do not indicate how the community would monitor pH levels. Acknowledging any maintenance and potential difficulties upfront rather than minimizing it builds trust in your reader.

Early review feedback should focus on the soundness of the argument as well as the clarity of ideas—the content. At this stage, it helps to think of the document as a "pre-production" draft instead of an initial "prototype" draft.

19.3 Revising structure and organization

Like the pre-writing strategies that help to build your "prototype" draft, early revision activities often concern the core logic of the content—its fundamental concepts, arguments, and evidence. We often enhance this central content most effectively when working with the document's organization, because its structural features tell the reader so much about what's most important. Most of our work in this area addresses two related questions:

- Does the organization give obvious logical priority to the main ideas, clearly subordinating the rest of the content to them?
- Can readers grasp those main ideas without getting bogged down in the details?

When revising, you will likely work more effectively with a paper copy. Paper helps you to see relationships among sections and topics and the degree to which you've developed and supported them: Where detail is absent or excessive, sections or paragraphs will stand out as being too long or too short.

Headings

Headings enable a reader to skim through a document quickly and still glean its main points. To check their effectiveness:

- Skim through the document—reading your headings and topic sentences—to determine whether a hurried reader can understand findings and arguments with general accuracy based on these prominent elements alone.
- Check the hierarchical organization of the headings. A heading expresses the main claim of a section, while subheadings signal supporting points that define subsections. Ideas within a particular heading level should be comparable in importance.
- Unify headings in length and complexity, and make sure that they are grammatically parallel.

Paragraphs

Paragraphs are the building blocks for an effectively organized document. To check your paragraphs for coherence:

- Scan through the document and check to see whether the paragraphs are of relatively uniform length. An unusually short or long paragraph may signal a step in your argument or exposition where you need to elaborate or reorganize.
- Identify a single focused idea for each paragraph. If you cannot easily do so, revisit the organization of the paragraph, and divide it into multiple paragraphs if needed.
- Make the topic sentences of your paragraphs explicit and easy to identify.

19.4 Revising and editing for clarity

Later in the revision process, clarity should become the primary focus. Reading a document aloud can allow you to hear abrupt transitions and awkward language: If you trip over a sentence when reading it aloud, mark it as likely needing revision. The tips below highlight some of the most common weaknesses in undergraduate writing. The modules in the *Components* section provide a more in-depth explanation of these topics and other ways to improve your writing.

Strengthening transitions

Transition words show the relationships between ideas. Avoid connecting ideas with vague pronouns and ineffective and wordy phrases ("this means that . . ."). Transitions connect ideas but can also signal a shift in logic, so select a transition phrase that indicates the appropriate relationship between ideas:

Agreement (and)	*in addition, also, too, likewise, moreover, furthermore*
Contrast (but, or)	*however, rather, instead, although, while, whereas, despite*
Cause and effect	*because, as, since, if . . . then, therefore, consequently, as a result*
Illustration	*such as, including, for example, particularly, especially, specifically*
Time	*previously, afterwards, immediately, simultaneously*

Refining word choices

Check sentences for these constructions—which are wordy and imprecise—and then revise:

- Sentences beginning with "there is" and "there are"
- Sentences beginning with "this is" and "that is"—often making unclear pronoun references. Make sure that each pronoun has a clear antecedent (i.e., that the pronoun "this" refers back to a specific noun in the previous sentence, not to the entire sentence).
- Broad, vague main verbs, such as *to be, to get, to have.* Use precise verbs—and active ones when possible.

Proofreading for errors

Finally, proofread for the smallest, most localized errors, such as misspellings and punctuation errors. Sentence fragments, comma splices, and fused sentences are most frequent in many engineers' writing and are discussed in some detail in Module 24: *Sentences.*

After revising and editing your document, you'll recognize improvement from the initial drafts. You may see the document as perfect and ready for submission, but ask for one final read-through by someone who hasn't already been involved in the revision. You've invested significant time and thought into making your writing more effective, and one more look at the document can catch the very last grammar mistake or weak transition sentence.

Benefits of vermicomposting

Vermicomposting increases the capacity of a pit latrine because the waste is being processed while the latrine is in use. With each individual community member producing 0.5 kg per day of waste, the pits can remain open for 2.5 years before being closed for 0.5 year to allow the worms to decompose the waste fully. Using a dual pit latrine system allows one side of the latrine to be operational while the other side is closed for decomposing.

[...]

Selecting the earthworm population

Gomoa Gyaman hosts an ideal earthworm species for vermicomposting. *Eudrilius Euganiae* ("the African nightcrawler") is native to Ghana and is particularly abundant in Accra [3]. These earthworms require wet soil with a neutral pH and a temperature above 10° C. The average soil temperature on site exceeds 10°C, and nearby rivers provide a sufficient source of water for wetting the soil during the initial set-up of the pits—leaving only the pH levels of the soil to be monitored by the Gomoa Gyaman community.

To decompose waste as quickly as possible, we recommend 150 adult earthworms per cubic feet for every pit [4]. Increasing the number of earthworms in the pit above this ratio will not decompose waste more quickly, and immobile earthworms in an overcrowded pit are less likely to reproduce. At the optimum production volume, the earthworms will reach sexual maturity thirty-five days after hatching.

[...]

In the original, the major benefit was not clearly stated in the heading.

This paragraph's topic sentence has been revised to reach its main point more quickly. Evidence then shows that *Eudrilius Eugeniae* is appropriate for the location and imposes a small monitoring burden on the local community.

Elaborating on the benefits of the general approach (vermicomposting) provides the reader with the context needed for the discussion of individual factors, such as the selection of an appropriate earthworm species.

After making the appropriate revisions above, this section of the report more clearly communicates the most compelling arguments being made.

Summary

From revision to editing to proofreading

- Revision focuses on improving the substance of writing, ensuring that the correct pieces are present and that the overall document fits them together optimally.
- Editing focuses on polishing the writing—improving clarity at the level of sentence structure and word choice.
- Proofreading focuses on the correction of errors in spelling, punctuation, capitalization, and the like.
- Some movement back and forth among these tasks is likely, but follow a deliberate sequence of stages focused first on revising, then on editing, and finally on proofreading.

Revising content and argument based on feedback from experts and peers

- Peers and experts will bring an outside perspective to your document and notice things that you will miss.
- Ask reviewers to focus on revising the substance while you focus on editing only after addressing their concerns.
- Expert reviewers can improve your document by identifying weaknesses in the argument, places where more evidence is needed, and sections that are confusing or inadequately explained.

Revising structure and organization

- Headings should capture the main points and hierarchical organization of the document.
- Headings at the same level should be comparable in importance and be grammatically parallel.
- Paragraphs should be of similar length, with each focusing on a single idea stated clearly in its topic sentence.

Revising and editing for clarity

- After the content has been revised, focus on increasing clarity by reading the document aloud and listening for abrupt transitions and awkward language.
- Strengthen transitions, signaling the relationships between ideas.
- Refine word choice by clarifying pronoun references and using precise and active verbs when possible.
- Proofread for typographic and punctuation errors: sentence fragments, comma splices, and fused sentences.

20 Collaborating

Teams in engineering workplaces commonly practice collaborative writing. A typical project requires the expertise of numerous colleagues, much of which needs to be collected into a single document. Documents with multiple authors pose twin challenges—to coordinate the authors' work purposefully and to present it in the single voice of the organization or team.

Objectives

- To distinguish tasks best accomplished as a team from those best accomplished individually
- To assign roles that leverage the strengths of the team members in formulating content, producing draft prose, revising, editing, and project management
- To unify a document with several authors, optimizing its consistency in organization, voice, style, format, and handling of content
- To manage the collaborative writing process in a way that minimizes unproductive conflict and perceptions of unfairness

20.1 Avoiding the "divide-and-conquer" approach

Like an individual author, a writing team proceeds through the stages of researching, planning, drafting, and revising. For collaboration to be effective, however, it is important to coordinate multiple individuals' involvement in all of these steps.

Student teams often attempt to complete collaborative assignments by reducing them to separate sets of tasks to be completed by individual members before the final compilation of the document—the "divide-and-conquer" approach. When writing, the simplest way to "divide-and-conquer" is to assign separate sections to each author. Assigning individual team members to write the introduction, methods, results, and discussion sections—with no shared planning or coordinating beyond merging the files—is not collaborating; it is merely compiling individual work. Moreover, the fundamental problems of "divide-and-conquer" exceed mere stylistic discrepancies and impair the underlying content. Documents prepared in this way often contain contradictions, inconsistencies, redundancy, logical omissions, and sudden leaps in reasoning that result from the absence of transitions. When readers encounter such issues, their understanding is impeded and the team's credibility is undermined.

Teams that avoid these problems typically use two strategies to do so:

Shared planning process	The entire team plans the document together. Ideally, this process is facilitated by a *project manager*.
Lead writer	A designated lead writer takes charge of the document's organization and style after collecting drafted material from the other members of the team.

Sharing the planning process

A more effective model of collaboration includes a shared planning process, preferably conducted in face-to-face conversations. The early stages of the process will require more time than the divide-and-conquer approach because the team will have to discuss issues of content, organization, and style that are often neglected as individuals tackle sections separately. However, this investment of time up front ultimately creates a greater savings of time during the revision stage. For example, it takes more time rewriting many headings to meet a common style than it does to decide together on a style before drafting. In short, the collaborative model produces better drafts and a satisfactory final product more quickly, making it more rewarding for everyone involved.

In her book *Team Writing: A Guide for Working in Groups*, Joanna Wolfe suggests that the team designate a *project manager* to oversee the collaborative planning process. Wolfe carefully distinguishes the project manager from a "boss" who gives orders to teammates. Rather, the project manager contributes to the project by dedicating his or her attention to the big picture rather than to an individual part of the document or content. Wolfe assigns the following tasks to the project manager; several of them are specified more completely in Module 21: *Meeting*:

- Creating and maintaining a *task schedule* that defines responsibilities and deadlines
- Enforcing the terms and rules of the team charter
- Organizing team meetings and their results using *agendas* and *minutes*
- Initiating the drafting stage by creating a *straw document*—a first preliminary draft (see Module 18: *Drafting*)

To bear these responsibilities successfully, a project manager will likely need to have other team duties reduced or waived. Placing a single individual in charge of both management and a specific content area typically means that neither task benefits fully from the person's capabilities and attention. (It is, of course, possible to divide the project management tasks among the team members, although most teams find such complex divisions of labor harder to coordinate and sustain.)

Assigning a lead writer

A project manager fully dedicates his or her time and effort to the coordination of project tasks rather than to generating prose. Appointing a *lead writer*, on the other hand, means freeing up a team member to focus solely on that prose. If one team member has developed special skills in organizing and formulating ideas, then the team can marry those skills to those of the other members. Lead writers on successful teams must be able to trust their teammates, who are working equally hard to make content decisions, gather research materials, compile data, and do all of the other necessary work.

A suggested process for working with a lead writer is detailed in the "Drafting by roles" section below. Working carefully through such a process can help to guarantee that all members of the team are carrying a fair share of the load and that their contributions are valued appropriately.

20.2 Planning a document as a team

Collaborative writing is a common professional practice because individuals bring different areas of knowledge to the document. Similarly, they can also bring their different perspectives to the planning of the document, producing a more effective document than a single author could. Collaborative planning consists of five major steps, undertaken as a team, in a fixed order, preferably in face-to-face meetings:

- Define your purpose.
- Profile your audience.
- Create an outline.
- Define a style sheet.
- Draft an introduction.

Define your purpose

The team members should begin by stating their goals explicitly and in detail to ensure that there is agreement. State every inference you want the audience to make and the effects that you expect your argument to elicit. In addition, identify the document's genre and list ways in which the genre affects the document design and content.

Profile your audience

Analyzing the audience as a team will enable you to quickly sketch out a well-developed profile. Is the primary audience an engineer, an executive, a client, a government agency, etc.? Is there a secondary audience? What background information does the audience need? What actions or decisions will these audiences face after working with your material? Likely, each member will have different insights into the needs of your audience.

Create an outline

It is crucial to complete the entire outline as a team before drafting. This step requires the largest time investment but produces the greatest rewards in the quality of the resulting drafts. As a team, one section at a time, and for every section:

- List every topic that belongs in a section.
- Arrange the topics to support the logic of your argument.
- Restate the topics as assertions in complete sentences.
- List the evidence that supports each topic. If you don't yet have the evidence, keep a list of actions required to obtain it.
- Note sources for each item that will need to be cited.
- Review the overall outline repeatedly, questioning every step in your argument as you plan it: Is the assertion necessary? Is it correct? Is it in the right place?

Creating a detailed outline benefits the team in two ways. First, it reduces the number of contradictions, inconsistencies, and unnecessary repetitions to which team-authored drafts are prone. Second, it asserts the team's expectations for the first draft. The more detailed the outline, the more likely it is that each individual writer will meet his or her teammates' expectations.

Team authors take responsibility for the outline by including their initials.

This outline was developed in a team meeting to coordinate the main points of the analysis section of an experimental report.

The main steps in the analysis of the results are decided by the team and listed on the outline.

Supporting evidence about each step appears as bullets. Notes to the team highlight when a reference is made to another section of the report—avoiding repetition of information.

Because this outline will never be shared externally, jargon, acronyms, and deep technical language are appropriate if understood by everyone on the team. (Variable names will be replaced with appropriate math notation in the final version.)

Required team decisions are documented so they aren't forgotten.

The outline concludes with a summary statement that ties this section to the overall goals of the report.

Mass-spring-damper parameter identification experiment

Team 4: TEO, VEM, JW
6/7/2013

Analysis section outline

Use data to compute overshoot Mp

- Show the definition of Mp from the book (needs reference)
- Refer to the table in the results section showing measured variables xp, xss, and x0. (get table number from results author)
- Substitute values and calculate Mp.

Having calculated Mp, continue calculation to find the period Td and gain K.

- Refer to the previous graph in the results section to show that calculated Td is reasonable. (Make sure this graph exists!)
- Show the gain calculation.

With these values calculated from the experimental results, initialize the simulation and show the simulation results.

- Show the math model that drives the simulation.
- Show full Simulink model in an appendix.

Display the simulation results and the experimental data together

- Show that parameters have been tweaked to create a best match (eyeball fit is sufficient).
- Should we show one representative graph here? (Team decision needed).

Conclude with: The parameters developed above are our best estimate of the unknown parameters that make the model fit the data. Connect this result to the report goals.

A detailed outline of the analysis section of an experimental report. The team meets in person to prepare the section outlines that coordinate individual work.

Define a style sheet

Many technical professionals are familiar with *style sheets* from web publishing, where cascading style sheet (CSS) files are used for encoding the details of the format in which a webpage is displayed—fonts, sizes, colors, spacing, and the like. (The same kind of code for managing document style is also used in markup languages like TeX.)

Even the simplest documents, of course, require management of these parameters:

- The fonts and sizes to be used in the document
- Formats for headings, lists, tables, and figure captions
- Placement of figures, tables, lists, and other formats that define the document design
- Spacing at all levels of the document—within paragraphs, between paragraphs, and between sections
- Margins, headers, and footers
- Citation format (see Module 17: *Researching*)

Novice writing teams are often tempted to dismiss these items as cosmetic, leaving them to be resolved at the end. However, great gains in efficiency result when these decisions are resolved up front. Use these techniques to create a style sheet that all team members can use throughout the writing process:

Styles menu	Define the formats that you want to use in the Styles menu of Microsoft Word, Apple Pages, or other word processor.
Document template	Many teams work by inserting content into a template that implements all of the document design features. The template starts with generic content, and authors overwrite their content into paragraphs, lists, tables, and figures that are already set up in the team's desired style.

Starting with well-defined style menus and document templates saves time in the later stages, prevents formatting errors that can be hard to spot, and even encourages authors by allowing them to work within an existing document rather than facing a blank page.

Draft the introduction

Delegating one person to write the introduction is ineffective because it cannot be written without coordination with all of the other parts of the document. After the team has outlined all of the document in detail, return as a group to your shared purpose and audience analysis. As a team draft the introduction before drafting individual sections of the document. Developing this draft together increases the likelihood that the individually drafted sections to come will be consistent with and support the work's specific context and goals.

20.3 Drafting a document as a team

Ongoing communication between team members during the drafting stage is important. Team members should keep the shared complete and detailed outline in front of them as they begin drafting their sections individually. As you draft, keep your collaborators in mind. Although you may notice places where the original outline was inadequate and needs to be revised, confer with your team before changing the outline. In addition, share with your team any findings that affect their assignments and defer any decisions that might affect the shared expectations established in the planning stage: These should be addressed together in the team's next meeting.

The team may draft *by sections* or *by roles*, selecting the approach that suits them best. Members of professional teams generally write to their strengths. To save time, members of student teams usually do the same. However, students should consider drafting content that challenges them to learn. It may seem expedient to allow teammates to cover content that you do not yet know well, but this approach may also prevent you from acquiring important knowledge and skills:

Drafting by sections	Assign sections to team members, who follow the collaborative outline and style sheet. Integrating each section can be time-consuming.
Drafting by roles	Assign roles to team members, who develop the entire document according to their roles. More experienced teams will find this method produces very high-quality results.

Within either approach, the team should negotiate to reach a general consensus about the fairness of the assigned work. The goal of fairness should not extend, though, to an insistence on perfect equality according to narrow measurements of productivity (such as counting the pages written). As long as members are fulfilling their commitments, the most successful teams are able to look past slight inequalities.

Drafting by sections

The simplest way to define individual contributions is to assign each section of the document to one member of the team:

- Assign sections to individuals, using experience, interests, and skills as a guide. Distribute the workload fairly (but do not assume that each writer must contribute the same number of pages or words). Do not at this time assign summaries such as abstracts or conclusions.
- Draft sections individually. Follow the outline meticulously, turning its assertions and evidence into paragraphs.
- As you draft, perform research necessary to support the assertions made in your section.

Drafting by roles

In this approach, rather than taking separate sections, individuals take turns adding to the document as a whole. The advantage of this approach is that the shared ownership of the planning stage is retained through the drafting stage. In this model, each team is assigned a role—lead writer, editors, or subject matter experts.

The *subject matter experts* compile their content. Generally, they are knowledgeable about the material, but they may need to research and collect additional content. The entire team agrees that the content is sufficient before the lead writer begins drafting.

The *lead writer* can record the shared process of the team, transcribing spoken ideas in real time, or can take the content produced by the team and work alone on drafting the document. The lead writer integrates all of the team's ideas into a document draft, and at this stage should be less concerned with polished prose than with capturing the content and reasoning.

The *editors'* role requires much more than merely correcting errors: Done correctly, editing is just as difficult, and requires just as much skill and effort, as lead writing. The editors shape the draft produced by the lead writer, adjusting it to maximize its clarity, coherence, and sound organization. (Verifying the correctness of the content and the soundness of the technical reasoning will require the effort of the whole team, including the subject matter experts.)

On the most successful teams, the coherence and clarity of the final product depend equally on the work of the editors and the lead writer. Modules 18: *Drafting* and 19: *Revising* present a more detailed guide to the stages of the project: The team may use this sequence of stages to decide exactly when the lead writer ought to pass the draft document on to the editors.

When drafting by roles, monitor the process carefully to ensure that the following conditions are met:

- Work is distributed in a generally fair way, without obvious shirking or exploitation of one person's labor.
- Each drafting and revision stage preserves the team's decisions, without any unilateral changes that override a team consensus in favor of one person's position.
- Everyone on the team is sufficiently involved in the process that they are both learning the technical content and developing their writing skills.

20.4 Integrating and unifying a document

If team members have drafted individual sections separately, the team must then take several steps to integrate and unify individual work into a complete and coherent whole. The team focuses on the substance, clarity, and correctness of the argument:

- Merge the files. If the team drafted in separate sections, select one team member to electronically merge all files and share the merged document with the entire team. A modest attempt at unifying the document format (font, headings, etc.) is permissible at this step. Keep it simple.
- Identify and fill in gaps. Individually read the complete document and identify gaps in the argument. As a team, decide what work needs to be done to fill the gaps.
- Assess coherence. As a team, review the organization within sections and as a whole. List the revisions needed to eliminate

inconsistencies, unnecessary repetition, and ambiguous or unsupported assertions.

- Draft the summaries. As a team, review the introduction and draft the summarizing elements of your work. Depending on the genre, such summaries include abstracts, executive summaries, and conclusions. Write conclusions corresponding explicitly to the goals described in the introduction.
- Unify the style, tone, and voice. While editing for mechanics, usage, grammar, and syntax, make the document appear as if it has been written by one person. Paragraphs should generally be of consistent length, headings should all be complete-sentence assertions, and the level of formality should be consistent throughout the document.
- Unify the format. Make final changes to page layout details only when revisions are nearly complete. Select fonts, spacing, margins, and heading styles per audience requirements. Finally, adjust figure locations, column breaks, and page breaks and revise the text to eliminate textual orphans and widows.

Though these integrating and unifying steps are presented as a sequence, you will probably move back and forth between the steps. For example, while drafting a conclusion, you may identify a needed revision to the introduction and goals. You may have to revisit elements of the argument in light of new findings. Be prepared to jump back to an earlier step in the collaborative process at any time and to repeat the steps that follow it.

Writing collaboratively in the cloud

Teams will often need to supplement in-person meetings with electronic aids for collaborative work but should think strategically about when and how to use these resources. They should not replace in-person meetings entirely. Online tools, such as Google Docs or Trello, facilitate the sharing of electronic resources (documents, files, data, links to web resources, etc.) but lack the spontaneity of face-to-face meetings. Fully synchronous systems for team conversation (telephone conference calls, Skype, etc.) may preserve spontaneity but allow participants to disengage.

If electronic collaboration is needed, try to use the virtual meeting time to work together on documents: Whenever possible, reserve important decisions for face-to-face conversations.

Summary

Avoiding the "divide-and-conquer" approach

- A teams isn't really collaborating if it merely merges a collection of individually composed sections. Such a process leads to contradictions and inconsistency, impeding reader understanding and eroding the authors' credibility.
- The team should work together on the planning of the document. This planning may be led by a project manager, whose role focuses on coordinating the work process rather than on writing prose for a particular section.
- If one team member is especially skilled at organizing and formulating ideas, he or she may serve as a lead writer, working with content developed by teammates.

Planning a document as a team

- Define your purpose by stating the goals of the document and the intended effect on the audience.
- Profile your audience by thinking about primary and secondary audiences. What can you assume they know and don't know about your content?
- Create an outline for each section of the document, defining topic sentences and required evidence that build upon each other. A well-considered outline will clarify expectations for each section, making the individual writing more effective.
- Develop a style sheet to define the document's design—fonts, sizes, spacing, and formats for tables, lists, and figures—using your word processor's Styles menu. Consider creating a document template that implements the design.
- Finally, draft the introduction together so all team members know the overall goals of the document as they write each section.

Drafting a document as a team

- Negotiate a generally fair division of the work, but do not insist on perfect equality according to narrow measurements of productivity (such as counting the pages each individual has written).
- If drafting by sections, work from the outline and overall goals of the document. If the outline was well developed, writing each section should come easily. Contact the team when you need information from a teammate or if you have a question that only the team can answer.
- If drafting by roles, take turns with the entire document. Subject matter experts share material and content that they have collected. Only when the team agrees on the content does the lead writer integrate the ideas into a draft, focusing on capturing the argument. Finally, the editor polishes the prose as well as the logic and clarity of the writing.

Integrating and unifying a document

- If drafting by sections, integrating the individual work will require the entire team to consider the correctness and clarity of the argument.

- After combining sections into a single document, look for gaps in the logic and work on filling them.
- Eliminate repetition and resolve inconsistencies. If a reference is made between sections, make sure it is correct.
- Draft the summary sections as required for the genre—this may include an executive summary or abstract as well as a conclusion—and then verify the consistency of the style, tone, voice, and formatting.
- These collaborative processes work best in person, allowing for smooth discussion with everyone's full attention. If electronic collaboration is needed, try to work together on documents: Whenever possible, reserve important decisions for face-to-face conversations.

Meeting

21

Because engineers typically work in teams, meetings probably occupy more time in engineers' professional lives than any other activity. To move the team's work forward, the collaborators must not only interact effectively with one another but also document decisions and plans in a systematic, reliable way that maximizes productivity and mutual understanding. When skilled teammates meet, the team can determine a direction, consider actions that will enable progress, and begin to execute those actions.

While this description may make teamwork sound easy, veterans of team projects know that most teams will have to navigate through some disagreements and misunderstandings. As your interpersonal skills develop, you will be able to consider and resolve most of these disputes more easily. A good first step, though, is to make full use of agendas, minutes, and other meeting documentation. These documents are designed with the goal of keeping the meeting orderly and removing any uncertainty about what the team has decided or who is responsible for pending work.

Objectives

- To decide in an initial meeting on procedures, expectations, and roles to govern a team project
- To prepare agendas that reflect the team's pressing tasks and discussions, helping to keep meetings productive and organized
- To prepare minutes that reflect the team's action items and decisions, providing a definitive record of completed and assigned work
- To apply document design techniques to agendas and minutes, allowing team members to find and use content easily
- To adopt best practices for virtual meetings, minimizing disruptions and inefficiency

21.1 The first team meeting: Roles, responsibilities, and charters

When meeting for the first time, many teams spend their time discussing the mission and objectives that have brought them together, considering the talents and expertise that the various members bring, and the criteria for success. This can be done quite informally: The members of a new team of mechanical and electrical engineers might simply introduce themselves and tell their colleagues about recent projects on which they have worked. Assigned roles might not be needed: The team trusts that volunteers will come forward as work needs to be accomplished.

Assign roles to share responsibilities

Many expert teams, though, assign roles to individuals on the team. These roles might be defined by technical expertise, with individuals charged with responsibility for tasks that the team needs to complete. More often, though, those roles are defined by the communication tasks to be performed during meetings:

Team lead, chair, or facilitator	Runs the meeting, which usually entails preparing an agenda and leading the team through it. Accepts a higher degree of responsibility for the quality of the final product. Carries interpersonal responsibilities, trying to help team members contribute their best work. Mediates, if needed, when conflict occurs.
Secretary or recorder	Documents the team's decisions and compiles information into the meeting minutes. Distributes minutes and other project documentation to team members. May maintain team records on shared web space such as Dropbox or Google Drive.
Monitor	Keeps the team on task by alerting teammates when digressions are pulling the team off topic or when distractions (such as personal electronics) are interfering with the meeting's progress.

While some individuals may take on such special roles, the team of course has expectations for all of its members—to attend meetings regularly and to arrive promptly, to keep their focus on the team's business during those meetings, to give a fair hearing to all teammates' ideas, and so forth. Even such seemingly obvious matters, though, can cause frustration when teammates see them differently. Does the weekly meeting begin promptly at 10:00, or is there a five-minute grace period for latecomers?

Use charters to manage expectations and plan work

Engineers in a workplace typically come to understand the expectations of different types of meetings through experience: Perhaps the database team meeting often starts and ends late, while the user interface team needs to begin and end its meetings very promptly so that team members can attend to other commitments.

A student team, though, will often find it most effective to specify expectations in an explicit document—variously called a *contract*, *charter*, *code of conduct*, or *code of cooperation*. Disagreements are much less likely to result when teammates have considered the possibilities and specified procedures in advance. Even then, the team needs to take care to be explicit and precise: Everyone might agree that it's acceptable to miss meetings "in case of emergency," but the team members might have very different ideas about whether an exam the next day constitutes an emergency.

Charters typically cover the routine business of meeting and making decisions:

Meeting times and locations	If a single weekly meeting is sufficient, the team can schedule it at a regular time and location that work for all members' schedules. When team members need to miss meetings, the charter may specify how and when they must notify the others. The charter may also specify permissible reasons for missing a meeting.
Roles and responsibilities	The same individuals may run the meeting, prepare the agenda, and take the minutes each time, or those roles may rotate among the team members.
Correspondence	The charter may specify how quickly members should respond to email or text messages related to the project (within 12 or 24 hours, for instance).
Other expectations	Everyone enters team collaboration with expectations about procedures, goals, and roles. These expectations are guaranteed to mismatch unless discussed directly, creating the potential for conflict. The team will thus benefit from explicit discussion of goals. (For instance, conflict sometimes arises on student teams between those who expect A work and others who believe that a B is good enough.) It can also be useful to specify procedures for making decisions: Will the team pursue consensus, or can disputes be settled by calling for a vote?

Teams using a charter should revisit it at later meetings, perhaps after key milestones in the project: It's common to lose track of some important expectations or procedures, or to revise the ones that aren't working.

Team Expectations:

1. The team will reach a consensus on decisions before submission of the materials.
2. Members will actively participate in meetings and present topics of conflict for discussion.
3. Email correspondence will include the entire team.
4. Two days before each meeting, that meeting's facilitator will circulate an agenda containing date, location, time, and objectives.

Individual Expectations:

1. Emails must be replied to in a timely manner—preferably within twelve hours.
2. Tasks will be completed in a timely manner, by the set deadline and proofread beforehand.
3. Every member is required to attend every meeting, unless excused by the team as far in advance as possible.
4. Every member will be on time to all meetings, unless a valid excuse is given and approved by the team.
5. Every team member will keep current with the material being researched and presented by other team members.

Items on a charter are often designed to avoid problems encountered on past team projects. Here, students seem to have had problems with being left "out of the loop" while decisions were being made.

"Timely" correspondence may not mean the same thing to all team members: Specifying 12 hours provides valuable guidance. Concreteness is usually desirable.

Charters can encourage close collaboration rather than "divide-and-conquer" work.

Part of a team charter establishing some rules and practices. Charters not only establish expectations but also provide members of the team with a basis for discussing problems when they arise. © McGraw Hill, reprinted with permission (Smith, 2005).

21.2 Why meet? (agendas)

Don't meet without a reason

While it is beneficial to have meetings on a regular basis, each meeting should have a purpose. The objectives for each meeting can be included in an agenda that outlines items for discussions, decisions to be reached, and tasks to be accomplished. The team's facilitator usually prepares the agenda.

Schedule the meeting

Meetings should be scheduled far enough in advance that all team members can plan for them. A workplace meeting is often scheduled by an executive, your boss—guaranteeing that you'll be there. Student teams, on the other hand, require consensus on the decision to meet and on the scheduling of the meeting. Once you've agreed to be there, though, honor the commitment.

Distribute the agenda prior to the meeting so that everyone can be prepared for accomplishing the meeting's goals. An agenda should also include the time and place of the meeting and any materials that team members need to bring.

Review the agenda before starting the meeting

Use your agenda to keep the meeting focused. Review the goals of the meeting at the beginning of the meeting, and if needed, add or delete items on the agenda. In addition, review the roles of each member of the team, such as who will be taking minutes.

You might consider using a timed agenda, as it helps to keep everyone on task and the meeting progressing. A timed agenda allocates the number of minutes that should be spent on each item and gives you a goal for how long it should take you to reach a decision. Timed agendas are helpful in keeping a team organized and productive even when the initially allotted time ends up being inadequate for the tasks at hand. (Naturally, such estimates improve with practice.) Perhaps most importantly, explicit management of the team's time together can spare teammates the frustration of an apparently endless meeting.

Including explicit objectives keeps team members aware of overall goals and specific actions that must be taken.

If even one team member forgets a date, time, or location, the team has to backtrack.

Prompts like this bulleted list encourage team members to think through important issues both before and during the meeting.

Teams often perform best when they have an explicit goal to meet, such as recording an official decision.

Agenda: First Project Team Meeting

Objectives: To generate preliminary topics and team roles

Date, Time and Duration: April 28, 2014, 11:00am

Location: Mechanical Engineering Reference Room

Items for Discussion and Decision:

1. Assign facilitator and recorder roles for this meeting
   Allotted time: 5 minutes

2. Brainstorm possible topics for RFP project, evaluating them according to the following criteria: *Allotted time: 25 minutes*

   - Can we prove or document the problem's *existence*?
   - Can we make a case for urgent, meaningful *need*?
   - Is the problem *appropriate for the RFP and its constraints*?
   - Are there *a number of possible solutions* for this problem?
   - Is the project supported by *abundant expert-level research material*?
   - Has this problem been *exhausted by past projects*?

Goal: include in the minutes a ranked list of 3 possible topics

Partial agenda from a student team's first meeting. Team members benefit from some advance information about what's to be covered: This may be covered in the agenda or in reports submitted in advance by individuals or subgroups who have been working on particular tasks.

21.3 What happened? (minutes)

Effective minutes assist action

Minutes are a useful tool for keeping a team organized and communicating effectively. In most cases, however, the goal is not to record everything said in a meeting. Such detailed records are *transcript*-style minutes, and they are important in some settings—in public hearings, for example—when having a complete record of events may later be important.

Team Meeting (Dec. 3, 2014)

During this meeting the team met initially during lab hours (approximately 3 hours) in M203 to begin working on our WBS. The team worked out a WBS for the Fair Chair portion of the document and decided that Lindsey would complete two more WBSs for the Posture Assurance System and the Communication App. The team spent approximately an hour working on the WBS but were unsure if we were doing it correctly so we determined Ryan would show the completed Fair Chair WBS to Dr. Rogge to do determine if it was correct. The team then gathered up the supplies from M203 and went to the machine shop to look for additional supplies. These supplies included plywood for the seat and additional metal tubing to use for the handles of the Fair Chair. While looking for the tubing, an additional idea was proposed for attaching the handles to their frame. The idea is to create an inner threaded lining for the main bar of the handles out of some plastic and then connecting a threaded plastic pin to the inside of the handles so that they will screw into the main portion instead of connecting with push pins which are both small and difficult to use. The team then moved the supplies to our work area in the Branam Innovation Center and spent time looking around and familiarizing ourselves with the area and tool locations.

Recording of team notes during the meeting will almost always use the transcript style—resulting in either a long list of items or large, dense paragraphs.

A WBS is a work breakdown structure, a document used in project planning and management.

Both action items and ideas for future discussion are present—but if they're left embedded in the initial transcript, they may never be unearthed again.

This transcript of a student design team meeting records both work completed and the results of team discussion. As recorded during the meeting, team notes will follow the sequence of what happened during that meeting—not always the best arrangement for locating information later. © 2014 Jeff Elliott, Zac Erba, Ryan Seale, and Lindsey Watterson. Reprinted with permission.

Minutes should be brief and should make use of document design features such as headings and bulleted lists to make them easier to read. It is useful to have a template to be used at every meeting. Not only does this make the job of the minute-taker easier, but it also makes information easier to locate if everyone is familiar with the format.

Formatting enhances readability but is simple enough to be read in the body of an email, even on mobile devices.

A basic structure, encompassing work accomplished, discussions and decisions, and action items, is sufficient for this meeting. It's important for the minutes template to match the way the team really works.

Action items need to be explicitly assigned (with due dates) to those responsible. Attributing ideas to individuals is optional—some minutes do, and some don't.

Team Meeting
Date: Dec. 3, 2014 1:30-4:15
Location: Myers M203
Attendees: Jeff, Zac, Ryan, Lindsey

Work Accomplished

Began drafting WBS for Fair Chair portion of document

Retrieved supplies from M203 and machine shop, including plywood for the seat and metal tubing for the handles of the Fair Chair

Set up work area in the Branam Innovation Center

Discussion of Handle Attachment Design (to be continued)

Push pins are small and difficult to use

Idea: screw handles to chair body with threaded plastic pin

- Threaded plastic pin would require an inner threaded lining within main handle bar

Action Items for Dec. 5 Meeting

Lindsey: Draft WBSs for Posture Assurance System and Communication App

Ryan: Show completed Fair Chair WBS to Dr. Rogge to check correctness

When converted into minutes, the chronological transcript of what was said and done is reworked into functional sections for future reference.

Minutes record decisions that were made

Minutes document the process of project development—incredibly important in industry, since a lawsuit may require a company to produce a record of decisions. They can also help project teams work effectively, keeping a team from unnecessarily revisiting the same questions and concerns. Shared documentation is much more effective than individual notes at keeping teams organized and preventing miscommunications. Individuals can remember decisions differently if not recorded formally; minutes ensure that team members are truly in agreement.

Minutes record action items that were assigned

The section of action items is the most important section and serves as a checklist of tasks to be completed. If tasks are assigned, include who will complete them and by when. Minutes should also include when and where the next meeting is as well as what team members should bring back to the team at that time.

21.4 Optimizing virtual meetings

Teams work best face to face and in person. Collaborators working in different locations around the world, though, may need to conduct much of the team's business virtually. The traditional conference call has become somewhat easier to manage as software tools (such as WebEx or GoToMeeting) allow easier sharing of documents. The facilitator can display a budget spreadsheet on everyone's desktop, or conduct a vote easily. However, teams should still expect some inefficiency and awkwardness when they can't see one another in a real physical space: Team members may experience dropped connections or software failures and may have difficulty speaking over one another. A few best practices can help to minimize these difficulties:

- Whenever possible, log on early. Arriving late can be more disruptive virtually than in person, and it can be harder to catch up on what you've missed. Most importantly, extra time allows you to troubleshoot any technical difficulties.
- Virtual meetings benefit particularly from written documentation prepared in advance. A detailed agenda is crucial, and it may be worthwhile for participants to distribute written reports of work they've done since the last meeting.
- Keep background noise to a minimum: Everyone's microphones or handsets should be muted when they aren't speaking.
- Resist the temptation to multitask: The team's efficiency and decision making both suffer when participants are paying only partial attention.
- Try to meet in person when possible—spending time together can be hugely valuable to kick off a project, to approach a deadline, or just to build relationships with teammates. Virtual meetings can be much more effective when they're building on time spent together in person.

Summary

The first team meeting: Roles, responsibilities, and charters

- Team members may be assigned fixed roles for the duration of the project or may elect to rotate those roles.
- At a minimum, one team member should run the meeting using an agenda, and another should record decisions, actions, and assigned tasks in order to prepare effective minutes.
- Charters for student teams specify roles, expectations, and decision-making procedures in order to minimize potential conflict and uncertainty.

Why meet? (agendas)

- Meetings should be dedicated to specific tasks and topics determined in advance.
- Agendas allow team members to know exactly what the team will be doing and discussing at the meeting, helping everyone to prepare.
- The agenda should be distributed in advance but reviewed at the beginning of the meeting, with items added or deleted as needed.
- Teams with a lot of agenda items and teams that struggle to reach decisions may both benefit from timed agendas, which allot specific durations to each item.

What happened? (minutes)

- Minutes should clearly record decisions, assigned tasks, and deadlines; the minute-taker should not attempt to create a transcript of the entire meeting.
- Headings, bulleted lists, and similar document design features make minutes much easier to read and to use in managing the project.
- Minutes should include enough detail and precision to provide a definite record of the team's decisions—important in the frequent cases where team members recall or understand decisions differently.
- Action items should clearly specify the team members responsible for completing all tasks, the deadline for completing them, and the time and place of the meeting at which the resulting work will be discussed.

Optimizing virtual meetings

- Meet in person when possible: Virtual meetings should be used to coordinate across large distances and not to replace face-to-face meetings for mere convenience.
- Conference call and online meeting participants need to carefully minimize distractions and sources of "noise" between meeting participants. Keep your microphone muted when not speaking, and log in early to prevent technical problems from disrupting the meeting.
- Detailed agendas shared before the meeting have special value in keeping a virtual meeting on track, with the goals clear to all participants.

Components

...of your writing work systematically together, accomplishing the functions for which you have designed them. Paragraphs are assemblies in which sentences must fit together to achieve coherence and continuity; within those sentences, you should be choosing words as purposefully as you would choose the values of resistors in a circuit. Mastery of the components of technical communication enhances the overall effect of whatever is being constructed. Conversely, the most compelling structure will fall apart in the eyes of the audience if the underlying components are badly chosen or inexpertly combined.

Components

22 Headings

Headings are more than convenient labels for a document's main ideas. Because most readers start using a document by skimming it rather than reading in detail, headings are indispensable, performing far more work than most writers realize.

Objectives

- To design headings that communicate a document's substance (rather than general labels like "Introduction" or "Problem")
- To manage multiple levels of hierarchy in a document's organization, logically dividing and summarizing sections and subsections
- To revise headings for parallel structure and consistent style across a document

22.1 Communicating the argument to the hurried reader

Rather than reading sequentially from beginning to end, most readers skim a document—at least at first—by looking over its section headings, topic sentences, and figure captions. Readers who are especially pressed for time may rely completely on such skimming (and most others will closely read only those sections that relate directly to their own work). Being precise in these parts of a document not only greatly aids a reader pressed for time but also allows you to direct that reader's initial attention.

Organizing in hierarchical levels

Even without any headings at all, paragraphs and their topic sentences organize good writing so that readers know where to look for main ideas and topics. A heading is similar to a topic sentence in stating the main point of a particular unit of text, but works at the larger level of a section or subsection. Headings are required when:

- The scope of the document is much larger than that of most individual paragraphs—as in a report or proposal
- Readers may not dedicate the time or undivided attention required to grasp the most important content—as in memos or even longer emails

Some genres—such as specifications and engineering standards—and some workplaces typically organize documents in a highly hierarchical structure composed of several carefully delineated levels: Nested within Chapter 2 is Section 2.4, which begins with Subsection 2.4.1, and so forth. Use such numbering systems when you know that audiences expect them. When you are able to devise your own structure, though, avoid making it too elaborate. Larger documents can require more than two levels of headings, but numbering beyond the first decimal is often more of a hindrance than an aid to the reader.

Space your headings and subheadings carefully: Because headings need not duplicate the work of topic sentences, each paragraph should not have its own subheading. On the other hand, a document would lose the benefits of its headings if it were to extend for pages without a heading or subheading.

The overall set of headings and subheadings defines a hierarchical relationship, so each one should appropriately identify the level of importance of the point that it states. A supporting point should not be expressed in a section heading nor a main point contained within a subheading. After writing your headings, revisit them to make sure that they convey the main point of the document without missing key ideas or mixing minor details with key points.

Using extended phrases

In scientific journals and elsewhere, experimental reports traditionally use the very basic headings that define the IMRaD (Introduction, Method, Results, and Discussion) format. However, such section headings (or comparable labels, such as "Problem" or "Analysis") provide little information to readers.

As always, follow the practices expected by readers; If a required format for lab reports limits you to such headings, you can focus on your topic sentences to make the structure of your content as clear as possible.

Where possible, though, provide more specific headings or at least subheadings. Headings such as "Wastewater Treatment" or "Stormwater Management" at least supply the topic being addressed, but they do not indicate what is being said about the topic. An extended phrase informs the reader of the main idea of the section. For example, Section 2 in the Table of Contents below—"The twin-pit latrine is the best solution for the community"—clearly foregrounds the recommendations in the report. If, instead, the heading was "Two-pit latrines," it would not be clear whether the authors were recommending or rejecting those latrines as a solution.

Extended phrases can be complete sentences (as in the Table of Contents below), comparable to the Assertion/Evidence style of presentation slides. Descriptive phrases, though ("Pit dimensions for adequate capacity," "Three-phase construction process"), still convey much more information than a single-word label ("Dimensions" or "Construction").

Setting headings off with typography

Given their quick skimming pace, the readers who rely most heavily on headings also need to spot them as quickly as possible. For this reason, they should be distinguished typographically from body text:

- Use a complementary font in a different style—headings are usually printed in sans serif type (in a font like Helvetica, Arial, or Gill Sans).
- Use a slightly larger font size. Bold type works, but increasing the type size is sometimes better for readability, and different sizes can be standardized for different levels of hierarchy.

To some degree, capitalizing headings is a matter of taste: Generally, though, longer headings are like sentences, and only their first words and proper nouns need to be capitalized.

TWIN-PIT COMPOSTING LATRINES FOR GOMOA GYAMAN FINAL REPORT

Table of Contents

Writers might begin with simple labels for their sections ("Capacity" for Section 2.2, "Lighting" for Section 2.4, and "Materials" for Section 2.5), but building those labels into complete assertions maximizes clarity and precision.

In this Table of Contents from a student report, the headings and subheadings all make assertions. Simply reading through the Table of Contents provides an overview of the report's argument. © 2014 Barker, Billingsley, Heidlauf. Reprinted with permission.

22.2 Unifying style and voice

While the main purpose of headings is to guide readers through the sequence of content, they also benefit writers. Headings are especially important when several writers are collaborating: A carefully worked-out system of headings can help team members to divide their work in a way that makes sense, with each writer or subgroup developing its content in consistent, parallel fashion. To accomplish this effectively, though, the headings themselves must be consistent.

Using parallel structure effectively

Any time items are presented in a list, whether that list is in a single sentence or a bulleted list, parallel structure—beginning each item in the list with the same part of speech—is required. As a type of list, the entries in the outline of your document should also begin with the same part of speech. For example, the heading structures in this module all begin with *-ing* verbs. Because they all use the same part of speech, it is easy for a reader to skim the section headings. In the sample Table of Contents, the section headings and subheadings are all simple sentences. If, instead, the headings changed structure or person, the reader would likely be disoriented.

Early drafting will likely yield headings that use several different types of phrases. The student authors of the design report might have started with something like this:

- Use three phases for constructing the latrines
- Convincing the community of the value of composting
- Sanitary, safe to handle, and beneficial
- Include special materials when composting

This gives the writers a list of topics to consider while they write, but a reader who read these items could not yet understand how they're linked or why they're arranged in this order. Parallelism allows the heading to make the big picture much clearer.

Such revisions also serve the important function of reconciling the work of different writers who have taken charge of the document's assorted sections. As you revise the headings for parallelism, you may be able to identify and eliminate unintended redundancy or contradictory information; you may also find stylistic differences that can be revised in the body text as you continue to make the document's style more uniform.

Summary

Communicating the argument to the hurried reader

- Precise headings enable a reader to skim a document more quickly and effectively.
- A heading expresses the main claim of a section, while subheadings signal supporting points that define subsections. Ideas within a particular heading level should be comparable in importance.
- Headings should not precede every paragraph but should also not be absent for several pages.
- An extended phrase can articulate the purpose and message of a section much more clearly than a single-word label.
- Typography makes it easier for readers to follow the sequence of headings. Headings are normally printed in sans serif type in larger sizes than body text.

Unifying style and voice

- Heading structures should be of roughly the same length and complexity throughout the document.
- Headings should use parallel structure, with each one built around the same part of speech.

23 Paragraphs

Paragraphs are the building blocks of professional documents. As you construct and arrange those building blocks, your ultimate goal is *coherence*, where the presence and sequence of sentences within the paragraph—and of paragraphs within the document—makes intuitive sense to readers. We have all encountered paragraphs that string sentences together in ways that confuse us, creating misunderstandings or the need to reread. The techniques that make paragraphs coherent, on the other hand, help readers to focus their attention on the merits of the document's ideas and the subtleties of its technical content.

Objectives

- To limit each paragraph to one main idea and its required support, explanation, and interpretation
- To choose the appropriate location for each main assertion—at the beginning, following a transition, or at the end of the paragraph
- To arrange a sequence of sentences leading readers through a paragraph's chain of reasoning
- To signal logical relationships and steps of reasoning with transition words
- To create variety in sentence structures and lengths, maximizing ease of reading and comprehension

23.1 Focusing paragraphs on a single idea

The first task in arranging a coherent paragraph is to ensure that it is *focused* narrowly on a single main idea, stated in the topic sentence. Other sentences in the paragraph will then define, support, qualify, or interpret the idea, so that readers understand its implications. We recognize insufficient focus when a topic sentence weakly connects several ideas, or when a paragraph contains sentences that lack a clear connection to the topic sentence.

Focusing on a single calculation helps to make this paragraph coherent, leading a reader easily through the calculation's purpose, main finding, assumptions, and implications.

We performed a calculation to determine the extent to which column load capacity could be reduced by the corner cracks and localized bulging. We determined that if a section of the column developed full-thickness cracks at any two corners, the column would still have the required load capacity as the crack lengths do not exceed approximately 10 feet. The assumption of cracks at two corners would make the column section behave in the cracked region either like two separate single-angle sections or like a single channel section, as opposed to a tubular section.

The repair approach we recommend is depicted on the enclosed Figure 1. It involves welding collars around all six columns at seven locations per column. Each collar would be formed using four 3-inch by 3-inch angles welded to the column and each other. The purpose of the collars would be to hold the column walls together at sufficiently close intervals to safeguard against localized column buckling. This way, existing cracks and any possible future cracks or existing crack extensions could be accommodated while still maintaining the needed load-carrying capacity of the columns.

The logic of moving from a problem to a solution is transparent enough that it doesn't require an explicit transition between paragraphs.

The second paragraph focuses on the repair approach being recommended. While the paragraphs are closely related, each topic requires its own paragraph.

The first sentences in both of the above paragraphs state the main idea in the first sentence, and then the body of the paragraph further details the problem in the first paragraph and the recommended solution in the second.

State the paragraph's main assertion in a carefully placed topic sentence

The topic sentence—which most often opens the paragraph, although it may also follow a transition sentence—will most often be an *assertion*. That is, it will state a claim that you want your readers to accept or a recommendation that you want them to endorse. An effective writer guides readers through the argument, and a topic sentence at the beginning of the paragraph helps those readers comprehend the description, evidence, and examples that follow.

When reviewing a draft paragraph, however, it is not unusual to discover that a *final* sentence contains the strongest statement of the paragraph's main idea. It is most often appropriate to convert such a sentence into a conventional topic sentence by moving it to the top of the paragraph. However, you may want to leave a particularly strong assertion at the end of a paragraph if you believe that it may be controversial or difficult for readers to accept. This unusual placement allows readers to process some of the evidence and reasoning before they are confronted with a main idea that might otherwise surprise, confuse, or upset them.

Choose paragraph lengths based on audience and content

While your audience will not expect paragraphs to be of identical length, paragraphs should be relatively uniform in length throughout the document. The thought process seems uneven if there is a significant disparity between the lengths of paragraphs. In the above example, both paragraphs have only a few sentences—three in the first paragraph and four in the second. If readers needed to understand the authors' predictive model in greater detail, longer paragraphs would be needed to accommodate the additional content.

A paragraph's length should accommodate audience needs, based on the genre and setting in which you are writing. Industry documents most often use shorter paragraphs like the ones above, ranging from three to eight sentences (and often tending toward the lower end of the range). Such brevity is helpful to ease the demands placed on readers by highly technical content or complex sentence structures, and it keeps readers' focus on the main idea. Paragraphs in scholarly research articles may be longer, reflecting an emphasis on comprehensiveness and mastery of details. Such articles, generally targeting an audience of expert specialists, may include paragraphs of twelve or more sentences, especially when citing detailed findings from earlier researchers' work.

23.2 Moving coherently from one sentence to the next

Coherent paragraphs require not only a narrow focus on a single idea, but also a fluent, logical sequence in the sentences that present the authors' reasoning about that idea. Reading through a paragraph should seem effortless for your readers. Moving from one sentence to the next should not be jarring or disorienting for readers—you want your reader focused on your message, not trying to decipher your message.

Link the content explicitly to connect consecutive sentences

Providing clear logical links between one sentence and the next helps make the reading process seamless. Refer to information with which the reader is familiar at the beginning of a sentence before introducing new information. In addition, you can sometimes segue into the next sentence with the information provided at the end of the sentence, as this engineering consulting firm does in its letter to a client:

The repair approach we recommend is depicted on the enclosed Figure 1. It involves welding collars around all six columns at seven locations per column. Each collar would be formed using four 3 inch by 3 inch angles welded to the column and each other. The purpose of the collars would be to hold the column walls together at sufficiently close intervals to safeguard against localized column buckling. This way, existing cracks and any possible future cracks or existing crack extensions could be accommodated while still maintaining the needed load carrying capacity of the columns.

The third sentence begins by referencing the collars introduced in the second sentence. A different sentence structure—"Four 3-inch by 3-inch angles would be welded to the column to form the collars"—would weaken the link between sentences.

After introducing the welding collars in the second sentence, both the third and fourth sentences reference these collars before introducing new information that makes clear the relationship of the angles and the column to the collars.

In this lengthier paragraph from a dissertation, the author also connects each relevant study to a previous one through a shared area of study:

Many studies have been performed to explore nitrate dynamics in streams. One method used to estimate nitrate uptake over a stream reach involves determining the spiraling length from a linear regression of the nutrient concentration at incremental distances downstream from a constant-rate nutrient injection (Hall *et al.* 2002; Payn *et al.* 2005; Ensign and Doyle 2006). Spiraling, or uptake, length is a measure of the efficiency of nutrient use in a stream (Mulholland *et al.* 2002) and is defined as the mean distance that a nutrient atom travels in a stream before uptake by biota (Newbold *et al.* 1981). Nutrient uptake in streams is the net difference between solute transfer from the water column into the streambed and solute release from the streambed back into the water column (Stream Solute Workshop 1990). Water and solutes are exchanged between surface water and groundwater through the hyporheic zone, where important biogeochemical

After introducing the concept of spiraling length in the second sentence, the third sentence begins with the same idea.

Similarly, the fourth sentence begins with "nutrient uptake," defined in the latter part of the third sentence.

The sixth sentence begins with "hyporheic exchange," defined at the end of the previous sentence.

processes (e.g., denitrification) occur (Boulton *et al.* 1998). Hyporheic exchange is promoted through variations in physical characteristics of the streambed along channels (Tonina and Buffington 2009). As solutes move from the landscape into streams and downstream receiving waters, the composition of reactive solutes can change through processes occurring in the hyporheic zone (Findlay 1995). Quantifying nitrate uptake rates in natural systems is complex due to coupled processes of biogeochemical reactions, which are influenced by microbial communities, organic carbon content, and hyporheic exchange as affected by flow and substrate characteristics (O'Connor and Harvey 2008). Additionally, denitrification is a major biogeochemical process occurring in the hyporheic zone that removes nitrate from stream systems.

The purpose of this paragraph is to ground the current study in previous scholarly work, and it does so by making logical connections between the previous studies. Even describing the studies in chronological order would be less clear than the topically linked sequence established here.

Signal your moves with transition words

Transition words show the relationships between ideas. Avoid using vague pronouns and ineffective and wordy phrases such as "this means that" to connect ideas. Transitions connect ideas but can also signal a shift in logic, so select a transition phrase that indicates the appropriate relationship between ideas:

Agreement (and)	*in addition, also, too, likewise, moreover, furthermore*
Contrast (but, or)	*however, rather, instead, although, while, whereas, despite*
Cause and effect	*because, as, since, if . . . then, therefore, consequently, as a result*
Illustration	*such as, including, for example, particularly, especially, specifically*
Time	*previously, afterwards, immediately, simultaneously*

In the repair plan proposed by the engineering firm in its letter to a client, several transitions link and clarify the number of steps in the plan:

In addition to the collar installations, the repair plan also includes provisions discussed in the January 19, 2011 letter including drilling holes in the bottom of each column to drain any water that may enter through cracks and drilling holes in the column at each end of each crack to attempt to prevent the existing cracks from lengthening. As mentioned in the January 19 letter, the zinc coating would need to be removed in the weld areas prior to any welding and a cold galvanizing compound applied after welding. Also, heat input would need to be controlled and monitored during welding to ensure column strength is not temporarily impaired.

Transitional words and phrases often begin sentences, but they can be placed midsentence as well, particularly if they signal a shift or other key point in reasoning.

The underlined transitional words and phrases show agreement; all of the described actions are part of the repair plan in the letter to the client. © 2015 Strand Associates, Inc. Reprinted with permission.

In the analysis section of the dissertation study on nitrate uptake, the author needs to illustrate cause and effect, agreement, and contrast:

This study demonstrates that nitrate uptake and transient storage are detectably influenced by flow variability and geomorphic characteristics, including median grain size and longitudinal roughness. The geomorphic context of streams mediates physical characteristics of substrate condition and roughness, which influence amounts of hyporheic storage versus in-channel storage. In turn, nitrate uptake is linked to transient storage, as it behaves differently in each storage zone and varies with changes in discharge. Accordingly, it is important to measure nitrate uptake over various flow rates to adequately characterize its behavior in streams. Furthermore, much of the focus on F_{med}^{200} as a parameter describing transient storage emphasizes hyporheic exchange; however, in-channel storage is also included in transient storage values of F_{med}^{200} and nitrate uptake behaves differently in the hyporheic zone as compared to in-channel storage. Differentiating between in-channel storage and hyporheic storage, which is explored further in Chapter 5, is important in future studies to more fully understand nitrate uptake and other biogeochemical processes occurring within stream ecosystems.

"In turn" and "Accordingly" both place the paragraph's content in a causal chain.

"Furthermore" suggests agreement with the earlier stated results.

"However" indicates a shift in logic. Such mild qualifications of a main claim are often placed at the end of a paragraph.

The underlined transitional words indicate cause and effect, agreement, and contradiction as the author presents a nuanced conclusion. © Jennifer Mueller Price. Reprinted with permission.

Varying sentence types by length and structure

Maintaining a generally consistent length and structure is desirable from paragraph to paragraph, but from sentence to sentence, variety is preferable. Short, direct sentences are effective for emphasis but create choppiness when several of them occur back to back. Longer sentences, with multiple clauses, help to qualify ideas or to relate them to one another; several in a row, though, have a lulling effect that makes it harder to pick out the central ideas. It isn't necessary to alternate strictly between the two, but moving among different lengths and structures helps to maintain and direct your reader's attention.

The topic sentence uses a simple subject and verb but lists several factors in the object. (CCW has previously been identified as coal combustion waste.)

Short, simple sentences create emphasis in the middle of the paragraph. "However" shifts from one way of thinking (current technologies are effective for one purpose) to a contrasting one (the same technology may not suffice for other criteria).

The list form in the paragraph's last sentence is logically simple, enabling a longer sentence than usual.

The chemical composition of CCW generated at a given plant will depend on the type and source of coal, the combustion technology used at the power plant, and the air pollution control technology used. Recently, EPA has focused particularly on the potential for changes in hazardous constituent levels as a result of the increased use and application of advanced air pollution control technologies in coal-fired power plants. These technologies are expected to reduce air emissions of metals and other pollutants. However, they are expected to transfer those pollutants to the fly ash and other air pollution control residues. As part of its regulatory proposal, EPA discusses research and information it has gathered in an effort to more accurately characterize the toxicity levels of CCW; determine how those toxicity levels may change with changes in air pollution control technology; and more accurately determine the potential for those constituents of concern to leach from their deposition site.

This paragraph, from a Congressional Research report, uses longer sentences when it forecasts the report's topics and shorter ones when it summarizes overall findings. Courtesy: Congressional Research Service (Luther, 2010).

Summary

Focusing paragraphs on a single idea

- Topic sentences should state the main idea of a paragraph, usually an assertion that subsequent sentences will define and support. Including the topic sentence at the beginning of the paragraph enables readers to comprehend the evidence, examples, or description that follows.
- Paragraphs should develop ideas evenly, generally in three to eight sentences. A significant disparity between the lengths of paragraphs can create the impression of uneven reasoning.
- Paragraphs with greater sentence complexity or more technical jargon can maintain clarity by using fewer sentences.

Moving coherently from one sentence to the next

- Clear logical links create a seamless reading process by connecting the end of one sentence to the beginning of the next.
- Transition words show the relationships between ideas, such as showing agreement or signaling a shift in logic.
- Varying sentences by length and structure helps readers to move through content efficiently. Whether short and choppy or long and convoluted, sentences of uniform length slow readers down and impede their comprehension.

24

Sentences

Sentences are to writing what molecules are to matter. They are made up of the "atoms" of individual words, which are essential to their fundamental character. The sentence, though, is the fundamental level in which those "atoms" of language are linked and arranged to deliver information, express assertions, or pose questions to audiences.

While producing early drafts, engineers are likely to think more about technical content than about the nuances of sentence structure. Revising for maximum clarity, however, requires detailed attention to those nuances—and ultimately, that attention will help you to form effective sentences even when composing a first draft.

Objectives

- To build a solid sentence *core* by choosing a subject, verb, and object to reflect the sentence's most important content
- To choose among sentence-building strategies (such as subordination and coordination) based on logical relationships in the content
- To correct the most common defects in sentence structure—sentence fragments, fused sentences, and comma splices
- To increase conciseness, without losing clarity, by adjusting both word choice and sentence structure

24.1 Solidifying the sentence core

The core: subject, verb, and object

At the sentence level, the easiest route to clarity is to embody each sentence's main idea in a simple *foundation* or *core* on which the rest of its content is built:

- The *subject* to which we attribute an action or property
- The *verb* that names the action or links the subject to the property
- The *object* on which the action is performed.

When we think of engineering design work, we usually think of highly complex systems, but the best engineers also recognize the value of simplicity. Modern microprocessors, for instance, are logical packages of billions of transistors, and the code that runs on them can be no less daunting—but the mobile computing revolution is a result of computer engineers and programmers delivering simple, even minimal, interfaces to users.

As technical professionals, much of our writing reflects the complexity of the problems and concepts with which we're working. It sometimes takes some extra thought, though, to minimize the difficulty that our audiences—the "users" for whom we're designing our writing—will experience. Focusing on the sentence core gives us a deliberate site in which to place the main message that our sentence should carry. In technical writing, the sentence core serves two particularly important purposes:

- Explaining the main point for non-experts, separating it from technical details: "Cast aluminum alloys / best match / our required material properties."
- Highlighting the particular finding that is most important to the engineer's analysis: "A 3.3 pF capacitor / will handle / the expected variation in voltage."

In this approach to sentences, the labels of "subject" and "object" include not just the nouns but also their modifiers: adjectives (*required material* properties) and prepositional phrases (*in voltage*). As long as they modify the sentence core and don't add additional subjects and verbs, prepositional phrases—beginning with *in*, *of*, *to*, *by*, *through*, etc.—can be treated as part of the core.

Yield strength data was approximated graphically using a 0.2% strain offset. Error: Reference source not found shows the stress-strain curve and yield strength approximation for sample #1. The bolts were treated as if they were of uniform diameter equivalent to the nominal root diameter. The active length was approximated to be the 1.125" between the bolt head and nutbar. The sample mean YS was 177.6 KSI with a standard deviation of 3.4 KSI. Assuming a normal distribution of strength values and with a 99.9% confidence level, subsequent modified bolts will meet or exceed the strength of an unmodified bolt with a YS 163.6 KSI. This exceeds the specified strength of 140.9 KSI.

For structural analysis the modified ¼-28 SHCSs meet the original bolt strength specification.

This paragraph accommodates engineering peers with sentence cores in which the object is usually a precise quantitative finding.

The memo's authors have separated and emphasized the main finding for executive readers, but clear transmission of their message relies on a clear subject–verb–object core.

Subject: ¼-28 SHCSs [socket head cap screws—defined earlier in the document]

Verb: meet

Object: the original bolt strength specification

24.2 Coordinating and subordinating ideas

While the sentence core provides the logical site for a sentence's central concepts and assertions, your findings and arguments will often require embedding that core in a sentence form that can place them in the appropriate context. Sentences may need to incorporate more than one sentence core to capture the logical connections among them precisely and accurately.

Coordinate to weigh linked ideas equally

The simplest way to move beyond simple sentence cores is to *coordinate*, using such logical connectors as *and*, *but*, *or*, *yet*, or *so* to provide equal emphasis to both ideas (the product is a *compound sentence*):

- Each SHCS sample failed along the reduced diameter, *and* each exhibited classic ductile tensile failure markings.
- All 5 SHCS samples survived the lateral pull test to 1000 lbf without fracture, *but* sample 5 yielded.

Coordination is the best sentence strategy when two criteria are met:

- The two sentence cores state related ideas.
- Neither idea is obviously and significantly more important than the other.

Subordinate to clarify, support, or qualify a main idea

Subordinate rather than coordinate when two separate sentence cores aren't equally important—when one of them states the main idea and the other clarifies, supports, or qualifies it: "Assuming a normal distribution of strength values and with a 99.9% confidence level, subsequent modified bolts will meet or exceed the strength of an unmodified bolt with a YS [yield strength of] 163.6 KSI." Here, the sentence core delivers the primary message: The project engineers' "modified bolt will meet or exceed the strength" that they are targeting. Because readers must be able to quantify that strength, the core includes two prepositional phrases ("*of* an unmodified bolt" and "*with* a YS of 163.6 KSI"). The engineers here are guaranteeing performance of a part—a claim important enough that it makes sense for the sentence to include not only the finding but also the circumstances in which that finding is valid. This information is delivered in the subordinate clause ("Assuming . . .").

24.3 Avoiding sentence fragments, fused sentences, and comma splices

Like a building or a bridge, a sentence needs to be structurally sound, and structural defects can have consequences. Audiences sometimes judge writers rather harshly when their sentences exhibit common defects: Errors in sentence technique can create the impression that the writer hasn't paid attention to subtle details or hasn't taken the care to check his or her work. Obviously, those are crucial qualities that you will want to convey to your readers. Fortunately, you can eliminate the most common errors in sentence structure by checking for three basic problems: sentence fragments, fused sentences, and comma splices.

Avoid sentence fragments by completing the sentence core

Sentence fragments occur when any required component of a sentence core is missing. All sentences need subjects and verbs, and many need objects. If a group of words is presented as a sentence—that is, it begins with a capital letter and ends with a period—all of these elements must be present.

Avoid fused sentences and comma splices by inserting logical connectors

Joining two sentence cores within a sentence requires a logical connector that specifies the logical and structural relationship between the two. Logical connectors include the following:

- Coordinating conjunctions: *and, or, but, yet, so*
- Subordinating conjunctions: *before, after, until, when, where, although, because*
- Relative pronouns: *who, whom, whose, which, where, when, why*
- Colons (:) and semicolons (;)

Two structural defects can occur where two sentence cores are present in a single sentence:

- A *fused sentence* (or *run-on*) occurs when *no logical connector is present* to attach the sentence cores to one another.
- A *comma splice* occurs when the sentence cores are attached together *only by a comma.*

The comma can join sentence cores together only with a conjunction or a relative pronoun. If you want to join those cores together with punctuation alone, you must use a semicolon or a colon. (Put another way, semicolons and colons are complete logical connectors by themselves; commas can only form one part of such a connector.)

One tricky case occurs when the second sentence core is introduced with a *transition word* (see Module 23: *Paragraphs*) such as *however*, *moreover*, or *nevertheless*. These are useful at marking a logical step in your reasoning, but they aren't conjunctions or relative pronouns. Therefore, they must be preceded by *a semicolon, not a comma.*

24.4 Increasing conciseness while maintaining clarity

More than any other feature of writing style, engineering audiences tend to appreciate *conciseness*. As in so much engineering work, efficiency is paramount: If you can deliver more information in fewer words while keeping the content easy for audiences to understand and use, you should do so.

Because it measures density of information, conciseness differs from brevity, which is simply a measure of length (in words or pages or speaking time). Making a message briefer *can* make it more concise and save time for your audience, but not if that brevity comes at the cost of clarity or substance. Readers or listeners who are confused or are missing key information will ultimately have to spend *more* time figuring things out—the opposite of the outcome you wish to achieve.

Word choice for conciseness

Audience considerations should drive the way that you approach revisions for conciseness. In general, liberally using technical jargon allows sentences to accommodate more information with fewer words, but this technique can create documents that make sense only to specialist experts. When a document may need to reach audiences with little technical expertise, less technical language may be preferable even when that language is less precise than jargon would be.

Choosing precisely the right verb to convey the most information is at least as important as choosing the right nouns. Active verbs are more concise than passive ones (because they don't require *to be* as a "helping verb"), although passive voice remains the convention for experimental reports and similar forms of scientific writing.

Introduction: Unicompartmental knee arthroplasty (UKA) is a minimally invasive treatment for unilateral osteoarthritis in the knee, and long-term awith contemporary prosthesis designs and improved patient selection. Failure in UKA has been ascribed to progression of arthritis in retained compartments, polyethylene wear, and implant malpositioning leading to tibial bone collapse. This study was designed to isolate the effect of tibial component alignment on loading the proximal tibia.

Acronyms can make technical writing dense but are an important way of balancing precision and brevity—here, referring to UKA efficiently without any risk of it being confused with other types of arthroplasty.

Methods: Sixteen composite tibias were implanted with a metal backed, mobile bearing tibial prosthesis in one of four different sagittal angles with respect to the anatomical axis of the tibia: 0, 5 and 10 degrees of posterior slope, and 5 degrees of anterior slope. Ten rosette strain gages were distributed on the medial cortex of the proximal tibia. Each bone was axially loaded to 1500 N using an electrodynamic materials testing machine through a UKA femoral component affixed to the actuator of the testing machine.

Results: In preliminary data, highest posterior strains were observed when tibias were implanted with extreme posterior slope. Strain observed in posterior measurement regions increased between 16 and 800 microstrain when components were implanted with 10 degrees of posterior slope as compared to other alignments.

The degree of conciseness achieved here depends not only on careful selection of nouns and verbs but also on decisions about which quantitative details to include.

Conclusions: These findings correlate well with prior clinical studies which have suggested that posterior slope greater than 7 degrees significantly increases risk of posterior tibial collapse and reaffirm the need for careful tibial component alignment in UKA procedures.

The genre conventions of experimental reports, which mandate passive voice, take priority over the general stylistic preference for active voice.

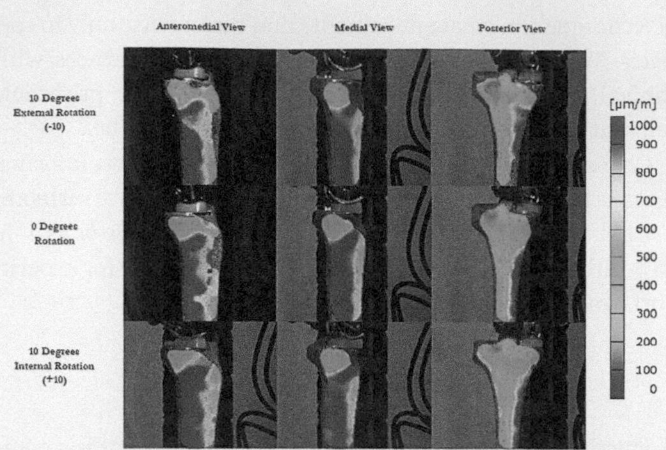

Conciseness becomes an especially important goal in documents like this one-page summary of a research study, written to a foundation to present results obtained with grant-funded equipment. ©2012 Joint Replacement Surgeons of Indiana Foundation, Inc. Reprinted with permission (Small et al., 2012).

Avoid expletives (there is, there are)

Vague, imprecise sentences are often built around a weak, imprecise subject–verb combination known as an *expletive* construction: *there is*, *there are*, or *it is*. While common in everyday conversation, expletives weaken the directness and clarity exhibited by good technical writing:

- *It is* necessary to calibrate all instruments before taking measurements.
- *There are* several problems occurring in the prototyping process.

Avoid using too many expletives: A simpler, more efficient sentence core can usually be devised by choosing a precise noun in the sentence to become its new subject:

- *All instruments require calibration* before taking measurements.
- *The prototyping process has encountered* several problems.

If the expletive is stating an instruction, the sentence can be reworked around the implicit subject "you": "*Calibrate all instruments* before taking measurements."

Summary

Solidifying the sentence core

- A sentence should house its main content in the core—the subject (performing the action), verb (the action), and object (on which the action is being performed).
- Well-crafted sentence cores can explain a main point for non-experts, separating it from technical details, or highlight the particular finding that is most important to the engineer's analysis.

Coordinating and subordinating ideas

- *Coordinate* two or more sentence cores, using such logical connectors as *and*, *but*, *or*, *yet*, or *so*, to provide equal emphasis to both ideas.
- *Subordinate* one sentence core to the other when one of them states the main idea and the other clarifies, supports, or qualifies it.

Avoiding sentence fragments, fused sentences, and comma splices

- Sentence fragments are missing a required component of the sentence core—a subject, verb, or object.
- Joining two sentence cores together in a single sentence requires a colon, semicolon, or conjunction (*and*, *but*, *or*, *yet*, *so*). Without a conjunction, a comma is insufficient to perform this function.

Increasing conciseness while maintaining clarity

- Jargon can always make sentences more concise, but this conciseness comes at the expense of clarity to readers whose expertise does not perfectly match your industry, specialty area, and technical interests.
- Avoid expletive clauses (*there is*, *it is*): To revise an expletive, find another word in the sentence to become its subject, and then pair the new subject with a verb.

25 Words

Engineering students sometimes become frustrated while writing when they see language as lacking the precision and accuracy that they associate with numbers. The fact that engineers privilege those qualities, though, can be a great strength of their writing.

Much of the technical writer's craft consists of arranging words in sentences and paragraphs, but that craft begins with the ability to choose exactly the right word with the right meanings—rather like choosing a resistor with exactly the right value for a circuit that you're designing.

Objectives

- To choose effectively between technical jargon and other available word choices based on the background, needs, and tasks of likely audiences
- To recognize and avoid word choices that impede clarity and create inefficiency, such as nominalization, affectation, and buzzwords
- To choose verbs that maximize the clarity of the sentence core
- To manage the emotive component of written language, avoiding unfavorable or inappropriate audience reactions

25.1 Achieving precision without needless jargon

Professions like engineering—and their specialized disciplines and subfields—are defined in large part by their specialized technical terminology: You can recognize your growth as an engineer by your ability to define and use that terminology. The technical terms of any intellectual field help practitioners to be certain that they're really talking about the same things, and having a convenient term for every concept helps us to be concise and efficient.

Deciding when to use technical jargon

If we think of *jargon* in this sense of a specialized vocabulary shared by a community of professionals, there's nothing wrong with it. In fact, we need such a vocabulary to communicate technical findings with any degree of efficiency. An engineer conducting materials testing would have a hard time speaking, writing, or even thinking precisely without terms of that field: *modulus of elasticity, ductile strength, force-deflection.* When speaking or writing to other engineers, or to technicians or managers with relevant experience, there's no reason not to use those terms.

At the same time, using these terms automatically in all contexts, without considering alternatives, can give an engineer a reputation as a poor communicator. This is why "jargon" has a bad reputation in general: We associate it with an inability to understand what's being said, or with excessive difficulty that goes "over our heads." In a worst case, jargon can alienate audiences, suggesting to them that a communicator is oblivious to the needs of the audience.

You might want to tell an engineering peer that a bolt "exhibited classic ductile tensile failure." For other audiences without experience in materials testing, though, you might want to state instead that the bolt broke when pulled with a particular force, and then determine whether readers or listeners will benefit from additional explanation of the fracture itself and how it happened. This strategy typically produces better results than introducing and then defining technical terminology.

Avoiding affectation and buzzwords

At its worst, imprecise and pretentious language can mask unethical business practices or hazards to public safety or to the environment (see Module 3: *Meeting your ethical obligations*). More routinely, flawed word choice prevents clear thinking by writers and audiences alike.

Buzzwords	Professional documents sometimes use fashionable language at the expense of clarity. In the right context with the right definitions, "paradigms" and "synergy" can be part of a meaningful argument, but these words have been damaged, and their meanings diluted, by overuse.
Affectation	When choosing words, writers in business sometimes put less thought into clear transmission of meaning than to sounding important and intelligent. Affected language favors the exotic (as in such Latin constructions as "per your request"), the multisyllabic ("utilize" rather than "use"), and broad clauses that approach the main point indirectly ("it appears to be the case that . . .") or modify it needlessly ("at the present time," "pending further notification").

Nominalization A particular variety of affectation works by creating complex nouns out of other parts of speech, so that "apply the sealant" becomes "initiate the application of the sealant."

These typical flaws in word choice are apparent in sentences like this one:

> After implementation of the finite-difference diffusion simulation described above, lubricant modulation profiles at increasing relaxation times are produced which enable the accurate prediction of the lubricant profile at any desired future state by the experimenter (assuming that the parameters are correctly determined).

Important verbs (*implement* and *predict*) have been converted into nouns that don't serve important functions within the sentence. Complex noun phrases (*desired future state*) only approximate the concepts required by the sentence. Perhaps because it's imprecise, the sentence lapses into redundancy (*the lubricant modulation profiles . . . are produced . . . of the lubricant profile*). A simple revision converts the first part of the sentence into a robust, straightforward subject–verb–object sentence core, reducing the risk of misunderstanding:

> Lubricant modulation *profiles* at increasing relaxation times *are predicted* with the finite-difference diffusion simulation described above.

The parenthetical comment makes an important point that is better served by placing in a separate sentence:

> The *accuracy* of the predicted profile *depends on* correctly determining the parameters.

Having more clearly explained the utility of the simulation, this writer is now well positioned to move forward, exploring the complexities of determining simulation coefficients from experimental data.

25.2 Selecting precise verbs

Actions and subject–object links

Interestingly, some of the simplest and the most complex verbs actually suffer from the same vagueness. We easily recognize the handful of broad, simple verbs—*to be, to do, to have, to get, to make, to go*—that populate so many of our everyday sentences. We often begin with one of these:

> In dynamometer testing, peak torque was 185 lb.-ft. at 2,500 rpm.

Recognizing a lack of precision here—the verb *was* doesn't add anything to this sentence—sometimes leads us to more complex, Latinate verbs—*to utilize, to operationalize, to implement, to facilitate.* However, these seldom deliver the precision and sophistication that they seem to promise. We end up with complexity in the sentence form, but the sentence itself doesn't perform any more work:

> Tests utilized a dynamometer, which indicated a peak torque of 185 lb.-ft. at 2,500 rpm.

Depending on what we want to say about this finding, though, we can build a much more potent sentence form around a more powerful verb:

- *The engine delivers* a peak torque of 185 lb.-ft. at 2,500 rpm. (simple cause and effect)
- The peak torque of 185 lb.-ft. at 2,500 rpm *suffices* for . . . (positive evaluation of the result)
- The peak torque of 185 lb.-ft. at 2,500 rpm *falls* short of . . . (negative evaluation of the result)

Active voice for responsibility, passive voice for detachment

Standard-issue writing instruction tells us that an active verb creates clarity, directness, and efficiency, but school lab reports also teach us a habitual reliance on the passive voice. We learn to write "the sample was placed in a calorimeter," not "I placed the sample in a calorimeter," even though the latter includes more information by telling the reader who performed the action.

Scientific writing developed this practice to emphasize objectivity and detachment: In an experimental report, it isn't supposed to matter *who* is performing a procedure. The validity of the findings depends instead on other experimenters being able to duplicate that procedure with similar results. As readers in a scientific community, we're supposed to judge the experimental methodology, not the person executing it, and so the authors of technical papers avoid mentioning that person.

Everyday professional correspondence, though, presents important reasons to be more candid about who's performing an action. Passive voice can seem evasive, as if writers are attempting to avoid responsibility. This impersonality can seem desirable—it's easier to write that "a decision has been made to discontinue the program" than to admit one's own involvement in that decision—but it's worth remembering that most readers will want and appreciate greater candor.

Most importantly, passive voice can contribute to inaction and poor planning. Your proposal may specify that "all oscilloscopes and other instruments will be inspected and calibrated annually"—but sentences using active verbs much more reliably state who exactly is responsible for conducting those activities.

Agreement with subjects

Verbs must *agree in number* with their subjects. A singular subject requires a singular verb, and a plural subject requires a plural verb:

Organizations	Use singular verbs with the name of an organization (even though that organization is obviously made up of many individuals): "Engineers Without Borders *addresses* many problems of daily life in industrializing countries."
Collective nouns	A few nouns are singular in form, even when they refer to multiple people (*team*) or things (*equipment*). These nouns almost always pair with singular verbs: *the team is*, not *the team are*.

25.3 Using pronouns precisely

Clear reference to antecedents

A pronoun (*he, she, it, they*) must refer to a noun, which is called its *antecedent*. Pronouns enable efficiency and variety in our sentences, preventing us from having to repeat nouns such as proper names and technical terms.

To avoid confusion, readers and listeners must effortlessly connect the pronoun to the noun that it replaces. *Unclear pronoun reference* results when that connection is muddied. For this reason, effective engineering writing uses very few pronouns, retaining clear descriptive nouns to prevent any possibility of unclear reference:

> In the case of Nunn Creek, a more active rehabilitation approach was taken with j-hook vane structures. *This restored reach* exhibited greater nitrate uptake than the unrestored reach, suggesting that *this rehabilitation technique* increases *nitrate uptake.*

In casual conversation, engineers might well replace any of these italicized phrases with the pronoun "it," but retaining or even repeating the noun phrases adds precision and clarity.

Agreement with antecedents

Like verbs, pronouns must *agree in number* with the nouns to which they refer. A singular subject requires a singular pronoun (*he, she, it*), and a plural subject requires a plural pronoun (*they, them*). Exceptions and special cases for verbs—to accompany collective singular nouns, the names of organizations, etc.—apply to pronouns as well.

25.4 Managing emotive language

Emotion won't usually be the primary focus of our word choices, but all writers and speakers need to consider the ways in which those word choices affect audiences' reactions, and this often involves emotional factors. We may associate engineering and other technical professions with rigor and objectivity, but they remain human activities practiced in human communities. Even when our writing concerns empirical findings or calculations, we're often delivering judgments or decisions about those findings, and those judgments can often prompt emotional reactions:

- The current environmental monitoring protocol *is a joke.*
- The current environmental monitoring protocol *is inadequate.*
- The current environmental monitoring protocol *creates an ongoing risk of crisis.*

As a matter of habit, writers should opt for word choices like those in the second and third sentences: The first sentence is openly derisive and therefore carries a high risk of causing offense. The alternatives are not completely value-neutral or objective, but we should probably not expect them to be: If the writer believes that a crisis is likely, a fairly high degree of concern is entirely appropriate.

The *fallacy of emotive language* occurs when communicators stir readers with emotionally loaded terms alone, in the absence of evidence. Because this fallacy is especially easy to commit when expressing a negative judgment, check word choice when calling colleagues' attention to a problem or expressing dissatisfaction with a situation:

- Inappropriate negativity can arise in nouns (*disaster*), adjectives (*pathetic*), verbs (*failed*) and even adverbs (*miserably*).
- If you want the reader to understand your feelings, negative words can describe your own state of mind (*disappointed, upset, anxious*) without pretending that those words express objective truths about the situation. Obviously, these should be used carefully.
- Word choices expressing urgency may be needed to arouse rapid or intense action or attention. When such cases occur, think of those word choices as claims that need to be supported with explicit evidence and reasoning.

Summary

Achieving precision without needless jargon

- Technical jargon will usually be appropriate for engineering peers whose backgrounds and training resemble yours. For those readers, highly specialized terms enable greater conciseness.
- For most other audiences and situations, clarity is most easily achieved by avoiding jargon (rather than defining it).

Selecting precise verbs

- Verbs can be too vague whether they are simple (*to be, to do, to have*) or complex (*to utilize, to operationalize*).
- Passive voice is the standard for reporting experimental procedures because it emphasizes detachment and objectivity.
- In most other cases, active voice allows for greater precision, especially in identifying the agent responsible for an action.
- Verbs must *agree in number* with their subjects. A singular subject requires a singular verb, and a plural subject requires a plural verb.

Using pronouns precisely

- Any pronoun (*he, she, it, they*) must refer to a noun, its *antecedent*. Writers should check to be sure that audiences can immediately connect the pronoun to the correct antecedent.

- Pronouns must *agree in number* with the nouns to which they refer. A singular subject requires a singular pronoun (*he, she, it*), and a plural subject requires a plural pronoun (*they, them*).

Managing emotive language

- Arguments commit the *fallacy of emotive language* when they rely on emotionally loaded words in place of evidence.
- In the workplace, emotive language creates the most problems when it is used to deliver negative judgments.
- Use emotionally evocative word choices only in rare cases to move audiences to action. When doing so, treat them as claims that need to be supported with explicit evidence and reasoning.

Summaries

26

Important audiences for engineering work will often need to understand key content but lack the time to read a complete document carefully. Summaries provide the context and the main points (or findings) of your argument without the supporting details of the longer document. Avoid "dumping data" into a summary: Instead, methodically craft the summary to meet the needs of the audience.

Objectives

- To enhance the decision-making role of executives by constructing executive summaries that are complete yet concise
- To describe a document's content in an abstract, allowing engineers to identify information relevant to their current task
- To persuade reviewers that a proposed project is well conceived when submitting an abstract as a proposal

26.1 Executive summaries

An executive summary is written *for* an executive, typically for decision making or authorizing. Executives will likely make decisions without reading the entire document, so the summary must be complete yet concise, enabling those decisions to be sound and sufficiently informed.

Meeting the needs of the executive audience

An executive summary facilitates decision making—for example, to act on your findings or to authorize your proposal. Because the executive is expected to make a decision without reading the entire document, the summary must accurately represent every major point of your argument. The sequence of points in the summary should preserve the order of the sections in the longer document.

Executives' needs define three overall objectives for a summary:

- Provide an overview of your technical analysis and recommendations.
- Avoid technical jargon.
- Focus on information important to a decision-maker: findings, recommendations, budget, timeline, lessons learned, or foreseeable problems.

Executive summaries also have secondary audiences. For example, reports or proposals submitted to city governments are usually made available to the public; in such a case, the executive summary should be written with that broader public audience in mind.

How to summarize a lengthy document

The length of an executive summary depends, of course, on the length of the document being summarized. Most executive summaries are 5% to 10% of the length of the full document (although very rarely less than a full page) and are placed after the document's front matter and before the main body of text. Adhere to any length and formatting standards specified by your primary audience.

After the longer document is completely drafted, begin the draft of the executive summary by reverse-outlining:

- Start with the section and subsection headings from the full document.
- Add the topic sentence of each of the full document's paragraphs.
- Review and revise the topic sentences, ensuring that every major finding is represented. (If your work has led to any "bad" results, include them: Even when seeking executive approval, a summary should reflect any reasons for doubt, pessimism, or caution.) This step also provides an excellent opportunity to review and revise the argument in the full document.
- Identify topics or ideas that need to be expanded beyond those topic sentences or that can safely be omitted altogether.
- Compile the outline into a complete draft with coherent paragraphs: Add transitions, and perhaps very brief supporting material, where needed. Including an informative visual is usually acceptable (although only summaries of very long documents will include more than one).
- Revise for completeness and clarity, using the audience's perspective—decisions that need to be made, and knowledge or views that the reader may bring to the document.

Summarizing very long documents will usually require you to design your sentences to summarize subsections or even whole sections, rather than individual paragraphs. Writing such a summary may also require you to omit sections of the larger document.

The executive summary is placed just before the main body text.

This feasibility study makes an argument for pursuing a particular course of action. The executive summary echoes the primary evidence, arguments, and recommendations made in the body text.

Table of Contents

This feasibility study's executive summary aims for maximal brevity, summarizing a fifty-page report in two pages.

EXECUTIVE SUMMARY

This report describes the results of a study sponsored by the Keck Institute for Space Studies (KISS) to investigate the feasibility of identifying, robotically capturing, and returning an entire Near-Earth Asteroid (NEA) to the vicinity of the Earth by the middle of the next decade. The KISS study was performed by people from Ames Research Center, Glenn Research Center, Goddard Space Flight Center, Jet Propulsion Laboratory, Johnson Space Center, Langley Research Center, the California Institute of Technology, Carnegie Mellon, Harvard University, the Naval Postgraduate School, University of California at Los Angeles, University of California at Santa Cruz, University of Southern California, Arkyd Astronautics, Inc., The Planetary Society, the B612 Foundation, and the Florida Institute for Human and Machine Cognition. The feasibility of an asteroid retrieval mission hinges on finding an overlap between the smallest NEAs that could be reasonably discovered and characterized and the largest NEAs that could be captured and transported in a reasonable flight time. This overlap appears to be centered on NEAs roughly 7 m in diameter corresponding to masses in the range of 250,000 kg to 1,000,000 kg. To put this in perspective, the Apollo program returned 382 kg of Moon rocks in six missions and the OSIRIS-REx mission proposes to return at least 60 grams of surface material from a NEA by 2023. The present study indicates that it would be possible to return a ~500,000-kg NEA to high lunar orbit by around 2025.

The first sentence states the scope and focus of the study, setting up a paragraph that arrives at the study's findings.

The summary includes the data most central to the study's conclusions but puts the numbers into context for non specialist audiences.

The opening paragraph's last sentence states the study's central conclusion—that an asteroid might be retrieved successfully by 2025.

Few executive summaries will include more than a single image—here, an illustration.

This executive summary provides a fairly comprehensive account of the document's first two sections, supporting the overall conclusion that an asteroid retrieval mission is feasible and advances important near-term objectives in space exploration. Subsequent sections are summarized only very broadly in the last paragraph, with one sentence often stating the main finding of an entire subsection. The overall report compares two types of asteroid retrieval ("Pick Up a Rock" and "Get a Whole One"); the executive summary omits these, choosing instead to focus on the feasibility of the project in general.

26.2 Abstracts

While an executive summary typically targets a known audience within an organization, an abstract is addressed to a somewhat larger external community. Engineers read abstracts looking for documents relevant to a technical problem on which they are working. To serve these readers, abstracts are short, stating just enough to describe the document's content without necessarily condensing its entire argument. Unlike an executive summary, an abstract assumes that readers will read the entire document when they need to know more than its general finding or argument.

Meeting the needs of a technical audience

Abstracts serve two main audiences in engineering—researchers and practitioners. Researchers are working at the boundaries of a discipline: Abstracts tell them the current state of their art. Practitioners are solving particular problems: Abstracts tell them what has already been solved. Both audiences can be expected to be familiar with standard technical jargon in the field.

Many abstracts are cataloged in searchable engineering databases, so they should contain key words and phrases to aid searching. Some journals require authors to submit a list of "keywords" following the abstract for just this purpose, helping readers to find the most relevant content within the large body of technical literature produced every year. Like technical journal articles, patent applications are searched by engineers who want to know what others have achieved; to assist in such searches, patent applications include abstracts that describe the purpose and operation of an invention.

How to write an abstract

Abstracts are short—usually one or two paragraphs, or around 200 words. (When submitting to a journal, authors will often need to write abstracts within a specified maximum word count.) Because abstracts are displayed in online databases, they include only text in standard paragraphs. Exclude visuals, equations, headings, and references. In evaluating whether to obtain the full document, readers typically look for a few sentences describing the substance and significance of the work that you have performed. For this reason, abstracts in scientific and technical journals often approximate the IMRaD format for reports:

- Introduction (purpose and scope): Why does the document exist? What questions does it address?
- Methods: What methods or procedures were used?
- Results and Discussion: What did you find, and what are the implications of those findings?

An abstract will seldom spend more than two sentences on any one of these questions.

Someone working in the field of laser excitation of subatomic particles would likely search a journal's database with a few keywords and then skim the resulting abstracts, saving articles that seemed useful and informative to read in full.

***Abstract*—Wedge-shaped holes are fabricated in the top mirror of proton-implanted vertical-cavity surface-emitting lasers (VCSELs). A radially symmetric fill factor approach is used to calculate the resulting transverse index profile. To investigate both the index confinement provided by the etched pattern and its effect on optical loss, continuous-wave (CW) and pulsed experiments are performed. Under CW operation, we show proper wedge design leads to improved fundamental-mode output power, decreased threshold, and increased efficiency. We report a significant decrease in threshold under pulsed operation for the etched device compared to an unetched device, indicating a significant reduction in diffraction loss to the fundamental mode due to strong index guiding. Single-mode output is maintained over the entire operating range of the VCSEL due to increased loss for the higher order modes.**

***Index Terms*—Distributed Bragg reflector lasers, semiconductor lasers.**

Targeting specialist experts, this abstract opens with the narrow technical objective alone.

As a whole, the abstract generally follows the IMRaD report structure: The first sentence introduces the study's objective, followed by two sentences on method and some generalizations about results.

I. INTRODUCTION

VERTICAL-CAVITY surface-emitting lasers (VCSELs) are an ideal source for short-haul optical telecommunication. VCSELs provide a number of benefits over edge-emitting lasers including low cost, high volume manufacture, the capability for on-wafer testing, inherent single longitudinal mode operation, and the ability to easily fabricate large arrays. However, VCSELs tend to operate in multiple transverse modes. Single-mode laser sources are desired for high bandwidth operation of fiber optic links, sensing, and atomic clocks.

The article still addresses the broader context that makes this technical advancement meaningful (although, targeting specialists, the abstract does not summarize it).

This abstract for a technical journal article demonstrates the typical IMRaD organization. In this case, all quantitative detail is reserved for the article itself.

26.3 Submitting an abstract as a proposal

While abstracts typically summarize completed technical work, researchers in engineering fields sometimes need to write abstracts for planned or proposed future work. This is common when applying to the program of a technical conference. In these cases, the abstract's audience and purpose will more closely resemble those of an executive summary because you will be seeking more formal and definite approval from decision-makers. Because your project will be judged on the abstract alone, you may have as little as two to three hundred words to make your case for acceptance.

When applying for a grant or a patent, the abstract is supplementary, as it is for a published article. The longer proposal or project narrative supplies opportunities to present and support your ideas in detail, but the abstract still needs to make a solid initial case, persuading readers to think of the project favorably from the outset.

Research grant applications for funding follow the usual abstract format but do not include the results (since the work has not been done). However, the expected implications of the results are often discussed, supporting the claim of the usefulness of the proposed research.

Design goals are given in sufficient detail for reviewers to understand what will constitute success for this project.

The authors describe benefits for the research community, focusing on "our understanding" of important technical topics.

Public agencies like NSF serve the public interest and require grant proposals to identify social benefits beyond narrow technical advancements.

ABSTRACT

The Principal Investigator (PI) will explore and test a number of hardware platforms and software algorithms whose goal is to facilitate sub-second human-robot synchronization. To this end the PI will utilize the medium of music, one of the most time-demanding media where accuracy in milliseconds is critical because asynchronous operations of more than 10 ms are noticeable by listeners. Specifically, the PI will develop up to three different kinds of robotic devices intended to allow a drummer, whose arm was recently amputated from the elbow down, to play and synchronize between his organic functioning arm and the newly developed robotic devices. In addition, he will develop and investigate the efficiency of novel anticipation algorithms designed to foresee human actions before they take place and trigger robotic actions with low-latency and in a timely manner. This research will advance our understanding in a variety of areas, including the biomechanics of limb operations, machine-learning techniques for the anticipation and prediction of human gestures, and highly accurate myoelectric robotic devices.

Broader Impacts: This project will ultimately benefit a large population of amputees whose quality of life could improve through the use of low-latency robotic limbs with sub-second synchronization. Facilitating such accurate sub-second human-robot synchronization could also improve efficiency and flow in other human-robot interaction scenarios where humans and robots must collaborate to achieve time-demanding common goals. The novel solenoid-based robotic device(s) created in this research should also benefit musicians in general (that is to say, who are not disabled), who will be able to explore novel drumming techniques and create novel musical results.

This abstract comes from a proposal to the National Science Foundation, seeking funding to investigate human–robot interactions. Like an abstract for a research article, this one covers the project's scope, purpose, methods, and anticipated significance—even though it describes future work yet to be completed. Courtesy: National Science Foundation. (Weinberg, 2013).

Summary

- An effective summary meets the needs of the audience while presenting the context and main points of the full document.

Executive summaries

- Executive summaries are written for decision makers or authorizers, so the summary must be complete yet concise, enabling them to make sufficiently informed decisions. Do not assume that the authorizer will read the full document.
- Readers of executive summaries will be looking for findings, recommendations, budgets, timelines, lessons learned, and foreseeable problems.
- Wherever possible, avoid technical jargon.
- An effective executive summary can be crafted by reverse-outlining the full document. Summarize paragraphs or entire sections by using the topic sentences found within.

Abstracts

- Abstracts are written for researchers and practitioners exploring what others have achieved in researching a technical topic or in solving a particular problem.
- Abstracts are one or two paragraphs long and include just enough material to describe the document's purpose and findings without necessarily condensing its entire contents. The reader will then decide whether or not to read the full document based on the abstract.
- Abstracts in journals use standard technical jargon in the field, often listing specific keywords to aid readers in finding relevant information.
- Readers of abstracts will want to learn the purpose and scope of the work being reported, the methods used, the results presented, and the significance of those results.

Submitting an abstract as a proposal

- Some abstracts are submitted as proposals, especially to seek grants or acceptance to technical conferences. Such abstracts resemble executive summaries in their aim to seek approval from authorizers.

27 Front and back matter

Many larger workplace and academic documents, such as reports and proposals, include supplementary components outside of the main sections that deliver the primary content. *Front matter,* such as the cover page, a table of contents, a list of figures, and the executive summary, precedes the primary document; *back matter,* such as the notes, appendices, and a bibliography or references, follows the main content of the document. These parts of the document assist readers in finding content and understanding its context.

Some parts of the front and back matter, such as executive summaries and appendices, require substantive treatment of *content* and are addressed elsewhere in this book. Other components of the front and back matter, though, are more about *form* than content. Yet making a favorable impression on your audience requires you to pay scrupulous attention to the details of conventional forms.

Objectives

- To create front and back matter sections that follow conventions for content and display of routine information to readers
- To help readers easily find desired information throughout a document
- To arrange information in a way that allows readers to follow up on points of interest and assess the credibility of your work

27.1 Front matter: Title page, table of contents, and list of figures

Title page

Title pages present fundamental information about the content of the document that a reader uses to assess the usefulness of the document and the credibility of the authors. The title should be the most prominent item on the page, usually appearing near the top in a large or boldface type. Make sure the title is descriptive—"Experimental Stability Verification of PID Controller" is a more useful title than "Experimental Report." The authors' names, titles, and affiliations come next, making clear who is responsible for the content. The date is also important to establish when the document was completed.

Wabash River Nutrient and Pathogen TMDL Development

FINAL REPORT

September 18, 2006

Prepared for
Illinois Environmental Protection Agency
Indiana Department of Environmental Management

Prepared by
Tetra Tech, Inc.

Contrast in the typeface is used to make the title of this report stand out.

Right-justified alignment is used here to make the page visually interesting, but it should be used sparingly.

Listing the organization that prepared the document increases credibility, but including the names of the primary authors would be even better.

The title page of this over-250-page report establishes the document's contents and the credibility of the work. © 2006 Illinois EPA. Reprinted with permission.

Table of contents

For long documents, a table of contents (TOC) helps readers locate relevant sections of the work. A table of contents repeats the section headings and subheadings that appear in the document itself. The TOC excludes any front matter that precedes it but might list the executive summary if one is included.

Indentation indicates subheadings that fall beneath the major section headings, showing the logical hierarchy that define the document's overall reasoning. Page numbers are included to indicate where each section begins.

When editing a document, particularly at the last minute, page numbers might shift, sections might appear in a different order, and figure numbers might change. Make sure the information presented in the TOC is up to date and reflects the actual content of the document.

The section headings and subheadings appear on the left with the associated page number on the right.

Notice that the page numbers are right-aligned while the headings are left-aligned. Using a table with hidden gridlines makes such complex alignment easy.

While extended ellipses are commonly encountered in a table of contents, careful attention to alignment and spacing can make them unnecessary.

The section headings include any appendices.

Wabash River TMDL Development Final Report

Table of Contents

i

The table of contents for a long report repeats the section headings and organization of the main document text with starting page numbers. © 2006 Illinois EPA. Reprinted with permission.

List of figures

After the table of contents, a list of figures and tables might be used. Figures are listed separately from tables. The organization of these lists follows similar rules for the table of contents. Figure and table numbers are listed in order and the caption is repeated exactly as it appears in the text, followed by the page number where the figure or table appears.

Wabash River TMDL Development — Final Report

Figures

Tables

ii

A separate list of figures and tables appears immediately after the table of contents.

When the caption of the figure or table runs to two lines or more, align the item number with the first line and the page number with the last line.

The list of figures and list of tables for a long report helps readers locate the items that interest them.

27.2 Back matter: Notes, appendices, and bibliography

While the front matter of a document helps readers quickly identify what they want to read, the back matter provides additional information that might be of interest to readers—either after they have read the main text or while they are exploring specific content within it in greater depth.

Notes	Clarifying information that might help the reader understand a point made in the primary document. Notes are short—often a few sentences and rarely more than a paragraph.
Appendices	Useful detailed information that some, but not all, readers might be interested in reading, but not critical to the understanding of the major points of the document. Appendices are longer than notes—they can be thought of as miniature documents.
Bibliography	Sometimes called "references," bibliographies present—in a consistent, standard format—the original sources for the claims made in the document.

Most of a document's back matter requires careful judgment; you're making distinctions of two essential types:

- Material that needs to be included in some fashion (vs. material that can safely be omitted altogether)
- Material that can be treated as a supplement (vs. material that needs to be present in the main body)

If the content with which you are working doesn't appear to be necessary in the main body but seems to be too important to be omitted altogether, it may be a good candidate to be presented in a *note* or an *appendix*.

Notes

When you are drafting a document and deciding whether to include information that you know is relevant but that you perceive as tangential, consider including it as a note. Notes should not contain any content essential to understanding a document's major argument or findings, but they can further clarify or qualify any of the document's content. One common use is to identify related research that is not referenced in the primary document.

Generally, an organization or a journal will stipulate whether notes are expected. If you are using endnotes, list them numerically both in your primary document and on the notes page in your back matter. In the primary document, add a superscript number (without a space) to indicate a note. Each note requires a new number.

Appendices

Appendices supply additional documentation that might be useful or even necessary for readers performing detailed analysis on the document's content, but that would be out of place in its main body. In an experimental report, tables of raw data or detailed calculation steps might appear as appendices. In a design report, an appendix might be used for transcripts of stakeholder interviews.

While the sections within a document's main body are usually numbered, appendices are traditionally labeled with letters, based on the order in which they are referenced in the main body. Itemize appendices carefully: One appendix gathering many supporting documents or data sets together is likely to be very hard for readers to use—especially if that material is arranged in chronological or alphabetical order through which the reader needs to navigate.

One-Way (existing)	Closer to Blair (Stop Controlled)	North	0	97	35.1	E
	Closer to Blount (Signal)	North	0	80	35.0	E
Two-Way (Scenario 3, 2-out 1-in)	1/2 way between Blair and Blount	North	1	15	36.2	E
			1	30	53.4	F
		South	2	1	74.3	F
	Closer to Blair (New Signal)	North	1	30	37.4	E
		South	2	1	68.8	F
	Closer to Blount (Signal)	North	1	15	35.9	E
			1	30	52.7	F
		South	2	1	130.5	F

Table 3.03-1 Delay and Level of Service for Left Turns from Driveways

The modeling indicates that for Scenario 1 (one-way) a high number of left turns (80 or more) can occur from a driveway during the PM peak hour before the delay results in LOS E conditions. For Scenario 3 (two-way with two lanes outbound in the PM peak), depending on where the driveway is located, LOS F is reached with a volume of 15 to 30 vehicles. In the most unfavorable locations, a single vehicle trying to turn left out of a driveway is predicted to wait an average of one to two minutes or more to do so. Detailed traffic modeling output is provided in Appendix C. This analysis also suggests that the one-way scenario provides more opportunity for pedestrian crossings than the two way scenarios.

C. Summary of Pedestrian Crossings

Both methodologies used to examine the length and frequency of the gaps occurring in East Johnson

The full detail of the traffic model yielding these results is excessive for the main body of the document but is referenced here as Appendix C for interested readers.

A traffic study involves detailed modeling of each scenario. When reporting the results of that modeling, an interested reader can investigate further by looking in an appendix for the detailed information that would unnecessarily clutter the main body of the document.

Bibliographies and References

If a particular claim made in the document is interesting to a reader, he or she can follow up with the reference source to explore the idea further. The most important aspect of the list of references is to use a consistent format that includes all necessary information to find the reference. Module 17: *Researching* elaborates on commonly used reference styles, but your employer or organization might have its own expectations. Use a single style consistently within each document.

A list of references is necessary for any document that does not include full reference information in the main body of the text.

Depending on the reference format being used, the references might be listed by consecutive reference number or by the author's last name.

A consistently formatted reference section demonstrates the credibility of the document.

Wabash River TMDL Development **Final Report**

7 References

Indiana Administrative Code Title 327 Water Pollution Control Board. Article 2. Section 1-6(a). Last updated November 1, 2003.]

Pennsylvania State University. 1992. Nonpoint Source Database. Pennsylvania State University. Department of Agricultural and Biological Engineering. University Park, Pennsylvania.

OEPA (Ohio Environmental Protection Agency). 1999. *Association Between Nutrients, Habitat, and the Aquatic Biota in Ohio Rivers and Streams.* OEPA Technical Bulletin MAS/1999-1-1. Columbus, Ohio.

OEPA (Ohio Environmental Protection Agency). 2006. Final 2006 Integrated Water Quality Monitoring and Assessment Report. State of Ohio Environmental Protection Agency Division of Surface Water. Final Report. Submitted to U.S. EPA: March 27, 2006. Approved by U.S. EPA: May 1, 2006.

Thomann, R.V., and J.A. Mueller. 1987. *Principles of Surface Water Quality Modeling and Control.* Harper & Row, New York.

USEPA (U.S. Environmental Protection Agency). 1986. Ambient Water Quality Criteria for Bacteria – 1986. EPA440/5-84-002. Office of Water. Criteria and Standards Division. Washington, DC 20460. January 1986.

USEPA (U.S. Environmental Protection Agency). 1991. Guidance for Water Quality Based Decisions: The TMDL Process. EPA 440/49 1 -001. U. S. Environmental Protection Agency; Assessment and Watershed Protection Division, Washington, DC.

USEPA (U.S. Environmental Protection Agency). 1997. Technical Guidance Manual for Developing Total Maximum Daily Loads: Book 2, Rivers and Streams; Part 1 - Biochemical Oxygen Demand/Dissolved Oxygen & Nutrient Eutrophication. EPA 823/B-97-002.

USEPA (U.S. Environmental Protection Agency). 2004. Total Maximum Daily Load (TMDL) for the Wabash River Watershed, Ohio. U.S. Environmental Protection Agency. Region 5 Watersheds and Wetlands Branch 77 West Jackson Blvd. (WW-16J) Chicago, Illinois 60604. July 9, 2004.

USFWS (U.S. Fish and Wildlife Service). 2005. Evaluations and Assessment of Fish Assemblages Near Electric Generating Facilities: with Emphasis on Review of Discharge Submitted Data, Development of Standard Operating Procedures, and Traveling Zone Assessment. December 2005, Department of Interior, US Fish and Wildlife Service, Biological Services Program & Division of Ecological Services, Bloomington Field Office, Thomas P. Simon, Ph. D. Fish and Wildlife Biologist.

USGS (U.S. Geological Survey). 1999. Review of Phosphorus Control Measures in the United States and Their Effects on Water Quality. By David Litke. U.S. Geological Survey. Water-Resources Investigations Report 99–4007. National Water-Quality Assessment Program. Denver, Colorado 1999.

Vallentyne. J. R. 1974. The Algal Bowl – Lakes and Man. Misc. Special Publication #22. Department of the Environment. Ottawa, Canada.

Wetzel. Robert G. 2001. Limnology – Lake and River Ecosystems. Academic Press, San Diego, California.

49

The list of references for a long report helps readers locate the background sources that informed the document. © 2006 Illinois EPA. Reprinted with permission.

Summary

Front matter: Title page, table of contents, and list of figures

- The title page presents information about the document—a descriptive title, the authors' names, titles, and affiliations, and the date completed.
- The table of contents helps readers locate relevant information by listing the page numbers for each section heading and subheading (excluding the front matter but including the back matter). Indentation is used to show the logic of the document's organization.
- The list of figures presents the figure numbers, captions, and corresponding page numbers. The list of tables presents the table numbers, titles, and corresponding page numbers. These lists are always separate, each with its own consecutive numbering.

Back matter: Notes, appendices, and bibliography

- Notes clarify or qualify a document's content without being essential to understanding it. Numerical superscripts are used to identify notes in the primary document, and they are listed numerically both in the document and on the notes page.
- Appendices are presented at the end of the document, elaborating on points for interested readers without distracting from the main argument. They are labeled with letters (rather than being numbered) consecutively and are presented like miniature documents.
- A list of references or a bibliography presents the scientific sources providing documentary evidence for the claims made in the main document. The entries should adhere consistently to a standard format.

Visuals

...such as graphs and illustrations make arguments, just as written prose does. Crafting an effective visual is as important as crafting an effective paragraph. A visual can evoke an immediate response in an audience, can be the most memorable aspect of a talk or report, and can actually be the deciding factor in making a decision. If poorly designed or purposely manipulated, visuals can also conceal unpleasant news, deceive an audience, or compel a poor decision. Like any other form of communication, then, visuals require deliberate and ethical design and conscientious, competent drafting, revising, and editing.

Visuals

28 Graphs

Engineers use graphs—visual representations of numerical information—to help audiences reason about content. Creating effective graphs requires careful consideration of the audience, the purpose of the display, and the context of the data. Carelessly created graphs should be scrupulously avoided because they can confuse the audience, hamper the purpose of the display, and obscure the truth.

In his book *Beautiful Evidence*, Edward Tufte defines the central imperative when designing data graphics:

> This is a content-driven craft, to be evaluated by its success in assisting thinking. . . . The first question is *What are the content-reasoning tasks that this display is supposed to help with?* Answering this question will suggest choices for content elements, design architectures, and presentation technologies (136).

Too often, visual displays are selected merely out of convenience; in particular, an engineer might simply settle for the default graphical output of software like Microsoft Excel. Tufte asks us instead to begin by *finding out what the content really has to say*; this will allow us to pin down the reasoning task and how the visual should enable it—how the data should be ordered, which comparisons should be made, and how causality ought to be suggested. Deliberate visual design decisions are needed to avoid obscuring data and conveying poorly considered messages to the audience.

Objectives

- To select the most effective type of graph for a particular data set, audience, and context
- For common graphical genres, to describe the conventions that support effective communication
- To caption graphs and label important features of the data for effective communication

28.1 Choosing the best graph for the task

Engineers sometimes create distinctive graphs that follow particular conventions in their specialized technical fields. When such specialty graphs are not expected, though, the most commonly encountered types are the following:

Bar chart	Encodes data using bar lengths. Suitable for comparing the quantities of one continuous variable for different levels of a category.
Pie chart	Represents quantitative data in the angle and area of the sectors of a circle. Like the bar chart, the pie chart is suitable for comparing the quantities of one continuous variable, often as a percentage.
Dot plots	Replaces most pie and bar charts to communicate the same comparisons more effectively. Can be expanded to the *multiway* form for displaying more than one category (Cleveland, 1993).
Histogram	Divides the range of a measured quantity into evenly spaced intervals and displays the number or percentage of observations in each interval, showing the shape of the distribution and revealing potential outliers.

Box plot	Displays a statistical summary of the distribution of one measured quantity, enabling comparisons between different distributions and revealing potential outliers.
Scatter plot	Displays data with paired values, with an assumed dependent variable graphed as a function of the independent variable.
Line plot	Displays data with paired values, with an assumed dependent variable graphed as a function of the independent variable and line segments connecting the data markers. Common in time-series data.

To decide among these possibilities, think about what you want the audience to do with the information, what assertions you are making, and what type of graph the audience expects to see. Familiarity with the common types of graphs makes this decision easier:

Expected uses	Does the graph need to support audience members as they complete a task—determining the nature of the relationship between variables, noting the maximum or minimum values, or comparing quantities?
Assertions being supported	What statements do you expect audience members to accept, or what ideas do you want them to understand, after viewing the graph?
Conventions among likely audiences	What types of graphs do your likely audience members know well, and which ones will they expect to accompany this type of content? Note that the expected graph is not always the one that most clearly communicates the information.

Showing the data that conveys your argument and purpose

You may be tempted to prepare a graph showing data exactly as it was recorded, particularly if you are rushing to meet a deadline. However, the data that makes the most convincing support for your argument may be a modified version of the original data. Thinking about the argument you are trying to make and the purpose of the graph can help you decide how to present the data.

The samples below show the same data in two forms: The first shows the original record (in the time domain) and the second shows the result of a mathematical transformation (to the frequency domain). Both graphs are truthful, but the second graph better supports the purpose of the display because it shows what frequencies are present in the measured signal.

The data measured are the sensor readout in millivolts and time in microseconds. As a time series, the conventional exploratory display would be the time series shown.

The graph shows the data truthfully but fails to meet the goal of the communication—displaying the value of the peak frequency in the signal—and thus is a poor choice of graph type.

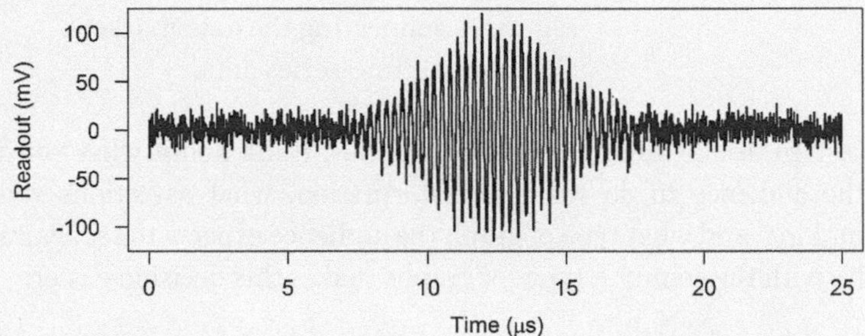

The LDV readout is characterized by high frequencies throughout and large amplitude during the 10–15 microsecond interval.

A laser Doppler velocimeter (LDV) is a sensor used to measure fluid velocity. This graph shows an LDV readout plotted as a function of time. This type of graph fails to communicate the information needed to complete the fluid velocity estimate.

This graph meets the communication goal by displaying the value of the peak frequency in the signal.

The display reveals that the LDV readout is composed of several frequencies and shows which has the largest amplitude (3.5 MHz), meeting the goal of the communication. Thus, this is a good choice of graph type.

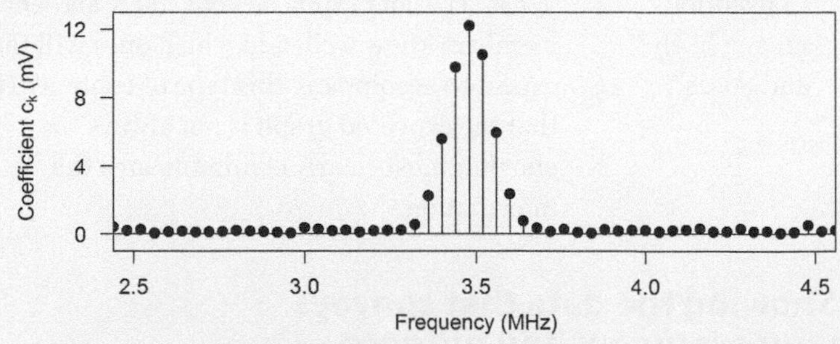

The peak amplitude of the LDV signal is at 3.5 MHz.

Computationally transforming the LDV data from the time domain to the frequency domain produces a spectrum of amplitudes plotted as a function of frequency. This graph provides exactly the information needed to complete the fluid velocity estimate.

Serving readers' and listeners' needs

Once you choose the type of graph, make sure it is prepared in a way that serves the readers' and listeners' needs. Because engineers prepare graphs for a range of audiences—including other engineers, technicians, and executives—it is important to consider the appropriate level of technicality for a visual display:

- For an audience of experts, the data should be represented in its full detail and complexity.
- For a non-expert audience, present the data in a way that guides the audience through the reasoning and presents the data truthfully, without distortion.

The most commonly used types of graphs—scatter plots, bar charts, and pie charts—are readily understood by an engineer's general audiences. However, if the data demands that a genre be adapted in a novel way, or that a new display type be devised, then you should be prepared to explain it to the audience.

28.2 Shared conventions of graphs

Tables are best for communicating exact numerical values. Graphs are designed for *visual* communication, and thus the conventions guiding graph design are about conveying meaning visually:

- Continuous data is encoded using a continuous line; discrete data is encoded using discrete data markers. Exceptions can be made to improve the clarity of an argument, but never "connect the dots" (the line plot genre) for data with no possible continuum between data markers.
- Three-dimensional visual effects are never used for decoration. Instead, use 3D graphs only for 3D data. Better yet, consider learning to create display types designed for multivariate data—for example, *scatter plot matrices* or *co-plots* (Cleveland, 1993).
- Graph elements should generally be monochromatic unless color is used to encode additional information. Be sure to meet the needs of audience members with color vision deficiency.
- Scale labels must be meaningful, units must be correctly identified, and all text must be legible for the intended audience.

- Legends or keys should be avoided. Legends force an audience to take additional time and effort to decode the information. You risk losing their attention entirely.

28.3 Bar charts, pie charts, and dot plots

Bar charts, pie charts, and dot plots all suit the same purpose: comparing values of one continuous variable among different levels of a category. The table lists examples of categories, levels, and variables:

Category	**Levels of the category**	**Example of a continuous variable**
Materials	metals, plastics, and ceramics	resistance to ultraviolet light
Human subjects	gender, socioeconomic status, and geographic location	life expectancy
Building type	industrial, commercial, and residential	building cost per square foot

Dot plots are less familiar to some audiences than bar and pie charts, but dot plots are more effective in helping audiences make comparisons. Because these three graph types all suit the same type of data, you should select the type that best suits your audience's needs and your goals.

Bar charts

Bar charts encode data using bar length. Bars typically have an arbitrary width. Color can be used to add an additional layer of categorical information to the display:

- Bars are usually horizontal, with category labels oriented horizontally for readability.
- Categories are ordered by the magnitude of the data values unless the categories have an implied order—e.g., high, medium, and low (called *ordinal* data). Avoid alphabetical ordering.
- The quantitative scale includes zero.

- *Stacked* bar and *grouped* bar charts, often used for additional subdividing of the continuous variable, should be avoided. Try to display such data in other, more effective, forms.

The advantages of bar charts are that they are easy to create with common software and they are familiar to many audiences. Their disadvantages are that they can easily be used to distort the message in the data (even unintentionally), their "stacked" and "grouped" variations communicate ineffectively, and they lend themselves to confusing and useless 3D (and other) effects.

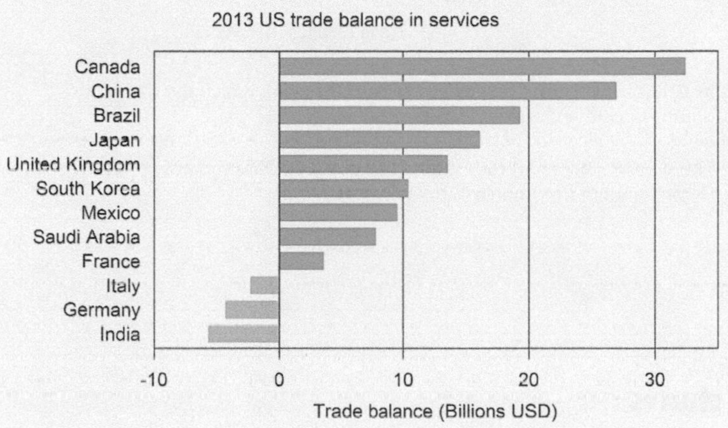

In 2013, the US had a positive balance of trade in services with most trading partners.

A bar chart of ordered categorical data. The category includes US trading partners, the levels are the country names, and the continuous variable is trade balance in USD. Data courtesy of US Department of Commerce (Census Bureau, 2015).

Category labels are most legible when oriented horizontally adjacent to horizontal bars.

Categories having no implicit order should be sorted by the magnitude of the data values.

The quantitative scale is horizontal.

To avoid distorting the audience's perception of the comparison, the scale must include zero.

Color is used to distinguish between positive and negative balances of trade.

Using a dot plot to replace a bar chart, a data marker is placed at the endpoint of the bar and the bar is removed.

Category levels are to the left of the display. The quantitative scale is horizontal.

Because the trade balance includes negative and positive numbers, zero is naturally included in the scale. Generally, however, dot plots do not require a zero on the scale.

Compared with the bar chart, the audience is much more likely to see patterns in the data in the dot plot.

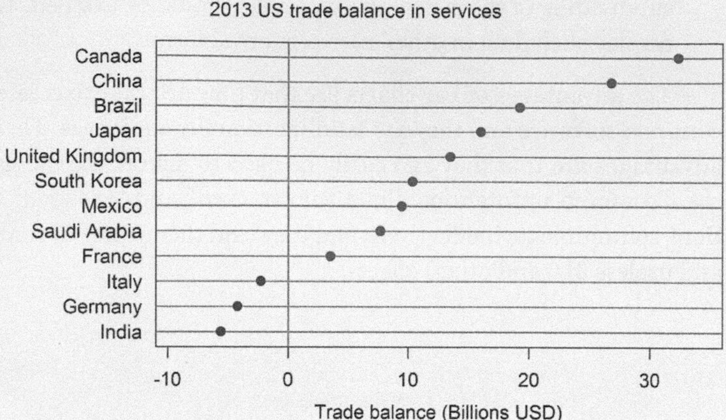

In 2013, Canada and China were the countries with which the US had the greatest positive trade balance in services.

The previous bar chart redesigned as a dot plot. Data courtesy of US Department of Commerce (Census Bureau, 2015).

Pie charts

Pie charts encode data using the angle or area of a pie slice (mathematically, a *circular sector*), often as a percentage of the whole, where the full circular area represents 100% of a variable. Color can be used to encode an additional layer of information or just to make each sector visually distinct from its neighbors:

- The area of the circle represents 100% of the quantity being subdivided by the sectors.
- The angle of the sector and its area both represent the numerical quantity.
- Sectors are ordered from largest to smallest, usually counterclockwise from the top. The location of the biggest sector can be adjusted to make labels easy to read.

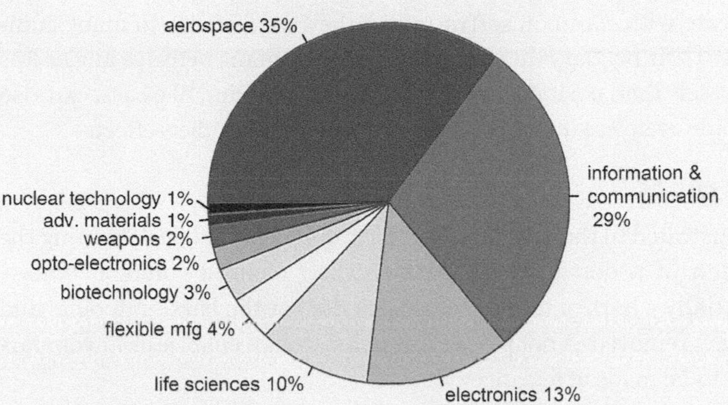

In 2013, approximately 75% of advanced technology exports were in aerospace, information & communications, and electronics.

A pie chart of ordered categorical data. The one continuous variable is percentage; the levels are types of exports in the category of advanced technology. Data courtesy of US Department of Commerce (Census Bureau, 2014).

The largest sector is approximately centered at 11 o'clock so that the labels for the smallest sectors are legible without callout lines.

Labeling the sectors with percentages makes the exact values available to the audience.

Color is used to visually distinguish the sectors.

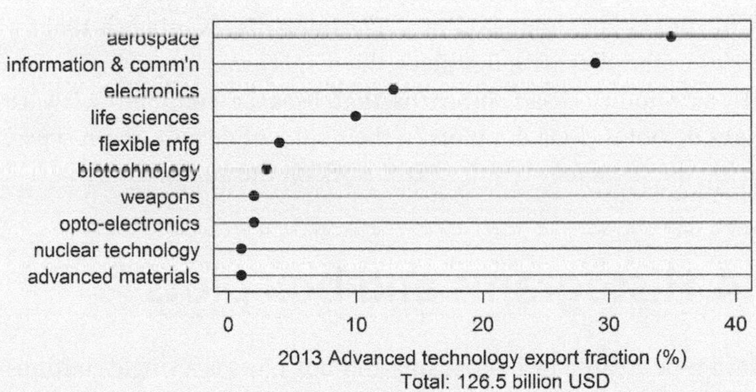

In 2013, both aerospace and information & communication groups had at least twice the export dollar value of any type of advanced technology.

The previous pie chart redesigned as a dot plot. Data courtesy of US Department of Commerce (Census Bureau, 2014).

The same data displayed in a dot plot removes the perceptual difficulty of assessing angles or areas and replaces it with the more effective comparison of position along a common scale.

The audience can distinguish between data values without the use of color.

Data clusters appear that are not as obvious in the pie chart at first glance.

Like bar charts, the advantages of pie charts are that they are easy to create with common software and they are familiar to many audiences. Their primary disadvantage is that humans perceive angles less accurately than position along a common scale. Pies, like bars, can also be made even less informative by adding 3D (and other) effects.

Dot plots

As illustrated in the previous examples, dot plots encode data using the position of a dot (or other data marker) along a common scale—essentially a horizontal bar chart with dots at the bars' endpoints and the bars removed. A dot plot's minimalist design enables clear comparisons to be made at a glance:

- The scale is horizontal with category labels on the vertical axis.
- Scales do not have to start at zero.
- Categories are ordered by the magnitude of the data values unless the categories have an implied order—e.g., high, medium, and low (called *ordinal* data). Avoid alphabetical ordering.
- A dot or other data marker encodes magnitude along the horizontal scale. Additional levels of information can be encoded using two dots on the same line.

The main advantage of the dot plot is that it addresses the perceptual difficulties of bar and pie charts. The main disadvantage is that for audiences unfamiliar with dot plots, the display may require some explanation. Another disadvantage is that because common software packages do not include dot plots in their suite of default graph types, preparing dot plots may require extra work on the author's part.

28.4 Histograms and box plots

The data type treated by histograms and box plots is a single continuous variable. These displays are used to visualize the variable's distribution. Histograms may be more familiar to the general audience because of their visual similarity to bar charts. Box plots might be less familiar to a general audience.

Histograms

When concerned with a distribution of a variable, a *histogram* is often the engineer's first exploratory data display. For example, samples taken from a manufacturing line may be weighed to check the process controls. Particular weights matter much less than how many samples fall into specific weight intervals.

A histogram has a horizontal scale that shows the range of the single variable divided into equal intervals or *bins*. A vertical bar height above each bin encodes the number of measurements in the bin, called the *frequency* of observation, or the number in the bin as a percentage of the total number of observations:

- Bins are usually intervals of equal length, though in some cases variable intervals may be required. Bins may need to be wider, for instance, to collect data across a distribution's long tail.
- The shape of the distribution is highly sensitive to bin width. Numerous rules have been developed for interval selection, though judicious trial and error often produces acceptable results. Be cautious of selecting intervals that produce falsely optimistic results.
- The more advanced user of statistical graphics might consider learning to create the *kernel estimator* graph, a smooth and continuous substitute for the histogram that is particularly useful when comparing several distributions (Silverman, 1986).

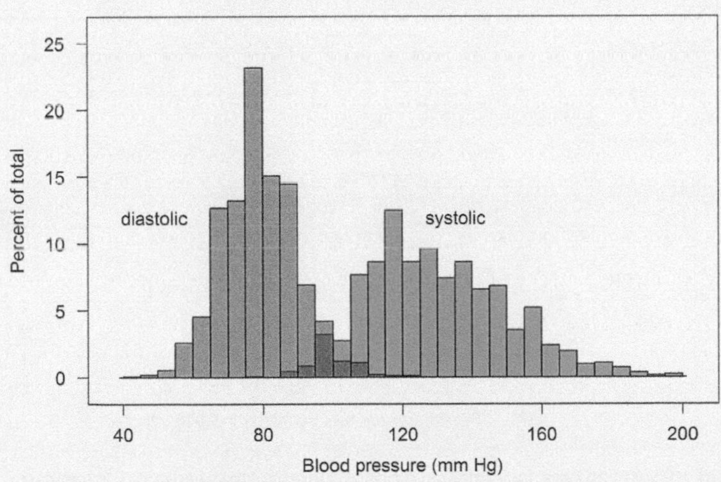

A blood pressure reading of 120/80 is the most frequent measurement. Systolic pressure has a greater range of frequent values than diastolic pressure.

A histogram shows the distribution of measurements of a single continuous quantity, sorted into bins of constant width. Data courtesy of the compareGroups R package (Subirana, et al., 2014).

Two histograms on one scale are used to compare distributions of two blood pressure measurements. Color is used to distinguish the two types.

Bin width is selected judiciously to reveal the characteristics of the distribution without creating a false impression of the data.

No gap is present between the bars, emphasizing that the bins capture a set of continuous data.

Instead of a legend, the two histograms are labeled directly to assist the audience in quickly decoding the display.

Box plots

Engineers will regularly encounter data summarized using a mean and standard deviation, a two-number summary of a distribution. Such minimal summaries, while easy to compute, can conceal important features of the data. Knowing this, you should summarize distributions using (at least) the five-number summary of the Tukey box plot (originally called a *box-and-whisker* plot). Up to nine-point summaries are possible.

Box plots display summaries of the distribution of a single continuous variable. For large data sets, box plots permit comparisons between groups of data. One definition of the box plot elements (definitions vary) is given by the following:

- The box shows the first to third quartile (25% to 75%), the interquartile range (IQR).
- The box is divided at the median (50%).
- The whiskers extend to 1.5 IQR.
- Numbers beyond the whiskers are potential outliers, drawn using individual data markers.

For audiences familiar with box plot conventions, the symbols shown here are meaningful.

The medians all lie on a fairly narrow range (15–25 mpg), but the distributions vary dramatically.

Potential outliers are identified by the statistical procedure, not the graph designer's intuition.

Rather than alphabetical order, the countries are ordered by the medians, descending from top to bottom.

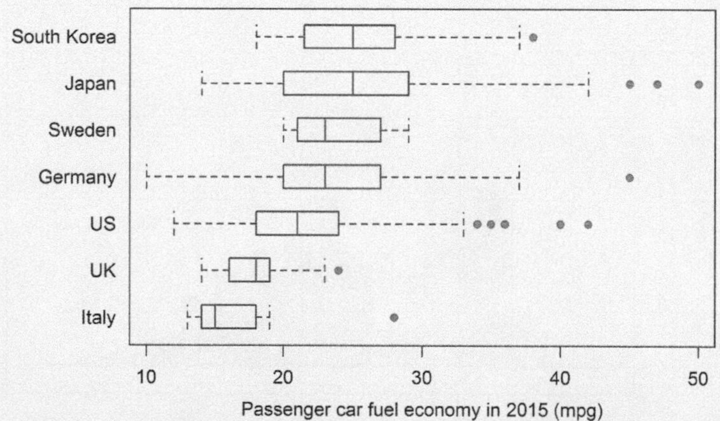

In 2015, Italian cars uniformly have poor economy, Japanese and German cars have by far the widest range, and only Japan, Germany, and the US produce cars with fuel economy greater than 40 mpg.

Comparing fuel economy using box plots communicates much more information than would be conveyed using a simple mean and standard deviation. Data courtesy of the US Department of Energy (Office of Energy Efficiency and Renewable Energy, 2015).

28.5 Scatter plots and line plots

Scatter plots and line plots display two continuous variables using the familiar Cartesian (x-y) coordinate system. The following guidelines apply to both types:

- The independent variable is displayed along the horizontal (x) axis. The dependent variable is displayed along the vertical (y) axis.
- Axes are labeled with the variable name (for instance, *heat flux*) and the unit of measurement (kW/m^2). Scale tick marks are usually shown at constant intervals.
- Data points must be shown unambiguously. In cases where the data markers are dense, they can be omitted and replaced by a line.
- Grid lines are used only when they serve a communication purpose.
- If more than one dependent variable is shown, the variables must be expressed in the same units or graphed in separate panels. If the graph is cluttered, use separate panels.
- If graphed in the same panel, different data sets must be distinguished using labels, different line types, or color.

Scatter plots

In *Trees, maps, and theorems*, Jean-luc Doumont explains that scatter plots reveal *correlation*:

> By encoding the data as positions along two orthogonal scales, a two-dimensional scatter plot reveals the shape and strength of the relation between two variables, together with outliers. It can easily identify subsets with different markers, although these must be well contrasted, lest the display look cluttered. When the form of the relation between variables is known, the scatter plot can show the corresponding regression curve (139).

Enclosing the scatter plot area in a complete rectangle is optional.

Scales are labeled, units are given, and tick marks aid in estimating values.

Data points are shown unambiguously. All data points are displayed, including what appears to be an outlier at 5 kW/m².

Based on the orientation of the axes, the audience assumes that heat flux is the controlled, independent variable and that the heat transfer coefficient is the resulting, dependent variable.

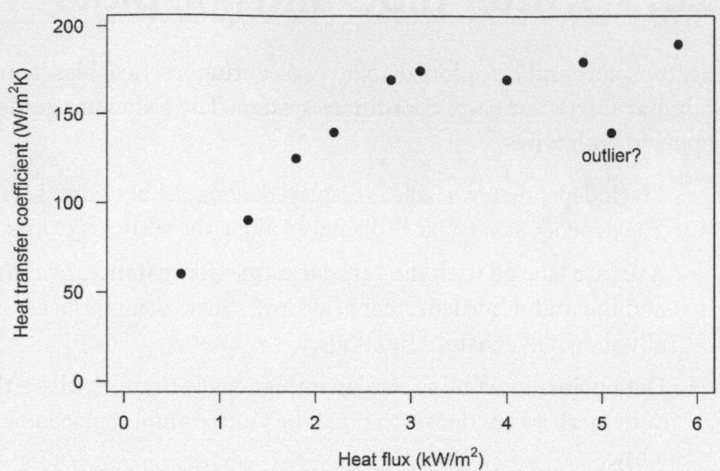

Coefficient of convective heat transfer increases as heat flux increases.

The visual shape of the data suggests a correlation between the two variables. Data courtesy of WITPress (Kriplani & Nimkar, 2008).

Scatter plots and curve fitting

Fitting equations to data (called *regression* or *curve fitting*) requires careful consideration of outlier data points to avoid misrepresenting the trend:

- Before fitting equations to data, check for outliers visually.
- Identify and omit outliers using established techniques (don't guess!).

A curve fit is correctly used in the following scatter plot showing the force-deflection characteristics of a nonlinear spring. Because the regression equation is the major result of this graph, it is included in the figure.

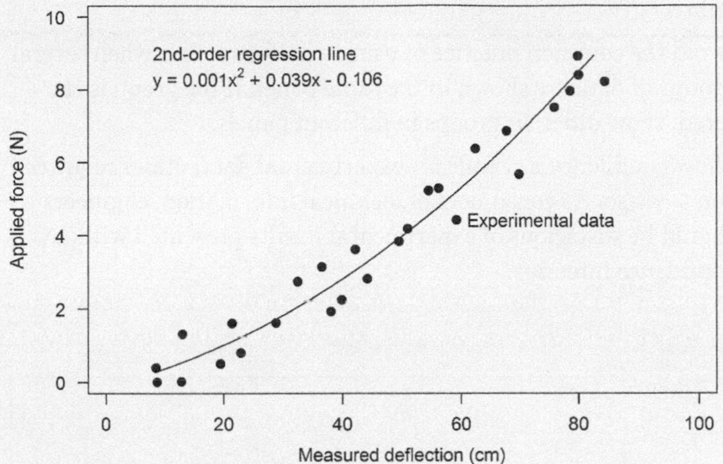

The spring exhibits second-order force-deflection behavior over the range tested.

This scatter plot with a curve fit shows that the second-order polynomial equation captures the trend of the experimental data.

A viewer can easily spot the important result of this graph—the coefficients of the polynomial fit.

Direct labeling of features (instead of using a legend) simplifies the audience's cognitive task.

While the force is applied (the independent variable) and the displacement is measured (the dependent variable), the convention for spring data is to display the variables as shown, violating the general rule of independent *x*-variable and dependent *y*-variable. To meet the audience's expectations, convention supersedes the general rule.

Scatter plots with confidence intervals

Confidence intervals (also called "error bars") give an estimate of the precision of a measurement. The best estimate of the true value is indicated by the data marker; a 95% confidence interval indicates that 95 times out of 100 that such intervals are constructed, the true value will lie somewhere on that interval.

Confidence intervals quantify the uncertainty in a particular measurement. When comparing different measurements, each with its own confidence interval, whether or not the confidence intervals overlap tell us something about the correlation between the two variables. For example, the mean values might suggest a trend, but overlapping confidence intervals mean that the correlation cannot be assumed without additional statistical tests:

- Avoid reducing the confidence level to obtain a better result; engineering results are conventionally reported with a 95% confidence interval.
- Display confidence intervals with a single solid line through each data marker.

- Keep the data prominent by making the error bar thinner or lighter.
- Avoid the common practice of using a half-error-bar when several groups of data are shown in the same panel. If the graph is cluttered, show different groups in different panels.
- Show confidence intervals for experimental data; ethics requires you to do so. As consumers of technical information, engineers should be suspicious of experimental results presented without confidence intervals.

95% confidence intervals are conventional in engineering results.

The linear trend downward suggested by the data markers cannot be assumed to be significant because the confidence intervals all overlap. Additional statistical tests are required to establish whether or not the means are statistically different.

Grid lines are not shown in this figure because the communication purpose does not require them.

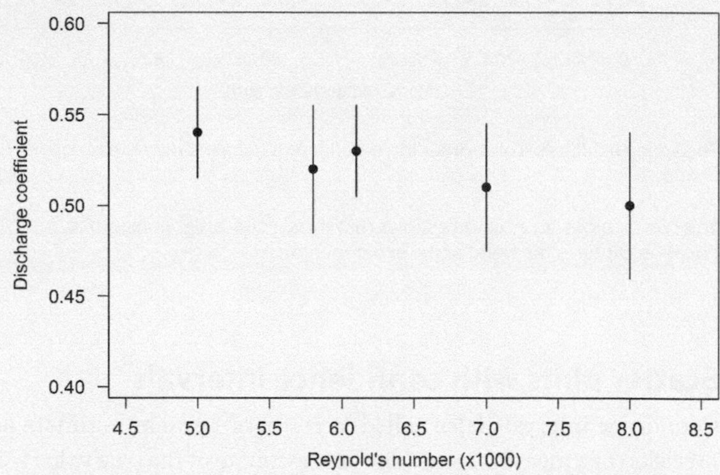

The experimental evidence is insufficient to claim that discharge coefficients change with increasing Reynolds Number.

Error bars prompt viewers to question apparent trends, helping to prevent them from making hasty conclusions that are not supported by the data.

Line plots

Line plots are scatter plots with "connected dots." Doumont recommends using line plots to visualize the *evolution* of a quantity:

> Line plots can readily display the evolution of two (or more) dependent variables versus the same independent variable. If all the dependent variables are expressed in the same units, they can be displayed along the same scale in a single panel. If not,

they must be displayed in distinct, juxtaposed panels, with different vertical scales and a common horizontal one (141).

Line plots are particularly suited to time-series data and for scatter plots comparing two or more groups of data. In a time series, the lines connecting the dots do not necessarily imply that the variable is continuous—the lines serve the communication purpose of helping the audience see which data values are part of the same data group.

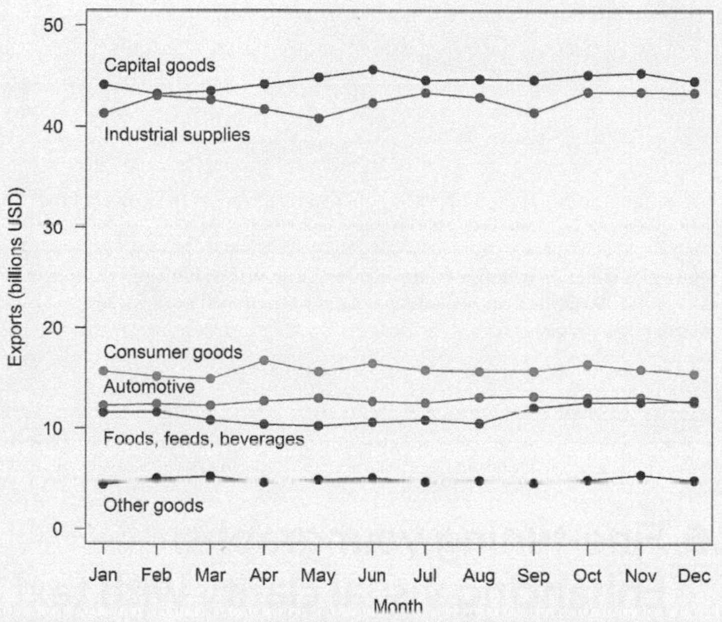

US exports of commodities and goods in 2013 remained fairly constant in each category.

Time series are often presented as line plots showing the evolution of the variables over time. In this particular case the evolution is not particularly interesting. Data: US census. Data courtesy of US Department of Commerce (Census Bureau, 2014).

Six groups of data are shown. Color distinguishes between groups, and each group is directly labeled.

The groups have little overlap, so all groups can be shown in one panel.

Intervals between data markers are large enough that all data markers can be shown unambiguously.

Data markers by month are discrete values, so the connecting lines indicate a data group, not that the data are continuous.

Typical of time-series data, the dots are connected by lines. Lines are particularly helpful when more than one group of data are shown in the same panel.

A line plot enables a comparison between two data sets by connecting adjacent data points.

In this case, the variables might be assumed to be continuous.

Two dependent variables are shown, using the same units on the scale. The color of the filled dots distinguishes the two.

Direct labeling of the lines reduces the amount of effort the audience needs to decode the information.

The experiment was performed with current at 2A intervals; thus, the tick marks are labeled at 4A intervals.

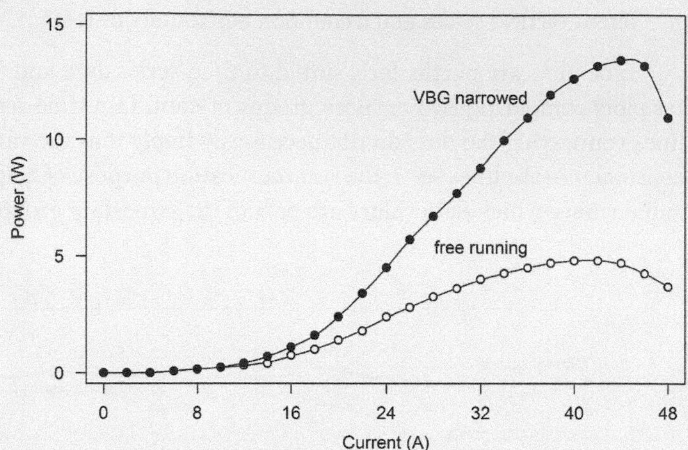

The volumetric Bragg grating (VBG) significantly increases the power of the laser system at higher current levels.

A line plot with two groups of data enables comparisons between the two curves. © 2006 Optical Society of America, reprinted with permission (Meng et al., 2006).

28.6 Fine-tuning your graphs: Enhancing visual clarity with text

When designing graphs, your goal is to lead viewers to visualize accurate and important conclusions. Text is used to help your audience reason about the visual information you are presenting, including captions, labels, callouts, and legends.

Captions

As a general rule, graphs and illustrations are given *captions*, while tables are given *titles*:

- Captions appear *below* a figure, while titles appear above it.
- Captions begin with a figure number. Graphs and illustrations are numbered in one series as "figures," while tables are numbered separately in their own series.

- A caption may include one or more sentences, while a title is ordinarily limited to a brief phrase.
- All of these rules are subject to change based on the established conventions of your company or the publication venue.

Captions and titles both benefit from being precise: Their job is to reduce ambiguity. While a simple descriptive phrase can identify the topic of the graph, a full-sentence caption can assert the main inference for the audience. Other important considerations—such as sources of data and complicating factors in the analysis—can also be discussed.

Graphs displayed in presentation slides or posters will likely benefit from the use of the assertion-evidence style (see Module 33: *Presentation slides*) developed by Michael Alley and Kay Neeley, in which each graph is preceded by a complete sentence or "headline" placed in the slide title frame. When executed well, this style improves on traditional figure titles because the grammatical completeness of the sentence helps authors and audiences alike to note the most important conclusions suggested by the data. To assist the audience when asking questions, figures and tables can still be sequentially numbered with short descriptive titles placed under the graph.

Labels, callouts, and legends

The labeling of axes is especially important: Both the quantity being measured and the units of measurement should be displayed clearly in the appropriate scientific or mathematical notation. If you're plotting data on a logarithmic scale or in another unusual way that readers might miss, be sure to emphasize the scale on the axis labels.

In small data sets, individual data points can be directly labeled to bring a feature to the attention of the audience.

In some instances—such as when you're using multiple colors to code regions on a map—you may need to use a legend. However, legends impose an extra "decoding" step on audiences, who must alternate their attention between the symbol and the legend that identifies its meaning. Consider instead simple text labels adjacent to parts of the figure.

If a label clutters the graph, use a callout, connecting the text label with the relevant visual area with a line or an arrow. Alternatively, different groups of data can be presented in separate panels.

Summary

Choosing the best graph for the task

- Choose graph types by considering the expected uses, the assertions being made, and the knowledge of the audience.
- Prepare graphs with a level of detail that suits your goals and the audience's expectations.
- Present the data truthfully. Avoid software defaults that distort the data, and always include all important data, even when some of it is unfavorable for your argument.

Shared conventions of graphs

- Graphs are designed for visual communication; tables are better for showing exact numbers.
- Axes are always labeled with the measured quantity and the unit of measurement.
- 3D visual effects are never used for decoration.

Bar charts, pie charts, and dot plots

- Bar charts encode data using bar length. Dot plots are a more effective alternative.
- Pie charts encode data using the sector of a circle. Dot plots are a more effective alternative.
- Dot plots avoid the visual drawbacks of bar and pie charts, but audiences are generally less familiar with this genre, and preparing dot plots may require extra work on the author's part.

Histograms and box plots

- Histograms display the distribution of a single variable. The shape of the distribution is highly sensitive to bin width.
- Box plots display a five-number summary of the distribution of a single variable and are useful for comparing distributions of different variables.

Scatter plots and line plots

- Scatter plots and line plots display two continuous variables using the familiar Cartesian (x-y) coordinate system. Scatter plots show correlation; line plots show evolution (often over time).
- When fitting equations to data, check for outlier data points first. If the equation itself is an important result, it should be added directly to the graph.
- Adding confidence intervals to a scatter plot is necessary to support a claim that measurements have a statistically significant difference.

Fine-tuning your graphs: Enhancing visual clarity with text

- Graphs and illustrations are typically given *captions*, which appear below the figure. A table's *title* appears above it.
- A caption may occupy one or more sentences, while a title is ordinarily limited to a brief phrase.
- When graphs are placed in slideware presentations, the descriptive caption should be placed as the slide title.
- Graphs are often improved by replacing a legend with direct labeling of the elements of the figure.

Illustrations

29

The familiar saying that "a picture is worth a thousand words" is true in certain senses. Text and images need to be coordinated carefully: Text is capable of stating and defining information with a degree of precision that images can seldom match. However, illustrating that information often makes it possible for audiences to understand it much more quickly, and to see relationships within systems and processes more clearly.

Like all other forms of technical communication, effective technical illustration involves many decisions based on the needs of the audience and the rules and conventions that govern different types of images. It is often tempting merely to select a convenient image from an online search, or to take a quick smartphone photo of a part. In practice, though, the time required to select and produce an illustration within a recognized style is almost always worth it for the clarity that it provides to audiences.

Objectives

- To make informed decisions about which features of a system, process, or part are sufficiently important to depict in an illustration
- To choose the appropriate type of illustration from those available (including photographs, line drawings, detail drawings, exploded view drawings, section view drawings, and flowcharts)
- To determine whether audience expertise makes it appropriate to use specialized genres such as circuit or process flow diagrams
- To label and caption illustrations and their important features in ways that maximize clarity and ease of use

29.1 Choosing the best illustration for the task

Engineers are usually good at recognizing when audiences need an illustration to help them to work with a system, process, or component. (Engineering *students* occasionally need to be careful not to include illustrations for the mere purpose of taking up page space.) It isn't always as easy, though, to tell which type of illustration fits the bill. Anticipate what your reader or listener may have to do, and design the illustration with that use in mind.

Expected uses	Does the illustration need to support audience members as they complete a task—assembling a device, performing a maintenance procedure, or reproducing an experiment?
Assertions or concepts being supported	What statements do you expect audience members to accept, or what ideas do you want them to understand, after viewing the illustration?
Conventions among likely audiences	What types of illustrations do your likely audience members know well, and which ones will they expect to accompany this type of content?

For instance, if you're writing a lab report and illustrating an experimental apparatus, you know to expect audience members who need to be able to set up the apparatus themselves (because the genre of experimental reports is founded on the goal of reproducing experiments and results). Therefore, audience members need to be able to identify all components and the connections among them. For specialists, you may be able to use specific notation that illuminates key features; an audience of engineers, for instance, will understand circuit diagrams.

29.2 The range of illustrations, from pictorial to schematic

An illustration is a "picture" of a system, process, or part—but the audience's needs and the conventions of your community determine which features need to be included in the picture. Different aspects of illustrations—*pictorial* and *schematic*—emphasize different kinds of features and properties.

Pictorial images resemble objects without encoding

Pictorial illustrations show what an object or event looks like (rather than detailing its specifications, explaining how it works, or revealing features invisible to the eye). Whether it's a photograph or a drawing, we judge a pictorial image primarily by its resemblance to the real-life object or event. (In a technical context, aesthetic appeal is secondary—although, of course, we always want images to "look good" enough to suggest that care and thought informed their creation.)

Needs like these call for pictorial images:

- Readers or listeners need to recognize the pictured object or event when they encounter it—for instance, selecting the correct part from a box of components.
- Readers or listeners need to observe properties that are visible in the object or event as it appears to the eye—for instance, seeing the appearance of a fracture when performing failure analysis.

On the other hand, purely pictorial images don't *encode* any information—specifications for parts, directions of movement, measurements taken—that isn't visible when viewing the pictured object itself.

Focus on real-world appearance means that viewing angle, lighting, and focus all impact what's visible to viewers.

Even an unedited photograph can show viewers some encoded information if the symbols are visible in the real-world system. For instance, resistor values are encoded in the pattern of colored stripes on each component.

It isn't necessarily helpful to capture all of an object's visual properties. It's difficult to see the printed trace on this circuit board because of its limited contrast from the green board.

Pictorial images like photographs work well when audiences need accurate guides to appearance—for instance, if a technician needed to begin assembling a device by selecting the correct printed circuit board.

Schematic images encode information without resemblance

The engineering field depends on many fundamental types of drawings that include schematic information, representing varying abstractions of the system being discussed or described.

When an object's or system's real-world appearance is less important than measurements and relationships, or properties not visible to the eye, use a *schematic* image with the necessary symbols to represent the information that audiences need to acquire.

Completely schematic images usually make sense only to expert audiences within a specialist community. For instance, chemical engineers can interpret process flow diagrams only because they have extensive specialized training to familiarize them with all of the symbols used in those diagrams.

A circuit diagram is completely schematic: The symbols for electronic components don't visually resemble the components themselves, and the physical layout of a circuit need not resemble the illustrated spatial layout showing their relationships. Courtesy U.S. Patent office (Cook, 2000).

Audiences require specialized knowledge to apply schematic or symbolic information to real-world objects—for instance, to identify a resistor or capacitor that matches the stated values.

Most of the notation in this image is standard to circuit diagrams. Some of it, though, labels parts of the circuit for the accompanying discussion in the text of the patent application.

Choosing pictorial and schematic elements

Illustrations range from purely pictorial to purely symbolic, with many including elements of both types. A photograph of a city street might help users recognize places when they arrive there—but pictorial images aren't designed to guide readers along the way. The only schematic information here is incidental—for instance, we recognize street signs that happen to show up in the photo.

Imagery © 2015 Google, Map data © 2015 Google. Google and the Google logo are registered trademarks of Google Inc., used with permission.

Like most illustrations, images from Google Earth combine pictorial and schematic information. We can see what the city of Bergen, Norway, looks like from above, but nearby roads and locations are also labeled.

Maps are more schematic still, ignoring most pictorial aspects to encode information in symbols—for instance, users can distinguish surface streets, ferry routes, and tunnels at a glance.

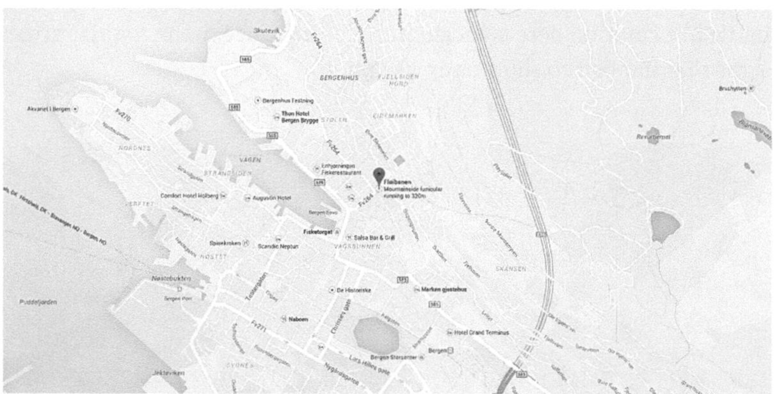

Map data © 2015 Google. Google and the Google logo are registered trademarks of Google Inc., used with permission.

29.3 Commonly encountered types of illustrations

Photographs

Photographs are the most purely pictorial images: In fact, we most often associate the word "picture" with photographs. Photographs are usually the easiest images to create: Smartphones and other devices make it possible to take photos at almost any time. This doesn't mean, though, that photographs are the right option for every instance. Generally speaking, photographs are called for only when users need to see an object's exact, lifelike appearance in exactly the conditions in which they'll observe it.

Photographs are appropriate when an audience truly needs to see the object or event as it appears in real-world conditions—for instance, to inspect a physical specimen from some process or experiment. An effective photograph can be taken by following these tips:

- Use sufficiently bright lighting so that fine detail is visible.
- Consider the view that most clearly shows the relevant features, and take the image from that perspective.
- Clear the background of clutter and elements that may distract from the specimen being photographed.
- If a close-up view is taken, also take photographs showing where the close-up view is located in relationship to the overall specimen.
- Add annotations to the photograph to point out important features.

The uncluttered background directs your eye to the details of the test specimen.

Because the details of the fracture surface are important in characterizing the material, a close-up is necessary here.

This photograph is taken from a memo from MIT Lincoln Laboratory describing material testing, showing the fracture surface of the test specimen.

Many distractors in the background obscure the information conveyed. If a drawing is not possible in this case, we can create a better photograph by removing objects not discussed in the document.

An important piece of the apparatus (the Petri dish) does not stand out against the rest of the image.

Some controls are seen, but their functions are unclear because the photo does not capture the labels on the apparatus.

A vague, generic caption does not convey useful information to the reader.

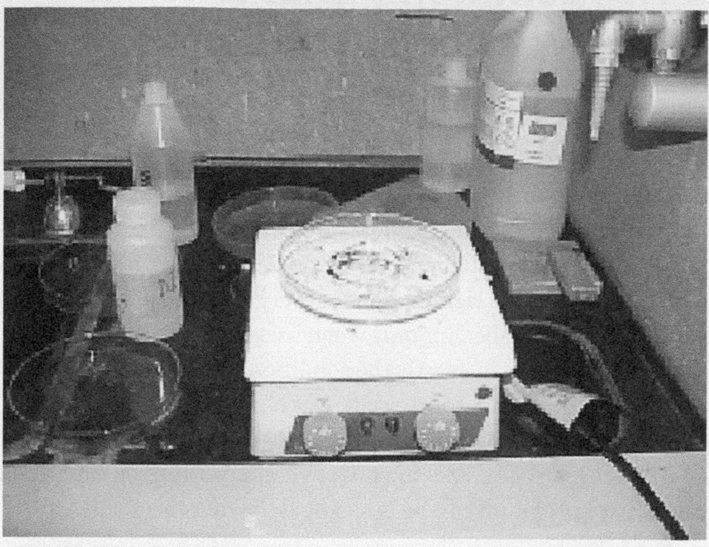

This photo, part of a laboratory instruction manual, shows an apparatus used to bake photoresist. Showing a photograph of the experimental setup is not the most effective way of describing the setup of the apparatus or the procedure being used. Because the image is meant to illustrate the nature of the process rather than the apparatus's exact appearance, a photograph works less well than a diagram might. © 2010 RHIT.

Line drawings

Drawings can present a *simplified* visual abstraction of a physical system. To clearly show three dimensions, an isometric view is often used (but is not the only projection possible). Line drawings allow us to remove irrelevant, confusing, or distracting elements of the system, drawing the audience's attention to the exact features, properties, locations, and relationships that we want to emphasize.

In many cases, we use photographs for convenience when other illustrations would actually work better. For instance, illustrations are often encountered in laboratory procedure manuals as a way of familiarizing the reader with the experimental apparatus. When writing such manuals, many engineers make do with photos of the apparatus. Drawings, though, better help users to operate the apparatus, limiting their detail to the features judged to be most important.

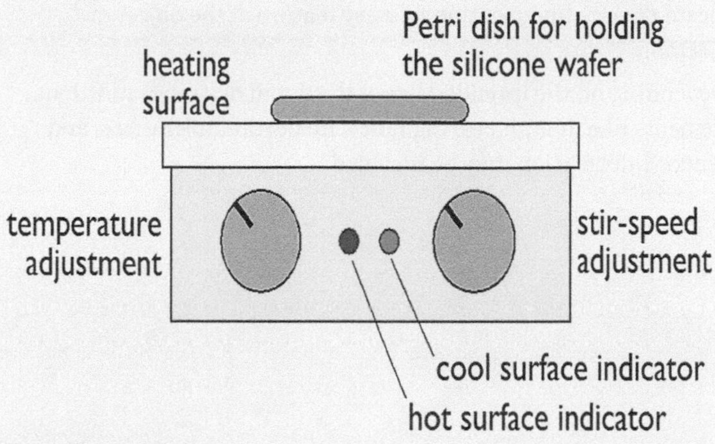

Silicone wafers are baked in a Petri dish using a hot plate with temperature and stir-speed controls.

A drawing distills the apparatus to its essential features, retaining only information needed by the user. © 2010 RHIT.

All important pieces of the apparatus are included in the illustration and nothing more.

Controls and indicators are labeled with their functions shown.

The caption starts with the noun–verb combination most important to the reason for the image.

Although it isn't a hard-and-fast distinction, more schematic drawings used to model or analyze a system are often called *diagrams*, such as the familiar free body diagram necessary for a force analysis.

Detail drawings

Detail drawings document the design process, the manufacturing requirements, and specifications to be examined during the inspection of a part or assembly. A design drawing focuses on specifications rather than on appearance, so it would not be used to display the final appearance of a design or to show the end user how to assemble a device.

More than any other individual type of illustration, detail drawings have specific conventions that engineers are expected to follow:

- All changes in the plane of a surface are shown as solid lines—unless the change in plane is obscured by the object itself, in which case a hidden (dashed) line is used.
- The geometric center and axis of circular features are indicated with crossed lines and center lines, respectively.
- Multiple views are used as needed to show all features of the object. Three principal views are usually sufficient, but if complicated features like internal geometry are not clearly shown, additional *detail*, *section*, or *auxiliary* views are used.

- Measurements of features, *dimensions*, are included to communicate the size and location of every feature of the object or assembly.
- Depending on the intended use of the detail drawing, additional elements like manufacturing notes, inspection tolerances, and vendor information may be included.

This detail drawing shows the critical dimensions for a component part.

Convention in the United States places the principal views (front, right, and top) in this orientation. The isometric view, here in the upper right, is often included.

The line types and annotations follow rules and conventions for the genre that are easily understood by other engineers.

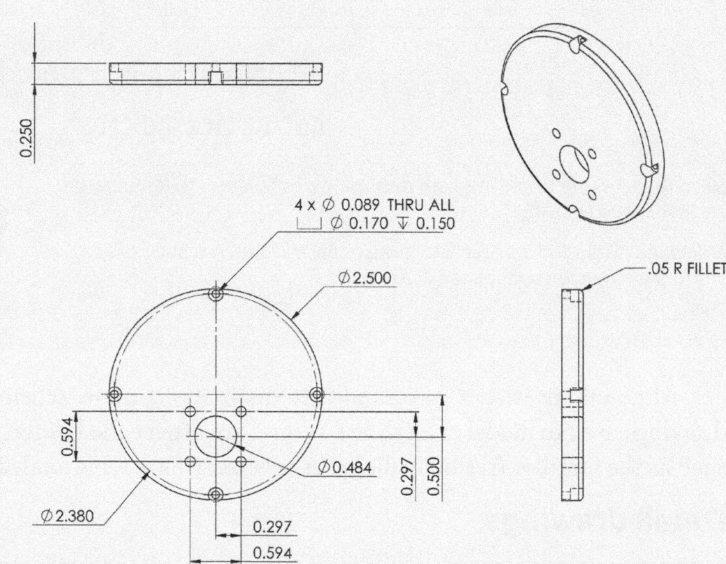

This engineering drawing shows the standard three principal views, along with an isometric view in the top right, of a component part. While the symbols and multiple lines may look confusing to novices, they convey meaning to an expert. © 2015 McCormack. Reprinted by permission.

Exploded view drawing

An exploded view drawing shows the component parts of an assembly separated in space but in relative position to one other, such that the assembly order is obvious. An exploded view drawing would never be used to show the physical appearance of the fully assembled system but would be ideal to communicate how the component parts fit together.

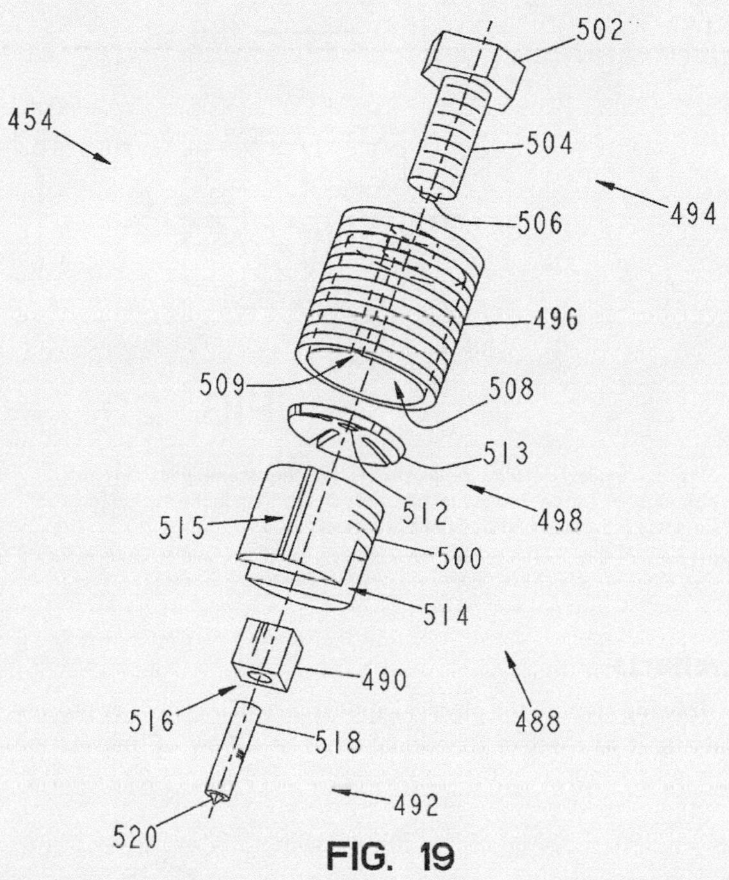

An exploded view shows the exact assembly order of the parts. Courtesy U.S. Patent Office (Stamper & Meek, 2003).

This exploded view drawing is shown to communicate how the component parts of an assembly are put together.

The component parts are separated by small gaps, and a dash–dot line indicates the shared axis that aligns the parts.

This drawing would not be very useful to show how the dimensions of the component parts work together, but it does work well as a guide to assembling the object.

Section view drawing

A section view, also called a *cross-section* or *cutaway view*, shows interior features of a part or assembly when a user could not see the relevant details as the object is ordinarily perceived. The object is most commonly depicted as if it had been sliced vertically, but some illustrations "cut away" an imagined hole in the object's surface to show just part of the interior.

This section view drawing is shown to communicate the orientation and location of component parts when the assembly is together.

The parts are distinguished by different shadings.

This drawing would not be very useful as a guide to assembling the object, but it does show how the dimensions of each component must work together for the overall object to function.

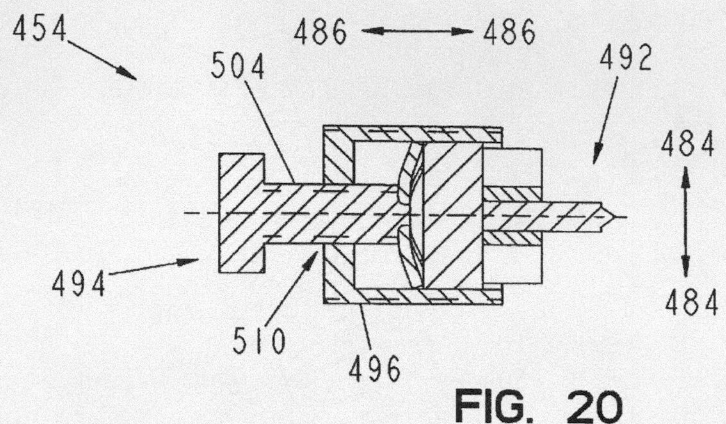

This section view drawing shows exactly how the assembly fits together, although we can see individual parts less clearly than in the exploded view. Courtesy U.S. Patent Office (Stamper & Meek, 2003).

Flowcharts

Some drawings ignore the physical appearance of a system or process, abstracting it to a set of functional steps or a flow of information.

A rectangle with circular sides indicates where a process starts or finishes.

A diamond indicates a process decision. The possible results of the decision (often yes/no) are added as labels on the arrows that exit from the diamond.

A rectangle indicates a step in the process.

A parallelogram indicates an input or output to the process.

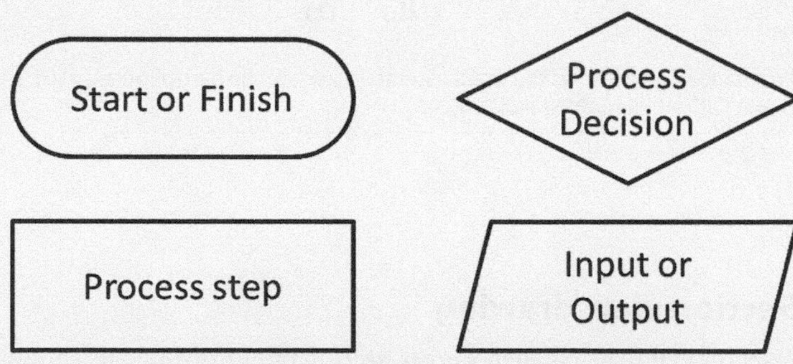

The four most common flowchart symbols are Start/Finish, Decision, Step, or Input/Output.

A flowchart depicts a physical or computational process, including the complete set of inputs, outputs, and procedures that define the system's possible states and operations.

Geometric shapes in a formal flowchart define the process's events and decisions while the flow of information is defined by directional arrows between shapes.

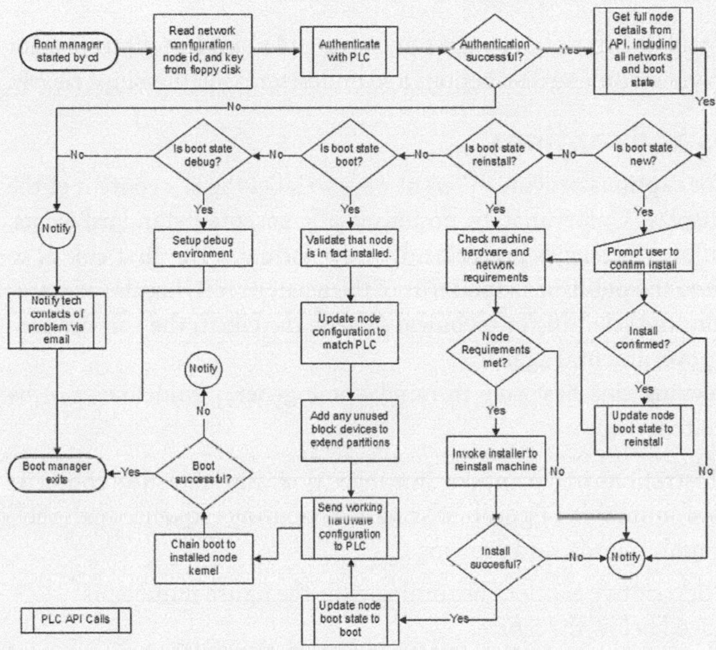

This flowchart follows the operations of the Boot Manager for PlanetLab, a system for coordinating large-scale network services. © 2007 The Trustees of Princeton University. Used with permission.

Vertical lines within the rectangle signify process steps that are predefined or automated.

Manual input ("Prompt user to confirm install") is represented with a quadrilateral shape. Like other specialty areas, software development uses a number of specialized flowchart shapes.

Decisions usually supply paths for "Yes" or "No" (sometimes "True" or "False"). Diamonds represent all process decisions with these possible outcomes.

A circuit diagram or a process flow diagram (used in chemical engineering) is essentially similar, although each type of system has its own specialized symbols. Many other kinds of illustrations also resemble flowcharts, since the basic pattern of shapes and lines can reflect many logical relationships. For instance, institutional authority and oversight are often formalized in organization charts, depicting senior executives, midlevel managers, and each worker, allowing a viewer to understand who's directly and indirectly responsible for different areas of operations.

Hybrid

Illustrations can be created that combine multiple categories, depending on the intent of the image. Modifying a visual display's features creatively may communicate your content more clearly than forcing that content into conformity with visual conventions that don't fit it well.

29.4 Fine-tuning your illustration

Almost all illustrations require some text, and many benefit from color to help viewers make distinctions and understand relationships clearly.

Titles and captions

Titles or captions provide different ways to label the key content of the data display. Unfortunately, no universally accepted standard exists, and different agencies prescribe different formats. The first rule is to adhere to the published standards of the agency receiving the communication: the scientific or technical society, the client, the corporation, or the governmental agency.

Keeping this first rule in mind, some general guidelines can be formulated:

- Illustrations in a print document are typically labeled as consecutively numbered figures, sharing a numbering sequence with the graphs, photos, and drawings.
- A descriptive caption, beginning with the figure number, is placed below the figure.

You will often see illustrations captioned with short sentence fragments that merely name the system, the event, or even the type of illustration being used ("Cutaway view of control module"). Captions like these seldom provide more information to viewers than they can see on their own; a caption accomplishes much more when it asserts the main inference that you want the audience to observe. Consider the reason for showing the illustration—what do you want the audience to learn?

Callouts, labels, and legends

When adding text labels to illustrations, the best option is to place the text label adjacent to the element being labeled, so that a viewer doesn't need to move back and forth between the illustration and a key elsewhere on the page.

If such a label would obscure or unnecessarily clutter the visual's content, use a callout, connecting the text label to the relevant visual area with a line or arrow. Additionally, when the exact location of the element being described is important, a connecting line or arrow is more precise than a textual label. Labels and callouts are particularly helpful for clarifying components of a system with many elements.

When text elements are added to illustrations, careful use of color will make the message clear. Particularly when adding callouts or labels to a photograph, there should be enough contrast between the text and the background so that reading the callout or label is not difficult. If the photograph is dark, use light-colored text, and vice versa. When annotating a photograph with both light and dark elements, medium-colored text will sometimes provide enough contrast; you might also use a text box that places a solid color behind the text.

Summary

Choosing the best illustration for the task

- Illustrations should be designed for audience needs—to enhance the ability to perform a task or to understand a concept.

The range of illustrations, from pictorial to schematic

- Pictorial images are designed to resemble an object or event, allowing audiences to identify important visible features or properties.
- Schematic images use symbols to encode information about a system or process.
- Most technical illustrations include both pictorial and schematic elements.
- Illustrations toward the schematic end of the spectrum typically require more expertise to interpret correctly.

Commonly encountered types of illustrations

- Photographs allow users to see an object's exact, lifelike appearance but may also introduce "noisy" irrelevant details that make it difficult to identify and focus on the most important ones.
- Drawings include features more selectively, allowing more deliberate decisions about exactly which figures audiences should see.
- Detail drawings document the design process, the manufacturing requirements, and specifications to be examined during the inspection of a part or assembly.
- An exploded view drawing shows the component parts of an assembly separated in space but in relative position to one other, such that the assembly order is obvious.

- A section, cross-section, or cutaway view shows interior features of a part or assembly that would not be visible as the object is ordinarily perceived.
- A flowchart uses labeled geometric shapes to define the states and operations of a physical or computational process.

Fine-tuning your illustration

- Figure captions are most effective when they assert the main inference that readers should observe or explain the central concept that they should understand.
- Text labels and callouts make illustrations easier to understand, especially when readers can easily read the text without having to shift their attention to a separate key or legend.

Tables, equations, and code

30

Tables, mathematical and chemical notation and equations, and computer code are *textual* components with *visual* conventions to consider. Authors have some control over the visual representation of these components (see Module 5: *Designing documents for users*) but should pay close attention to established rhetorical conventions lest they risk undermining their technical credibility.

Tables	Data is arrayed in tables when an audience expects to be able to read exact values. Table design should facilitate making comparisons.
Mathematical notation and equations	Mathematics is often described as a "universal language" for engineers and scientists. Like any language, though, its effectiveness depends on observing convention.
Chemical notation and equations	Chemistry, like mathematics, has a mature technical vocabulary and grammar that technical audiences would expect to see.
Computer code	When writing about software, lines of code are displayed using a visual design aesthetic—but the most important decision is whether audiences need to see the code at all.

Objectives

- To choose between graphs and tables for numeric displays based on the ways in which audiences need to reason about their content
- To design tables that minimize grid lines and other unneeded design elements, clearly displaying data in a meaningful arrangement
- To follow standard notation practices (including the use of subscripts, superscripts, and Greek and Latin typography) for mathematical and other formally notated expressions
- To document analytical computations, derivations, and computer programs with explanatory language that helps audiences to learn and interpret what's presented
- To integrate such expressions into clear prose by treating them as parts of speech

30.1 Designing tables

The first decision to make in table design is whether the data should be displayed in a table or in a graph. Graphs have the advantage of telling a story visually. Tables have the advantage of displaying exact numerical values and providing flexibility in ordering information. The decision depends on the needs of the audience:

- If an audience needs numerical data, use a table (or consider adding numbers to a graph).
- If an audience needs a visualization, use a graph (or design a table that assists in the visualization).

Use rows and columns to organize the information

Like any data display, tables should make a point that supports an author's argument. Rows and columns should be deliberately ordered to support that argument.

The leftmost column is the audience's guide to the table organization. Subsequent column names subdivide the information into categories useful to the audience, following a logical progression from left to right. The rightmost column is the "punchline," displaying information that supports your argument.

Rows are ordered top to bottom by the information in the most important column, with the "best" in the first (top) row.

Table 1: Comparing the mass of filter configurations.

Engine model	Filter type (in-out)	Mounting dia. (mm)	Length (mm)	Mass (kg)
B	end-end	246	920	36.0
	side-end		915	36.0
	end-vertical		940	36.2
	side-side		935	36.2
L	end-end	284	920	43.7
	side-end		915	43.7
	end-vertical		940	43.8
	side-side		935	43.8
X	end-end	322	994	54.3
	side-end		1015	54.4
	end-vertical		982	54.8
	side-side		1003	54.9

This table lists the mass of different configurations of diesel particulate filters (DPF). Mass is the critical characteristic. Printed with licensed permission from Cummins, Inc. © 2015 Cummins, Inc., all rights reserved.

The first column organizes the information. Horizontal lines (rules) and white space reinforce the visual organization.

The important information, mass, is in the rightmost column and is used to order the rows with the smallest mass (best) at the top.

Entries that are identical by model are entered once to declutter the display.

Numerical columns are right-aligned. Text entries are left-aligned.

The number of decimal places shown does not exceed the audience's needs.

White space between rows and columns groups the information.

Use white space and text alignment to group information

When creating a table, the designer typically begins with a rectangular grid of outlined cells. The gridlines (called vertical and horizontal *rules*) simplify data entry but should be replaced before showing the display to an audience. Outlining every cell distracts a viewer by "activating the white space"—a human perception phenomenon in which the white space between a line and data seems to lift off the page.

White space alone is generally superior to rules for visually organizing information (see Module 5: *Designing documents for users*).

Alignment is the second tool for formatting tables. Columns are generally left- or right-aligned, depending on the type of information; rows are generally top- or center-aligned. General guidelines are:

- Left-align text; right-align numbers.
- If numbers have significant decimal values, align the decimal points.
- Eliminate vertical rules (lines). Avoid stripes of shading; use white space.
- Reduce the number of horizontal rules; use white space to organize the information visually. Horizontal rules may be used judiciously to separate information groups.

A specific table entry can be given prominence by using bold or italic typefaces. A specific row or column can be given prominence with light shading.

Observe conventions for table numbering and captions

Tables are numbered and captioned according to conventions expected by the audience. Firms may have one standard, government agencies another, professional organizations yet another. In the absence of specific guidelines, the general guidelines are:

- Number tables consecutively (Table 1, Table 2, etc.) and separately from figures and illustrations.
- Use a descriptive caption that supports the argument.
- Locate the caption above the table.

30.2 Writing mathematics

Mathematics has a mature technical vocabulary and grammar. Its basic rhetorical components are symbols, expressions, and equations. Different engineering disciplines have strong conventions of mathematical mechanics, usage, and grammar that must be observed if you wish to establish your technical credibility with an audience:

Symbols — *Symbols* are the "nouns" and "verbs" of mathematical expressions. Variables and constants, the "nouns," are represented by symbols such as a, x, π, ∞, and Re. Operations, the "verbs," are represented by symbols such as $\surd$ (square root), Σ (sum), and $\int$ (integrate).

Expressions — An *expression* is a collection of symbols arranged in a mathematical phrase, such as "$x + y$", that makes sense in context.

Equations — An *equation* is a mathematical assertion that equates two expressions. The defining characteristic is the equality symbol "=".

Observe the conventions expected by your audience

Writing mathematics correctly takes time, even with the help of computer software. Accordingly, an appropriate amount of time *must* be allotted for drafting and revising or you risk presenting a sloppy approximation of the mathematical typography and syntax expected by the audience. Ignoring mathematical details makes an audience wonder what other details you ignored, undermining (yet again) the credibility of your argument.

General guidelines include:

- Observe symbol conventions expected by the audience; for example, g for the acceleration due to gravity, c for the speed of light, or m for mass. Be aware that different technical communities interpret the same symbol differently. For example, the letter I can represent sound intensity, electrical current, or moment of inertia, depending on the physical context.
- Use subscripts and superscripts; for instance, write x_2, not *x*2 or *x* <2>; write d^2, not *d*^2; write 2.5×10^{-3}, not 2.5E–3. In brief, use math syntax, not computer code syntax.
- Use Greek letters, not English equivalents or transliterations; for instance, write α, not *a* or *alpha*.
- Use italics for letter symbols, such as a, x, π, and Re.
- Use correctly typeset operator symbols. For example, the minus sign (–) is not a hyphen; the product symbol (×) is not the letter x; numerators and denominators of a quotient are shown with a horizontal line, not a slash /.

Treat mathematical components as parts of speech

Whether speaking or writing, the basic unit of technical communication is the sentence. A component of mathematical speech—a symbol, expression, or equation—conveys meaning best when presented as part of a sentence that is itself an element in your larger argument.

Symbols representing variables or physical qualities are treated as nouns. However, sentences should not start with a symbol; the written meaning is used instead. Variables are defined the first time they are used; symbols can be used thereafter, though the convention varies depending on the audience.

Use conventional words and phrases to treat mathematical components as parts of speech, such as "where," "substitute," "given by," "yields," or "follows from."

The variable names and typography in the figure are identical to those used in the equations that follow.

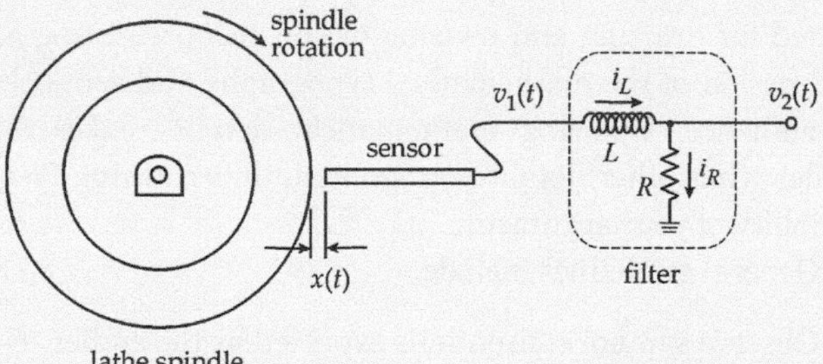

In a derivation, identify the physical principle by name and use symbols that match those in the figure.

Explain the process used to reach the next step.

Equations are punctuated as parts of speech.

Components are "substituted," not "plugged in."

Equations are referenced by number.

"Yields" is the conventional verb for walking through a derivation.

Applying Kirchhoff's current law at the electrical node in the filter yields

$$i_L = i_R , \tag{1}$$

where i_L is the current through the inductor and i_R is the current through the resistor, assuming a high impedance at v_2. Using complex impedance, the inductor current is given by

$$i_L = \frac{\Delta v_L}{Ls} , \tag{2}$$

where Δv_L is the voltage drop across the inductor and Ls is its complex impedance. Similarly, the resistor current is given by

$$i_L = \frac{\Delta v_R}{R} . \tag{3}$$

Substituting (2) and (3) into (1) yields the model

$$\frac{v_1 - v_2}{Ls} = \frac{v_2 - 0}{R} . \tag{4}$$

[...]

The model has the form of an ideal, first-order, ordinary differential equation. It can be rewritten in the form

$$\tau\frac{dv_2}{dt} + v_2 = Kv_1\,, \tag{8}$$

where $\tau = L/R$ is the time constant and $K = 1$ is the static gain. Equation (8) is the final result of the derivation.

The convention for defining terms in an equation uses the word "where."

Adding natural language reasoning and presenting the equations as parts of speech explains the reasoning of a derivation.

Numbering and referencing equations

Like tables, equations are typically assigned their own sequence of numbers. In the absence of specific guidelines, the general guidelines are:

- Number equations consecutively: (1), (2), etc.
- Align equation numbers to the right margin.
- Use a number in a sentence to refer to an equation presented previously—for example, (1). Use the construction "Equation (1)" only if the phrase starts a sentence. If the equation has been named (e.g., the conservation of energy), the name may be used instead of the number.

The model is given by (8), where $\tau = L/R$ is the time constant and $K = 1$ is the system static gain. The parameters of the filter are given by $L = 4$ mH and $R = 10$ Ω. The input voltage v_1 is a sinusoid with a 5 V amplitude and a 100 Hz frequency.

List parameter values early.

To determine the unknown steady-state output v_2, the input frequency is converted from Hz to rad/s,

$$\omega = 2\pi f \tag{9}$$
$$= 200\pi \text{ rad/s.}$$

Multiplication is implied for adjacent symbols—for example, 2πf. A product symbol is not required.

With zero initial conditions, the Laplace transform of the model is

$$\tau s v_2 + v_2 = v_1\,. \tag{10}$$

Observe conventional usage of symbols. For example, frequency in rad/s is represented by the Greek letter ω, not the letter w.

[...]
The magnitude of the transfer function is the magnitude of the right-hand side evaluated at the input frequency w.

$$|v_2| = \frac{1}{|\tau j\omega + 1|}|v_1| \tag{15}$$
$$= 0.97|v_1|$$

Before substituting numbers in an equation, show the equation in symbolic form first.

The equals signs are aligned and the numerical substitutions are made below, not to the right of, the symbolic equation.

The phase angle of the output is the angle of the TF evaluated at input frequency ω, assuming the input angle is zero,

$$\begin{aligned} \angle v_2 &= \angle 1 - \angle(\tau j\omega + 1) \\ &= 0 - 14\,\text{deg} \end{aligned} \tag{16}$$

Thus the steady state output is given by

$$v_{2,SS} = 4.85\sin\left(200\pi t - 14^{\circ}\right). \tag{17}$$

Using equations as parts of speech explains the reasoning of the analysis and defines the nomenclature.

30.3 Writing chemistry

Many of the guidelines for writing mathematics apply to writing chemistry: observing the conventions expected by the audience, treating chemical components as parts of speech, and numbering and referencing chemical equations.

Specific components of chemistry's rhetoric are formulas, names, equations, and structures. Engineers should pay close attention to these rhetorical conventions lest they risk undermining their technical credibility:

Formulas	Familiar symbols such as H_2O and CO_2 represent *molecular formulas* of chemical compounds.
Names	Compounds have *names*; for example, the name of H_2O is water; the name of CO_2 is carbon dioxide, the name of CH_4 is methane.
Equations	A chemical equation describes a reaction involving inputs (reactants) and outputs (products). For example, methane combines with oxygen to produce carbon dioxide and water: $$CH_4 + 2O_2 \rightarrow CO_2 + 2H_2O$$
Structures	A schematic, or *structural formula*, shows the individual bonds between atoms. For example, the structures of water and methane are shown as:

water

methane

Use standard symbols in formulas

The symbolic representation of a compound is its molecular *formula*, such as H_2O, CO_2, or CH_4. Formulas are the "words" used in chemical expressions. The conventions for formulas are:

- Elements are represented by standardized combinations of uppercase and lowercase letters (e.g., Na, Cl, and Pb). Unlike symbols in mathematics, chemical formulas are not italicized.
- Use *subscripts* to denote the number of atoms in a formula (e.g., write CO_2, not CO2).
- Ion charges are noted with *superscripts* (e.g., beryllium Be^{+2}, fluorine F^{-1} or F^{-}). As in mathematical writing, use a proper minus symbol (–), not a hyphen (-).

Translate nomenclature for the audience

Chemical *nomenclature* is a set of rules for naming chemical compounds. Fluency with nomenclature is somewhat analogous to fluency with languages—it involves an ability to *translate*. The most common types of translations are between names, formulas, structures, and common names, illustrated as:

abbreviation, common name, or brand name ⟷ name ⟷ molecular formula ⟷ structural formula

Engineers should be able to translate *from* any form *to* any form to suit the needs of their audience. For example, the nomenclature for a well-known nonstick coating used on cookware is shown below. One might use "Teflon" if presenting at a public hearing, "PTFE" if presenting to a group of materials engineers, or the formula $(C_2F_4)_n$ if presenting a technical paper:

brand name	abbreviation	name	formula	structure
Teflon	PTFE	polytetrafluoroethylene	$(C_2F_4)_n$	$\left(-\underset{F}{\overset{F}{C}}-\underset{F}{\overset{F}{C}}- \right)_n$

The *common* (or traditional) name of a compound is preferred in situations where a systematic name might confuse an audience. For example, the compound H_2O would usually be named *water* rather than deuterium oxide or dihydrogen oxide.

Names include roman numerals when an element can form more than one type of positive ion. For example, $FeCl_2$ contains Fe^{2+} ions, so the name of the compound is iron (II) chloride.

Use equations to convey meaning precisely

If compounds are chemistry's "words," chemical equations are its "sentences," expressing relationships among compounds in the same way that mathematical equations express relationships among expressions. Chemical equations have these attributes:

- Reactants are placed on the left, products on the right.
- The one-way arrow ($\rightarrow$) indicates that the reaction proceeds only in the direction shown.
- A special double arrow ($\rightleftharpoons$) indicates a reaction in chemical equilibrium. *Do not* use ($\leftrightarrow$).

The symbols for elements are treated as nouns.

The authors assume the audience will recognize common names.

Superscripts are used to indicate ions.

Equations are balanced. One-way arrows indicate one-way reactions.

Based on the average levels of Zn, Fe, Mn, and Cu in plants [3], the weight ratio of Zn : Fe : Mn : Cu was chosen as 3 : 1 : 0.5 : 0.25.

Synthesis was done with commercial grade chemicals, namely, hematite, pyrolusite, zinc ash, copper sulfate, roasted magnesite, and phosphoric acid.

Zinc ash, copper sulfate, and magnesite were analyzed for their content of Zn^{2+}, Cu^{2+}, and Mg^{2+} by atomic absorption spectroscopy (AAS).

[...]

Subsequent (partial) polymerization to the polyphosphates may be represented as

$$\begin{aligned} n\mathrm{Zn(H_2PO_4)_2} &\longrightarrow \mathrm{Zn}_n\mathrm{H}_{(2n+2)}\mathrm{P}_{2n}\mathrm{O}_{(7n+1)} + (n-1)\,\mathrm{H_2O} \\ n\mathrm{Fe(H_2PO_4)_3} &\longrightarrow \mathrm{Fe}_n\mathrm{H}_{(4n+2)}\mathrm{P}_{3n}\mathrm{O}_{(11n+1)} + (n-1)\,\mathrm{H_2O} \\ n\mathrm{Mn(H_2PO_4)_4} &\longrightarrow \mathrm{Mn}_n\mathrm{H}_{(6n+2)}\mathrm{P}_{4n}\mathrm{O}_{(15n+1)} + (n-1)\,\mathrm{H_2O} \\ n\mathrm{Cu(H_2PO_4)_2} &\longrightarrow \mathrm{Cu}_n\mathrm{H}_{(2n+2)}\mathrm{P}_{2n}\mathrm{O}_{(7n+1)} + (n-1)\,\mathrm{H_2O} \\ n\mathrm{Mg(H_2PO_4)_2} &\longrightarrow \mathrm{Mg}_n\mathrm{H}_{(2n+2)}\mathrm{P}_{2n}\mathrm{O}_{(7n+1)} + (n-1)\,\mathrm{H_2O} \end{aligned} \tag{1}$$

Chemical symbols and equations are treated as parts of speech. Reprinted under CC license 3.0 (Bandyopadhyay, Ghosh, & Varadachari, 2014).

- The reaction must be *balanced*; that is, the number of atoms of every type on the left side of the equation must equal the number of atoms of the same type on the right side.
- The *state* of compounds is clarified if necessary using the notation: solid (*s*), liquid (*l*), gas (*g*), or dissolved in water (*aq*) (i.e., an aqueous solution).

Use structures to convey information about bonding

Schematic or stick-figure *structures* (also called *structural formulas*) are used to convey information about how elements are bonded in a compound—who's tied to whom. To the knowledgeable audience, a structure can convey information about reactive site locations, stability, potential solvents, and its 3D shape. For many engineers, looking up the structure in a *reputable* source and copying it or redrawing it will suffice for most audiences. Example structures are shown below:

Formula	Name	Structure
CO_2	carbon dioxide	 O=C=O
$(C_2F_4)_n$	PTFE or polytetrafluoroethylene	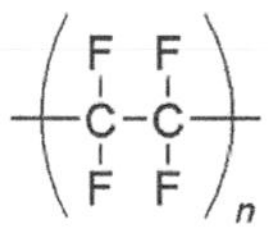 F F –C–C– F F n
$(C_3H_6)_n$	polypropylene	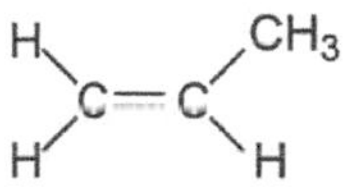 H CH3 C=C H H

30.4 Writing computer code

If you are using software to automate data processing and are explaining your method, most audiences don't need to know the exact syntax and series of commands used. Instead, present the process as a flowchart so the audience knows what steps are being taken without seeing the actual computer code. If you're not directly discussing the code itself, you probably don't need to include it. (If you're uncertain, including code as an appendix or attachment is less intrusive than pasting it into the body of a document.)

However, engineers will sometimes need to share programming code directly with an audience that does need to learn about it in some detail. You might be teaching colleagues how to use a programming language, sharing a particularly creative solution, or helping to debug an existing program.

Display code like a block quote

Programming code should stand out from the surrounding text because it serves a different purpose. You are presenting something verbatim—"quoting" a program—so indent the code from the body text. Like equations, each segment of code can be integrated into the syntax of a sentence that precedes or follows it.

Use a monospaced font

In most programming languages, individual characters are significant, particularly spaces and punctuation like periods, parentheses, and commas. A monospaced font—a typewriter-style font such as Courier—makes these typographical units easier to notice, decreasing the chance of a misunderstanding by the audience.

Indentation and white space set off the programming code as if it were a block quote.

A monospaced font (e.g., Courier New) displays each character clearly and emphasizes the syntax of the programming language.

Given the standard deviation and a vector containing distances from the mean

```
stDev  = .75;
distSD = [2.08  -1.07  1.28  1.26  -0.59]
```

outliers could be counted using a "for loop," as in

```
countOuts = 0;
for i = 1:length(distSD)
   if abs(distSD(i)) > 3*stDev
      countOuts = countOuts + 1;
   end
end
```

The same task is accomplished in a single line of code by taking advantage of array operations in MATLAB,

```
countOuts = sum(abs(distSD) > 3*stDev);
```

This excerpt from a tutorial incorporates code fragments as parts of speech in the sentences describing the work.

Summary

Designing tables

- To best support your argument, arrange the rows and columns of a table deliberately.
- Information is grouped in a table through careful use of white space and text alignment, with a minimum number of horizontal grid lines.
- Bold and italic type can be used to highlight single important cells in the table.

Writing mathematics

- Using correct mathematical notation for symbols, expressions, and equations—including the use of subscripts, superscripts, and Greek and Latin typography—establishes your credibility.
- Incorporate equations into sentences by treating them as parts of speech, making clear the reasoning behind their use.
- Use an equation editor or other type settings to execute standard notation correctly.

Writing chemistry

- Correct chemical notation follows similar rules to mathematical notation—with defined roles for uppercase and lowercase letters, subscripts, and superscripts—but chemical formulas are not italicized.
- A chemical name might be translated into an abbreviation, a molecular formula, or a structural formula to best communicate with a specific audience.
- Relationships among compounds are expressed through chemical equations that communicate the direction of a reaction, the state of compounds, and ions.

Writing computer code

- Because the exact sequence of commands used is rarely as important as the higher-level procedure, computer code can frequently be displayed as a process flowchart or list of steps.
- When code must be displayed, indent the code like a block quote and use a monospaced font like Courier to make individual characters easier to notice.

Media

...transmit your ideas to your audience in various material forms: A proposal might be printed on paper, delivered aloud, projected on slides, or displayed on a poster. Many fundamental tasks and criteria apply equally to these cases: You must formulate your ideas clearly, support them with compelling evidence, and address the most important priorities of your audience. Nevertheless, each medium emphasizes specific skills, has its own conventional rules for design, and requires some practice before you will feel comfortable. This training will help you to develop excellent results in the media in which engineers work most often today but will also help you to become adaptable as new digital media emerge.

Media

Print pages

31

For most professional documents, page layout and design are secondary to the writing itself. Nevertheless, a page's typography, line spacing, and margins can impact the reader's task, making it easier or more difficult to comprehend and use the material.

Attention to page design principles can create a satisfyingly professional impression, making a document appear not only attractive but also orderly and systematic. Many engineers prefer to finalize documents in the TeX typesetting system, which allows for the control of many subtle parameters (and also handles scientific and mathematical notation admirably). Even in a standard-issue word processor, though, the clarity and credibility of workplace reports and correspondence can be enhanced by applying just a few basic rules of thumb for *layout* and *typography*.

Objectives

- To set up a page layout using design principles to enhance readability and clarity
- To follow typical conventions for the layout of print documents
- To place visuals alongside text on a page in a way that maximizes their value and minimizes their disruption of the surrounding text

31.1 Creating a page layout

A beautiful layout will not salvage sloppy writing, but a poor layout distracts the reader from your ideas. An effective layout, on the other hand, guides readers through the information without being noticeable to readers, enabling them to focus exclusively on the content.

To achieve this subtlety, avoid redundant visual cues. For example, if you use white space to set off a visual from the surrounding material, adding a border line is redundant. If you distinguish section headings with a different style of type than that used for body text, underlining the heading is redundant.

Setting the margins and line spacing

Like any other space in which information is displayed, a page amounts to a grid, with invisible lines separating its components. The simplest pages have empty "cells"—margins—surrounding a single central cell full of text; the composition of the page becomes somewhat more complex if it needs to include additional elements, such as visuals and footnotes.

Most of us begin with the default settings of our word processor, often 1 inch on each side and a line spacing of a bit more than 1.0 (or 6 points between lines). Ultimately, though, margins and line spacing should make the text of a document easier to read. If a page is cluttered with information, readers are likely to become overwhelmed and to miss important information.

Top and bottom margins	Set top and bottom margins to a minimum of 1 inch (25 mm) to leave room for text in the page header and footer.
Left and right margins	Adjust left and right margins for a line length of 50 to 80 characters: More than this is difficult to read comfortably. For example, 75 characters in 12-point Times New Roman produce a line of text approximately 6 inches (152 mm) wide. Margins on an 8.5-inch (216-mm) page, then, can be set to 1.25 inches (32 mm).
Line spacing	Use line spacing slightly greater than 1.0 to improve readability, unless your audience requires a specific spacing: 1.5 and 2.0 are common.

Resource-conscious engineers will sometimes try to save paper by reducing margins (or by shrinking the font). Avoiding waste is laudable—but this objective can usually be achieved in more reader-friendly fashion by trimming less important sentences from the text itself.

Documents may benefit from more complex rules for the size of margins if they are to be bound in a book or pamphlet with pairs of facing pages, where the inside margins are usually smaller than the outside margins and the top smaller than the bottom. Such page layouts should be composed using publishing software (such as Adobe InDesign or Microsoft FrontPage) rather than a word processor.

Integrating visuals

Visuals should be integrated into the document so that the visual and the text relevant to it can be read together:

Location	Locate a visual near its accompanying text. If placement results in a few lines of text between a visual and a margin, align the visual to the margin.
Size	Design visuals to be readable at small sizes. Smaller visuals are easier to place on a page adjacent to the most relevant body text.
White space	Border a visual with sufficient white space to avoid crowding. If white space is sufficient, visible grid lines or borders are seldom needed.

Text wrap

Avoid narrow columns of text next to a visual. Avoid wrapping text around visuals while drafting documents.

III. Mission Overview & Safety Considerations

A basic Asteroid Retrieval mission concept is illustrated in Fig. 1. The spacecraft would be launched on an Atlas 551-class launch vehicle to low-Earth orbit. A 40-kW electric propulsion system would then be used to reach the NEA in about 4 years. Once at the NEA, a 90-day operations phase is divided into two phases. During the first phase, the target would be studied thoroughly to understand its size, rotation, and surface topography. In the second phase the spacecraft would capture and de-spin the asteroid. To accomplish this, the spacecraft would match the target rotation, capture it using the capture mechanism described in Section VI, secure it firmly to the spacecraft, and propulsively despin the combination. The electric propulsion system would then be used to depart the asteroid orbit, return to the vicinity of the Moon, and enter a high-lunar orbit. After reaching lunar orbit the spacecraft would stay attached to support human activity, which is anticipated to include the development of NEA proximity operational techniques for human missions, along with the development of processes and systems for the exploitation of NEA resources.

The ACR spacecraft concept would have a dry mass of 5.5 t, and could store up to 13 t of Xe propellant. The spacecraft would use a spiral trajectory to raise its apogee from LEO to the Moon where a series of Lunar Gravity Assists (LGAs) would be used in concert with SEP thrusting to depart the Earth-Moon system. This initial leg of the trajectory would take from 1.6 to 2.2 years to reach Earth escape. From escape it would take roughly 2 years to reach the target asteroid. The return time would range from 2 to 6 years depending on the actual mass of the NEA. The concept system could return asteroids with masses in the range 250,000 kg to 1,300,000 kg, to account for uncertainties in size and density.

White space creates a sufficient border around the visual, so no border line is needed.

The visual is large enough to read the text but not so large as to take over the entire page.

The visual is pushed to the edge of the page, leaving no orphans.

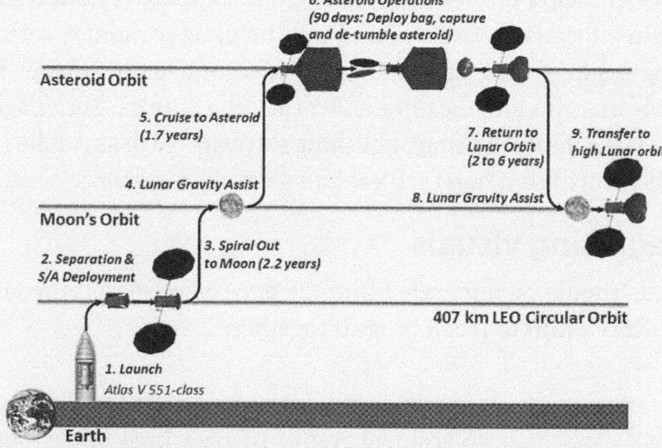

Figure 1. Asteroid return mission concept. Return flight time of 2 to 6 years depending on the asteroid mass.

The size and placement of this diagram are optimized so that all details remain readable, but the reader can easily use the image and the body text together.

Numbering pages, headings, and visuals

Technical and scholarly documents follow a general set of standards set by publishers for numbering pages and page elements. Within those standards, simplicity is desirable, as more complicated numbering systems tend to make content harder to find:

Pages	Single-sided pages are usually numbered on the upper right. Double-sided documents use upper-left and upper-right numerals on facing pages. Use lowercase italic Roman numerals for front matter, such as the table of contents.
Headings	Numbered headings add clarity only in longer documents. Multiple levels of section or subsection numbering (1.2, 1.2.1) quickly become overwhelming to readers, and you will rarely want to use more than two levels.
Visuals	Number visuals consecutively by type. Each type of visual has its own sequence: Table 1, 2, 3 and Figure 1, 2, 3.

31.2 Managing the appearance of paragraphs

Many of the rules of good writing amount to applications of the principle of *chunking*—deciding which small sets of information and ideas belong together, and then placing them together. Persuasive, articulate sentences and paragraphs are constructed according to this principle; their placement on a page should be, too.

Aligning text

Text generally looks best and is easiest to read when it is aligned *flush left* (also known as "ragged right"). Professionally typeset publications usually use text that is *justified*—aligned at both the left and right margins—but this format can create peculiar spacing that impedes readers' progress. In general, justify text only to satisfy a specified formatting requirement. (For instance, engineering professional societies often specify a format for conference papers that requires two justified columns per page.)

Alignment also requires decisions about how to mark the transition from one paragraph to the next. Books, periodicals, and many academic essays typically indent the beginning of a new paragraph. Workplace documents, however, most often double-space between paragraphs instead, allowing every line to remain flush left. This typographic style is sometimes called "block" format, especially for correspondence.

Avoiding widows and orphans

Near the top or bottom of a page, a transition between paragraphs can create another issue—a *widow* or *orphan*, a single line of text separated from the rest of the paragraph to which it belongs. A paragraph's last line is "widowed" if it carries onto a new page by itself; a first line is "orphaned" if it is "left behind" on the page before the rest of the paragraph. Most word-processing software will automatically control widows and orphans, ensuring that at least two lines of a paragraph appear together on a page, but minor editing is sometimes required—for instance, inserting an extra line space to move a heading onto a new page to prevent it from being "orphaned."

Another form of typographic widow occurs when a paragraph ends with a single word alone on its own line. Many writers try to avoid such widows; doing so usually requires editing the paragraph text.

31.3 Selecting typefaces

Typefaces are discussed at greater length in Module 5: *Designing documents for users*. For most documents that will be used chiefly in hard copy, body text should be displayed in a traditional font with serifs, such as Times New Roman, Palatino, Georgia, or Garamond. For headings, use a sans serif font; Helvetica, Arial, Calibri, Futura, and Verdana are the most popular choices. One font from each of these two categories will suffice for most documents; adding more than that to the mix can be difficult to execute well. When more variety is needed, try using multiple sizes of the same font (for instance, in different levels of headings).

A font's on-screen appearance is at best a general guide to how it will look on the printed page, so it's a good idea to print a test page at some point while you're working on a document.

Summary

Creating a page layout

- Scrupulous attention to page formatting is an important part of the impression of rigor, care, and intelligence that engineers and other technical professionals need to maintain.
- Margins and line spacing should make the text of a document easier to read. This is typically achieved with a line width of 50 to 80 characters and a line spacing slightly greater than 1.0.
- Visuals should be integrated into the document so that the visual and the text relevant to it can be read together.

Managing the appearance of paragraphs

- Text generally looks best and is easiest to read when it is aligned flush left.
- Alignment also requires decisions about how to mark the transition from one paragraph to the next.
- Avoid widows and orphans—single lines of text appearing alone on a page, separated from the paragraphs to which they belong.

Selecting typefaces

- Serif typefaces are the default standard for body text in print documents.
- Sans serif typefaces work well for headings; use multiple sizes of a single font rather than multiple fonts for this purpose.

32

Talks

When we think of "public speaking," we usually imagine a formal speech delivered to an auditorium full of listeners, and we admire the charisma of stars like Neil Degrasse Tyson, who command our attention while on stage. In fact, though, this is merely one setting in which engineers might need to present—and like writing, speaking to others in a professional role is a craft, founded on techniques and principles that reward study and practice. Many engineers are highly accomplished presenters who have honed their skills both in their engineering courses and in the workplace.

Giving an effective and engaging talk requires mastery of your content. Once you have achieved that mastery, you can take up the matter of clearly communicating your message.

Objectives

- To use personal experiences with voice, body, and space both to read your audience and to improve your own confidence when giving a talk
- To plan a talk by drawing on the conventions of specific genres of oral communication, such as elevator talks or keynote addresses
- To modify the content and delivery of a talk based on the audience, setting, purpose, and expected result
- To rehearse the talk—targeting improvements in both content and delivery—while approximating the setting and audience as closely as possible

32.1 Overcoming stage fright and connecting with listeners

Stage fright afflicts many performers. Even world-class athletes, actors, and musicians—who spend most days in front of huge audiences—routinely feel some anxiety or pressure. (Your professors feel it every time we stand up in front of a new class for the first time—and sometimes more often than that.) When you learn that you will need to speak in front of a class or present at a research conference, you will likely feel it, too. Even after you become a seasoned presenter, working "on stage" in front of a group may accelerate your pulse slightly and make you sensitive to perceived reactions on listeners' faces. Effective public speakers gradually develop these parts of the experience into assets, helping them to be engaging and personable while responding in real time to their audiences.

This process of development doesn't happen overnight, but it is possible to become calmer and more confident in short order by deciding to adopt two key beliefs about speaking in front of others:

- Have a *conversation* with your audience.
- Treat *voice*, *body*, and *space* as tools of the presentation medium, and make conscious decisions about how to use them.

Deciding to adopt these strategies may not eliminate nervousness, but it can reduce it and make it almost invisible to your listeners.

Establishing a conversational connection

We often think of presentations as a one-way channel from a presenter to his or her audience. Engineers know that in such a channel, our message (or "signal") will be degraded in the presence of noise, and we tend to be trapped in a mindset in which we're increasingly anxious about what might be going wrong. A conversation, on the other hand, obviously features channels running in both directions. The feedback you get from your audience can both correct your approach when needed—for instance, if listeners don't understand a technical term you're using—and reassure you when your message is clear and persuasive.

Many presentation settings that induce anxiety are actually much more conversational than you may think. Presenting a project to a meeting of executives in your company is hardly a casual affair, but you can work on initiating a back-and-forth discussion with your listeners—asking questions as well as answering them—rather than delivering a lecture to them about your content. Most public speaking tasks can be steered in this direction: In a sales meeting or in many class presentations, you can pose questions to your listeners or create some informal interactions that can engage your audience and relieve some of the pressure on you. (When you plan such moments in your talk, be sure that you're aiming for authentic interaction: This tactic can fall flat if it seems forced or gimmicky to your listeners.)

To be sure, some settings require a more formal address—if you're giving a keynote address to a large gathering of experts, for example. Conversational skills for interacting with an audience, though, will ultimately help you to gain confidence and flexibility in these situations as well: You'll develop the ability to adjust your performance in response to what you perceive in your listeners.

Monitoring voice, body, and space

Everyday conversations also offer us a stress-free opportunity to reflect on all of the subtle behavioral cues that shape the responses that we give to one another. We "read" those around us and communicate our own thoughts and feelings with a few basic tools:

- *Voice*: Vocal qualities, especially volume, pitch, and tempo, tell us about a person's *attitudes*.

- *Body*: Posture and movement usually show *degree of comfort and engagement.*
- *Space*: Physical proximity and frequent eye contact show *personal closeness and feelings of attachment.*

Whether you're talking with an individual or addressing a large audience, compelling speech relies on *deliberate*, *controlled* use of these tools:

- When you move, move deliberately. Remaining largely still is better than shuffling or pacing without purpose or awareness, because stillness can convey calm and control of the situation.
- You don't have to move too much to make an impact. Try moving closer to your listeners, or using deliberate hand gestures, just to emphasize one or two key points.
- Practice keeping your body open (shoulders back, feet apart) and your muscles relaxed.
- As you move through the content, carefully maintain a moderate tempo. Rushing not only makes the content harder for your audience to follow but also creates the impression of panic.
- Excessively slow delivery, on the other hand, can produce monotony. Varying your pitch and rhythm is also important in keeping your voice engaging—especially for speakers with low voices that can naturally tend toward a monotone delivery.

32.2 Analyzing audience and setting

Speakers who deliver compelling talks have a very deep understanding of their audience and the setting of the talk—although possibly not a conscious understanding. Taking the time to explicitly consider the following questions will improve your talk, and you will begin to internalize these considerations over time:

Audience	Who will listen to your talk? Are they co-workers (students), direct supervisors (teachers), high-level managers (administrators), or the public? Will several different types of audience members be present?

Setting — Where will you give the talk? Will you be able to walk around the room, or will you be limited to standing behind a podium? When will the talk be given? Is this a routine meeting (suggesting definite expectations for the talk), or are you doing something out of the ordinary?

As in a written document, listeners who don't know much about the topic may need a thorough and honest overview before they can understand the main content. Audiences who are thoroughly acquainted with the topic can engage that content with much less background and context.

Setting includes both the physical space in which you're speaking and the occasion for your talk, which helps you to understand the expectations that your audience is bringing. If you're delivering a status report in a regular meeting, attendees will expect your content and format to resemble the status reports that others have given. (Similar expectations will exist in other standardized situations, such as a technical talk at a particular conference. Rhetoricians call such a recurring setting a *forum.*) Thinking about these expectations allows you to follow them exactly, creating a "standard" talk—or to bend or break the implicit "rules" in a careful, strategic way in order to create a more memorable talk. Such a creative approach can be undertaken with less risk if your talk is not an everyday part of a well-known forum.

CSER 2013, March 20, 2013, Georgia Tech—Victoria Cox

Transcript of keynote presentation by Victoria Cox

Thank you, Tom. Let me also start by thanking Chris Paredis and the CSER Program Committee for the invitation to speak here today.

Keynote talks begin when the speaker thanks the person who introduced him or her to the audience.

I hope that my remarks will serve as an appropriate backdrop to the work to be presented over the course of the next couple of days. Systems engineering concepts and practices are key to success in project delivery, and I am sure that you will see many new and innovative ideas, systems, and processes presented along with interesting case studies outlining challenges in the delivery of complex systems.

A conference keynote is expected to explore core themes or problems in a way that can be discussed in later presentations and conversations.

[. . .]

I have had the great fortune of working on some very exciting, technically challenging and complex systems over the course of my career: the Hubble Telescope; Missile Defense—known then as the Strategic Defense Initiative or Star Wars; and, most recently, the Next Generation Air Transportation System (NextGen).

Listing major projects quickly establishes credibility with the audience. This could also be accomplished in the introduction that precedes the talk.

Cox explains NextGen in jargon-free terms that still capture its complexity—important for an audience of systems engineers.

This talk did not include presentation slides. At a typical conference, the keynote might have slides, or it might be the only talk without them.

[. . .]

For those of you unfamiliar with NextGen—it is not a single program; rather, it's a comprehensive initiative that integrates new and existing technologies. It represents the complete transformation of our national airspace system.

NextGen is intended to move today's mid-twentieth century system into the 21st century resulting in operations in low visibility conditions where aircraft cannot operate today; shorter, more efficient flight paths; reduced emissions and noise; and a better passenger experience—thus ensuring the long-term viability of this vitally important economic sector.

[. . .]

Conference keynote address "Challenges with deploying complex systems of systems: Some perspectives" given by Victoria Cox, former FAA Assistant Administrator for NextGen.

32.3 Identifying the genre, purpose, and desired outcome

Engaging and worthwhile talks need to be crafted with an eye toward the outcome that you and your audience want to achieve. Often, you can identify exactly what type of talk you're being asked to give. Any of the genres covered in this book might be delivered as a talk (which might either supplement or replace the written version): You might give a conference talk based on an experimental report that you've written, or make a proposal in a meeting rather than writing up a formal proposal document. Some genres, though, are always delivered as talks; these are covered below.

Elevator Talks

An "elevator talk" imposes a very tight time limit—a few minutes at most—in which you are to present a new idea, your previous work, or your own credentials, usually to someone you don't know well. The idea is that the talk might be delivered during a chance encounter in an elevator at an office building—giving you no more than two minutes.

The classic elevator talk is a "pitch" to a venture capitalist or other investor, but its format has been adapted for many other short presentations. In most of these, it's important to avoid the overt salesmanship of a "pitch."

Audience	Often a potential client, sponsor, or employer. Such listeners will have at most a broad acquaintance with your topic; when in doubt, assume no background knowledge.
Setting	Usually somewhat informal. Limited time defines the elevator talk: anything longer than a five-minute poster talk doesn't fit the genre (although five minutes is less restrictive than the time you'd have for the prototypical elevator talk).
Purpose	Convince the audience that your new idea is worth pursuing, that your previous work is well considered, or that you're the right person to hire. Don't expect to discuss the finer points of your content.
Desired outcome	Hope for listeners to show interest during the talk, ask follow-up questions, or arrange for a more formal meeting at another time.

Rules of thumb for elevator talks are as follows:

- Maintain focus on the big picture—your main arguments or conclusions, or the merits of your idea—and don't be sidetracked or led into a digression.
- Use precise evidence, but include only a few key pieces of information, especially when dealing with quantitative data.
- Avoid technical jargon, keeping the entire talk as friendly as possible to non-experts.

Project talks

When you're summarizing your previous work on a project and the time limit is not so short to be an elevator talk, you're likely giving a project talk.

Audience	Usually interested in your work in some way—supervisors, administrators, and colleagues from the same company, or students taking the same course.
Setting	Usually a meeting of some sort, which may include a management group, your own work group, or a larger audience of colleagues. Formality will vary with the space and tone of the meeting.

Purpose	To share progress on the project, showing what has been accomplished, what work remains, and what results can be expected.
Desired outcome	Hope for listeners to understand what's been going on with your project, to regard its progress as reasonable, and to support its overall goal. You may be asking for resources (time, equipment, personnel) to advance the project and need decision-makers to approve those requests.

Here are rules of thumb for project talks:

- A full project talk can incorporate more detail than an elevator talk. Still, emphasize the project's purpose, meaning, and direction over exact details of what's been done.
- Plan the talk around the concerns that you think will be most important to listeners—often deadlines, budgets, and other factors that affect the larger organization and its plans.
- Avoid defensiveness. Treat listeners' questions and suggestions as sincere attempts to contribute to your ongoing work.

Conference presentations

In the industrial and academic worlds alike, engineers attend conferences to share the latest research results, to learn about the current state of the art, and to make connections with other professionals in related fields. At many conferences, presenters are delivering content that they have already published in technical papers in the conference proceedings; the talk presents the paper's main ideas to listeners who might otherwise be unlikely to read it.

Audience	Usually technical experts of some kind. Listeners' levels and types of expertise depend on the breadth of the conference: At a smaller conference with a more specialized focus, you can expect the audience to have more direct expertise in your own content area.
Setting	Usually a session featuring several separate presentations on related topics, held in a conference room at a university, hotel, or convention center. You may or may not have a podium with a microphone.
Purpose	To share results, solicit feedback, and initiate ongoing conversations with others performing similar work.

Desired outcome	Hope for wider awareness of your own work and leads that might help you take the next steps in the work. A successful conference presentation might also lead to invitations to submit your work to a technical journal.

Rules of thumb for conference presentations are as follows:

- Expert audiences often discuss finer points of the data being presented. Be ready for this, especially with accurate, easy-to-read graphs and other visualizations of your data.
- Most conference talks use presentation slides, but consider distributing a short handout that displays the most important findings. If everyone has this information at hand, you can minimize the need to scroll back and forth through slides.
- Try to ignore audience members coming and going during your session: Such listeners are likely trying to visit several talks during a single time slot. A sudden departure isn't a reflection on the quality of your presentation.

[. . .]

It has been my observation that the greatest challenges are seldom limited to the integration of cutting edge technologies or the development of those technologies. Rather, other challenges are often the most difficult to plan for and overcome.

I am interested in exploring this topic because it highlights a lot of issues that I have dealt with over the course of my career. I would like to say that I am going to unveil my solutions to these difficult challenges; but, sadly, that is not the case.

However, I can say that I have examined a number of case studies; read and conducted some informative analyses; and learned from personal experience; and I have observed some commonalities across disparate projects; noted some commendable successes and, likewise, some stupendous failures.

I'd like to share some thoughts about these with you this morning.

[. . .]

The purpose of this keynote talk is to share "lessons learned," based on personal expertise and study, that would be broadly applicable to everyone in the audience with knowledge of the basics of the subject of the conference (systems engineering).

Based on these introductory remarks, the speaker is making a claim that the greatest challenges in a complex project are not technological.

The remainder of the talk gives multiple examples where large-scale projects failed.

The conclusion is a recommendation for personal attributes for "successful complex systems delivery."

So, the best advice I can leave you with is—whatever your level of responsibility for a project—whatever stage your project is in—consider these 5 things: Think Critically; Seek appropriate Competencies; Communicate; Consider the Culture; and be Courageous.

Conference keynote address "Challenges with deploying complex systems of systems: Some perspectives" given by Victoria Cox, former FAA Assistant Administrator for NextGen.

32.4 Rehearsing and preparing the talk

Until you have years of experience giving a particular type of presentation, full rehearsals of your talk should be a standard part of your preparation. Such rehearsals will help you to improve your talk in several ways:

- Adding, removing, or revising content (to substantiate a claim or to fit within a time limit)
- Crafting specific parts of your delivery (how you move while speaking, or how you formulate a key point)
- Gaining fluency (reducing hesitation and minor speech errors) and smoothing out transitions

Rehearsal is more than review

Skimming over your talk notes or slides may be necessary before you speak—but to rehearse a talk, you must actually deliver it, preferably to an audience of friends or colleagues. Addressing listeners allows you to practice eye contact and speaking volume, and to test your content to see whether others can understand it easily. The more closely you can approximate the actual talk setting, the more effective the rehearsal will be. Delivering your talk alone in an office or bedroom should be a last resort but is better than not rehearsing at all.

- Time yourself, particularly if you tend to get more talkative when you're nervous. Most presenters do respond to nervousness in this way, and so often have to cut their talks short. When this happens, your conclusion or summary—and, in turn, the coherence of your talk as a whole—is shortchanged.

- If you'll stand while presenting, stand while practicing.
- Use a display technology similar to the one you will be using for the actual talk. Having to wrestle with technology in front of an expectant audience is sure to fluster even a well-rehearsed presenter. (For instance, a projector's maximum resolution will likely be lower than your laptop's; that mismatch is one frequent cause for presentation woes.)
- If you plan to use a laser pointer, use it while practicing. Pointers should be used briefly (less than 5 seconds) to draw attention to something being displayed and then turned off—a small dot moving around the screen is very distracting. Stabilize your pointer hand by placing your arm against your torso and holding that wrist with your other hand; a shaking pointer dot can be a distressingly visual reminder of your nervousness.

Rehearsal helps you prepare

Take time to assess the rehearsal after you've finished speaking. Ask for honest, constructive feedback from your listeners. If you are rehearsing alone and have a hard time noticing how well the talk went, you can record yourself with a webcam and play back the recording. You'll find areas for improvement in the rough spots present at the early stages of most talks:

- Missing or jarring transitions between ideas or topics
- Stumbling over difficult words and phrases
- Vocalized pauses—"uh," "um," "like," "you know"
- Posture or movement that signals anxiety (swaying, shuffling, fidgeting)

With notes from your rehearsal, begin to make some changes to your content or delivery. The more time you allow for revision (and additional rehearsal), and the more deliberately you consider your delivery, the more polished your talk will be.

When you have rehearsed your talk and feel confident about the content, prepare your "Plan B." If your talk includes slides, have a backup USB drive with both a software-specific file and a general file (.pptx and .pdf). Another strategy is to email yourself the finalized file(s) so they can be retrieved from any computer with an internet connection.

If you plan to use note cards during your talk, record only the most important high-level points. To minimize the temptation to read directly from the note cards, write them down as phrases, not complete sentences. Some kind of marking on a note card that reminds you when you're halfway through can help you better manage your speaking time.

Summary

Overcoming stage fright and connecting with listeners

- Establishing a conversational connection with the audience—by asking questions and watching the audience's reaction to your main points—can reduce the commonly felt anxiety of speaking to a group.
- Deliberate, controlled use of voice, body, and space—moving deliberately, keeping an open body position, and speaking at a moderate tempo with engaging variations in pitch and rhythm—helps to create a compelling talk.

Analyzing audience and setting

- The audience for a talk will influence both the content and delivery of the talk: the amount of background information given, the interaction level, and the evidence provided.
- The setting for a talk includes the physical time and place (which may influence the audience's attention) and expectations regarding content and delivery.

Identifying the genre, purpose, and desired outcome

- A clear understanding of the purpose of the talk—the *why*—can help you decide what to emphasize.
- Think about what would make the talk a success, in terms of the impact on the audience and on yourself. Knowing the expected result allows you to bring it about more directly as you prepare the talk.
- Commonly encountered talk genres have their own particular rules of thumb that should be considered.

Rehearsing and preparing the talk

- Rehearsal is necessary to deliver a polished talk.
- The most effective rehearsals closely approximate the environment and conditions of the talk itself.
- Expect to make changes to the content, ordering, or delivery of the talk based on feedback and practice.
- Having a backup plan can make you feel more confident during the talk, even if it's not needed.

33 Presentation slides

In most professional meetings and forums, audiences will expect to see presentation slides. However, presenters should always ask themselves whether slides are the most effective tool to deliver a message or to initiate a conversation with a particular audience in a given situation. Even where slides are the standard, expected delivery device for content—as part of a briefing in a recurring weekly or monthly meeting, for instance—it is sometimes possible to "break the mold" and command the audience's attention by forgoing slides.

When slides are the correct (or the expected) accompaniment to your remarks, you can serve your content most effectively by applying key principles of graphic and information design.

Objectives

- To identify the factors that limit presentation slideware as a communication medium, such as low graphical resolution and the difficulty of conveying complex content in brief textual fragments
- To orient audiences to the main ideas guiding your presentation by using a complete-sentence assertion as the heading for each slide
- To design each slide's graphical content to provide evidence for its assertion (minimizing the use of visuals for mere decoration or branding)
- To assess other presentation software for its ability in conveying particular logical relationships and organizing frameworks—such as Prezi's strength in showing spatial relationships
- To adapt slides into a platform for project documentation or written communication outside of the usual context of a talk, creating hybrid "slidedoc" texts

33.1 Recognizing the limitations of slideware

Microsoft PowerPoint is the dominant tool for slide presentations. Even presentations delivered with Apple, Google, and open-source equivalents typically follow the same conventions that have evolved with Microsoft's software. (Another presentation tool, Prezi, has unique features and is addressed at the end of this module.) In recent years, many typical features of PowerPoint slides—especially their bullet points, clip art, and animated audiovisual effects—have been criticized widely.

In 2010, for instance, *The New York Times* reported on a backlash against PowerPoint by senior military commanders, despite the program's ubiquity "in a military culture that has come to rely on PowerPoint's hierarchical ordering of a confused world." Marine General James Mattis lamented that "PowerPoint makes us stupid."

These criticisms have been most forcefully and influentially stated by renowned visual designer and expert on data graphics Edward Tufte in his essay "The Cognitive Style of PowerPoint." Tufte argues that PowerPoint's default output distorts the content of a presentation for the following reasons:

- Reflecting the limitations of low-resolution projectors, slideware typography and data graphics often push presenters to abbreviate and oversimplify their ideas.
- Bulleted lists and slide sequences favor long series of basic informational items—unlike sentences and paragraphs, which can convey the complex relationships among ideas.
- Slides' visual elements are often ornamental or even whimsical, seldom conveying the intellectual substance demanded by serious content.

Not all commentators agree with Tufte's position: In fact, most dispute his conclusion that these flaws should drive presenters to use handouts or other media in place of slides. However, these objections have driven many innovations in slideware design—by presenters and even by the software developers themselves.

33.2 Designing slides using assertion-evidence style

To address these criticisms with traditional slides, a number of scholars and visual designers have created alternate slide presentation formats. Among engineers, the most influential and important of these is the "assertion-evidence" style promoted in Michael Alley's book *The Craft of Scientific Presentations*, which draws on the principles of rhetoric and cognitive psychology to create "more readily understood, more memorable, and more persuasive" slides. (The assertion-evidence technique is also recommended for engineers in *Slide Rules* by Traci Nathans-Kelly and Christine Nicometo.)

Combined with the title of the presentation, the hip joint diagram provides a quick visual cue to the content of the presentation.

Identifying individual authors rather than just the organizational author increases individuals' sense of accountability.

Logos of sponsoring organizations make sense on an opening slide—but can be omitted on subsequent slides to avoid needless repetition and distraction.

Femoral Component Installation Monitoring

Cup
Ball
Stem

Deena Abou-Trabi–University of Houston
Mike Guthrie–University of Wisconsin
Hunter Moore–Virginia Tech
Dr. Phillip Cornwell–Rose-Hulman
Dr. Michael Meneghini–St. Vincent Center for J Replacement, Indianapolis, IN
Dr. Aaron Rosenberg–Rush University Medical Center, Chicago, IL

LADSS
www.toc-stl.com/ info/totalhip.htm

Los Alamos

The title slide should orient the audience to the subject of the talk. © 2014 Cornwell. Reprinted with permission.

Using full sentences as headings

Headings consisting of individual words or brief phrases can indicate subject matter but can't easily tell your audience what you're saying *about* that subject matter. A full-sentence assertion, on the other hand, can state the exact proposition that you want listeners to understand or accept. This is the difference between just indicating that a "Problem" exists and defining the nature and scope of that problem.

Complete-sentence assertions tell audiences exactly what you are saying about the content. Writing such assertions might require more effort from presenters at first, because it requires more critical and careful thought about a talk's message and purpose. The complete-sentence slide heading serves the same function on a slide that a well-executed topic sentence serves in a paragraph: Both define the claims or observations that are to be explained and supported with evidence. In both cases, multiple drafts may be needed to express the idea as clearly and accurately as possible—but this work will make it easier to arrange the rest of the content, and that content will in turn be clearer to audiences.

This presentation provides a method of detection for a fully seated prosthesis within a femur

Background Information

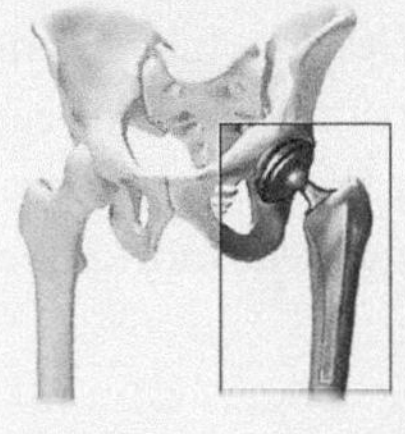

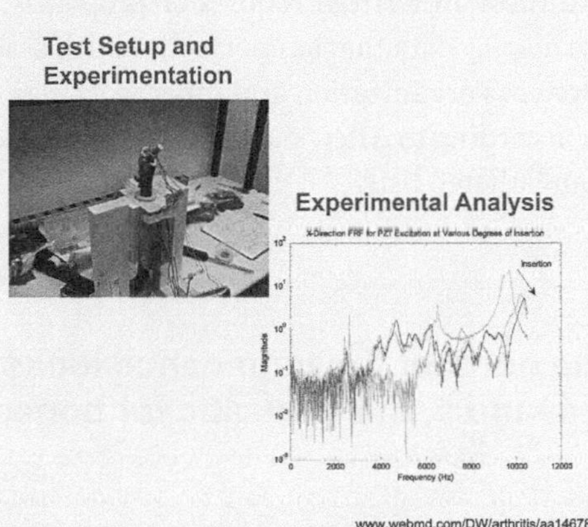

www.webmd.com/DW/arthritis/aa14678.asp

The outline slide maps the stages of the talk and provides images to make those stages memorable for the audience. © 2014 Cornwell. Reprinted with permission.

In the first slide, the assertion will often indicate the purpose of the presentation—providing much more information than a generic heading such as "Introduction" or "Problem."

This slide forecasts the presentation's organization without a full textual outline that will likely do little for listeners' comprehension.

Citations are necessary for images taken from a source—even if they come from a simple web search.

Once modal testing was completed, a test structure was constructed for the sawbones

The bolts and C-clamps were used to provide repeatability

The foam around and under the sawbone was used to simulate a human leg

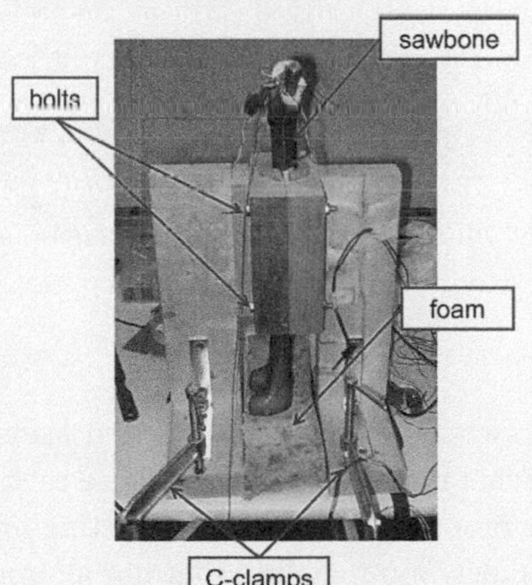

This sentence heading is more effective than a general "Experimental Design" heading because it informs the audience of the stage and purpose of the testing. © 2014 Cornwell. Reprinted with permission.

This slide's text varies in size and color, creating three levels to address different levels of detail. The sentence heading addresses the apparatus as a whole, while the explanatory sentences and callouts address specific components within it.

Callouts identify the parts of the testing apparatus—especially important in a photograph, where components might be harder to see than in a diagram.

Providing evidence

In the body of the slide, Alley recommends providing visual evidence that supports the assertion and that helps the audience remember the key points. The most effective visuals in technical presentations will be similar to those in written reports or proposals. Graphs, tables, and maps can display data that supports the slide's assertion; diagrams can show a system's organization and inner workings or the steps in a process. All, according to Alley, show connections among ideas more effectively than bulleted lists.

Assertions state major conclusions drawn from the quantitative data provided in a graph.

The graph fills the entire slide, making it easier for the audience to read.

Lines of contrasting colors differentiate the data series—but using different symbols for each series' data points keeps the graph usable even if printed in black and white.

The norm of the impedance looks good for bones 2, 4, and 5, but is erratic for bones 1 and 3

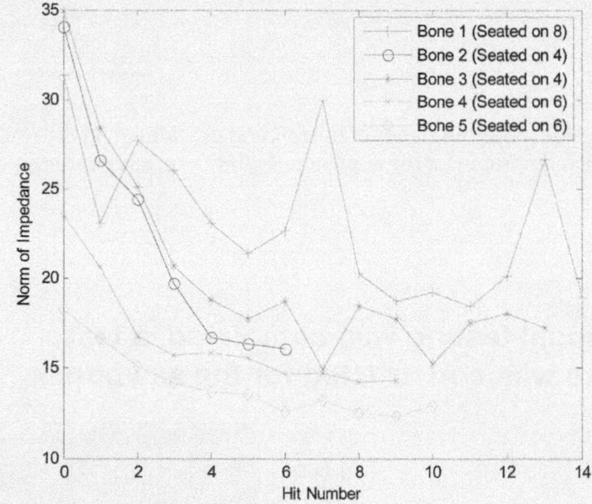

Visual evidence, such as this graph, enhances the presenter's credibility.
© 2014 Cornwell. Reprinted with permission.

Assertion-evidence style often leads to more precise slides with somewhat narrower scope, because each is dedicated to illustrating a single proposition (rather than listing multiple ideas). While creating your slides, you may find that not all important content will easily fit into your main assertion. Create separate notes for yourself—or for your co-presenters—to remind yourself of what you want to say without unnecessarily cluttering the slide.

In instances where your assertions are particularly complex, Alley suggests presenting your evidence before the assertion so that you can talk through your graph and data first. This strategy can also be effective when trying to persuade a skeptical audience.

At the conclusion of the presentation, including a "Questions" slide is unnecessary. Alley suggests a summary slide that emphasizes the main points of the presentation. Alternatively, an acknowledgments or references slide is appropriate. If there are too many references to include them on the individual slides, a bibliography slide should conclude the slide presentation. IEEE bracket citations can be included on the individual slides.

33.3 Using Prezi to illustrate spatial relationships

Prezi is a web-based presentation software tool that replaces a sequence of slides with a "canvas," a single virtual presentation space that a presenter prepares and then navigates for his or her audience. While the presentation "canvas" suggests painting, much of Prezi's appeal is cinematic: The presenter can zoom in and out and pan around as if using a movie camera, directing the audience's attention to different areas of the "canvas." Within a Prezi presentation, audiences will view a series of still "frames"—the program's equivalent to slides—in which users will organize content as they would in PowerPoint. (It's easy, for example, to create assertion-evidence frames.)

The "canvas," though, situates frames in a virtual "space" as well as sequencing them in time. As a result, Prezi lends itself to various kinds of spatial metaphors for organizing your content. Examples created for prezi.com take audiences around a game board, along a tree's branches, and so forth. In engineering, such a presentation might lead an audience through a circuit, one component at a time, or through the various spaces of a manufacturing plant. When designing a Prezi canvas, convey logical ideas with systematic ideas like these:

- Organize the presentation around a large frame that provides what film directors call an "establishing shot." Show your audience the overall scene, displaying the major ideas and how they relate to one another.

- Use proximity on the canvas to indicate the degree of connection between ideas. Pan further on the canvas when changing topics.
- Zoom in on specific details, letting viewers see that you've remained within the same topic.
- Zoom out periodically to your "establishing shot," reminding viewers of the big picture.

These features of Prezi have rich potential, but they should be used in a restrained, deliberate fashion. Try not to zoom or pan too far all at once. Especially on a larger screen, these effects can create audience discomfort—something like vertigo or motion sickness. Viewers should be thinking about the way that Prezi fits your content—not about its spectacular visual effects.

33.4 Adapting slide designs for other purposes

Assertion-evidence design should work well for most standard presentation settings:

- Briefings and reports during routine meetings
- Talks at scholarly or technical conferences
- Public forums and sales presentations
- Classroom presentations and other educational settings

In some organizations, though, presentation slides are used for other purposes as well. For instance, some organizations divorce slides from presentations, using PowerPoint files as substitutes for written reports, user manuals, or other forms of technical documentation. Many commentators believe that the slide format is inherently unsuited for such detailed documentation. At a minimum, though, presentation slides require a fundamentally different design structure if they are the sole vehicle for content (rather than a complement to spoken delivery).

In response to this need, Nancy Duarte has developed rules for the "slidedoc," filling the gap between a document (such as a report or proposal) and a slide presentation, drawing on the strengths of both. Like a conventional text report, a slidedoc depends on precise, carefully planned sentences that accurately record sound reasoning. Those sentences are laid out on slides, however, with the visual tools (headings, graphs, diagrams, and the like) ordinarily used to make slides easy to understand. The resulting document, Duarte argues, hits a "sweet spot" that lets audiences engage with its content with ease. The logic and evidence are neither buried in dense series of paragraphs nor deferred in favor of shallow slogans or vague phrases announcing topics.

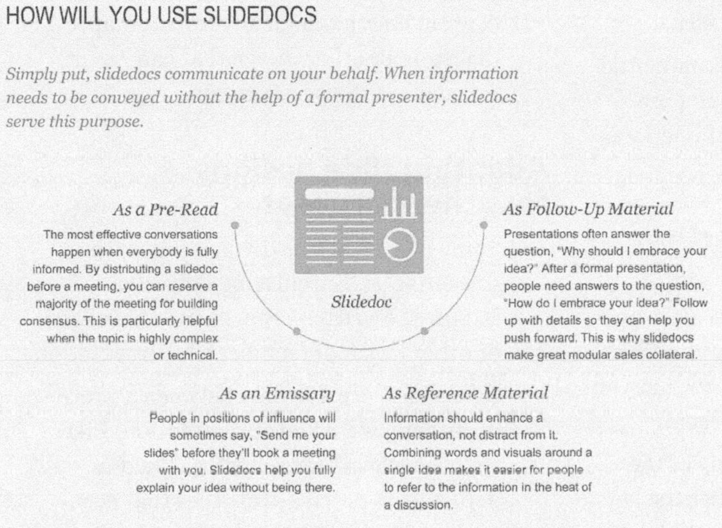

HOW WILL YOU USE SLIDEDOCS

Simply put, slidedocs communicate on your behalf. When information needs to be conveyed without the help of a formal presenter, slidedocs serve this purpose.

As a Pre-Read

The most effective conversations happen when everybody is fully informed. By distributing a slidedoc before a meeting, you can reserve a majority of the meeting for building consensus. This is particularly helpful when the topic is highly complex or technical.

Slidedoc

As Follow-Up Material

Presentations often answer the question, "Why should I embrace your idea?" After a formal presentation, people need answers to the question, "How do I embrace your idea?" Follow up with details so they can help you push forward. This is why slidedocs make great modular sales collateral.

As an Emissary

People in positions of influence will sometimes say, "Send me your slides" before they'll book a meeting with you. Slidedocs help you fully explain your idea without being there.

As Reference Material

Information should enhance a conversation, not distract from it. Combining words and visuals around a single idea makes it easier for people to refer to the information in the heat of a discussion.

This slidedoc includes headings at both the general-topic and complete-assertion levels.

Much like a poster, the slidedoc relies on dividing text into short chunks and placing those chunks into an arrangement that can be seen quickly. Here, each chunk shows a typical use of a slidedoc, and the four uses are arranged as a sequence in time.

Slidedocs provide more substantial content than a presentation slide but retain visual appeal and ease of reading. © 2014 Duarte Press, LLC. Reprinted with permission.

Duarte provides an extensive rationale of this design style and free templates on her website, http://www.duarte.com/slidedocs/.

Summary

Recognizing the limitations of slideware

- Slideware typography and data graphics often push presenters to abbreviate and oversimplify their ideas.
- Bulleted lists and slide sequences favor long series of basic informational items—unlike sentences and paragraphs, which can convey the complex relationships among ideas.
- Slides' visual elements are often ornamental or even whimsical, seldom conveying the intellectual substance demanded by serious content.

Designing slides using assertion-evidence style

- Use full-sentence headings so that audiences can easily follow the purpose of your content on each slide—the exact proposition that you want listeners to understand or accept.
- Provide visual evidence in the body of the slide to help the audience remember the key points. Graphs, tables, and maps can display data that supports the slide's assertion; diagrams can show a system's organization and inner workings or the steps in a process.

Using Prezi to illustrate spatial relationships

- Prezi replaces a sequence of slides with a "canvas," a single virtual presentation space that a presenter prepares and then navigates for his or her audience in a cinematic style (zooming, panning, etc.).
- Prezi often works well when a presentation's content emerges logically from a map, diagram, or other image that can recur throughout the presentation.

Adapting slide designs for other purposes

- Some organizations divorce slides from presentations, circulating PowerPoint files in place of written reports, user manuals, or other forms of technical documentation.
- "Slidedocs" are a hybrid between a prose document and a slide presentation. Like a conventional text report, a slidedoc depends on precise sentences that accurately record sound reasoning. Those sentences are laid out on slides with the visual tools ordinarily used to make slides easy to understand.

Posters

34

Many professional conferences and trade shows host poster sessions, which create a very different environment from the sessions in which speakers present with slides. In the typical poster session, many invited speakers—perhaps several dozen—are featured simultaneously, while audience members browse casually among the offerings. As in other settings, a knowledgeable, practiced presenter will be best able to engage and persuade his or her listeners. However, the design of the poster itself is crucial in helping the presenter to attract listeners and to guide them through the content—especially in a setting that can challenge an audience's ability to establish and sustain focus.

Objectives

- To plan a poster presentation for an audience of mobile listeners with varying interests and knowledge of the topic, beginning with the structure of the elevator talk
- To adapt the talk while speaking, adjusting the selection and sequence of content in the course of conversation with listeners
- To place major elements of a poster in a way that enhances audience comprehension, creating a path for the eye that follows the sequence of delivery that you anticipate
- To apply visual design principles that allow audiences to identify main ideas and to perceive major logical relationships among them

34.1 Understanding the audience for a poster presentation

Many conference organizers receive substantially more submissions than they can possibly assign to presentation times. Historically, poster sessions have provided an "overflow" for proposals that show enough merit to be accepted to the conference but aren't easy for organizers to place into a panel. It's a mistake, though, to think that a poster presentation can't be as rigorous or as valuable as any other conference presentation. When the 1991 ACM Hypertext conference diverted Tim Berners-Lee's proposed paper into the conference's poster session, it certainly didn't reduce the importance of his content—his recent invention, the World Wide Web (Wardrip-Fruin and Montfort, 2003).

The poster session may seem less formal than a talk in front of a seated, silent audience: Ultimately, though, posters and the accompanying talks simply define a slightly different style to which you'll need to adapt. A poster presentation's audience is defined by several key characteristics:

Mobility	Listeners can arrive at any time. Because poster presentations usually occur with everyone standing, there is little disruption when people come and go.
Varying interest	Depending on their interest level, some of your audience may wish to quickly skim over many posters while others may spend a significant amount of time with just a few.
Varying expertise	Listeners will not all be experts in your poster's area. Some may only be aware of the general context while others may have been working in the field for decades.

Walking and talking: Addressing a mobile audience

Expect listeners to arrive at any time, to stay for as short or as long a duration as they would like, and to leave at any time. Because you are not guaranteed their attention for any length of time, you must make the best of what you get. In some ways, a good poster presentation is like good advertising: If you "hook" the audience early—interesting them in the stakes or implications of your content—they may stick around longer. However, you must also respect the audience's time. They may wish to spend a few minutes with many posters, so you should be prepared, as they walk away, to thank them graciously for their time.

Audience variety: Commitment level, interest, and expertise

Some of your audience may have come specifically to see your poster; others may have paused on the way to the poster they really want to see. An effective poster presentation can address the deep interest of the former while also satisfying the idle curiosity of the latter. Similarly, you may speak to audience members with considerable expertise in your technical area as well as to novices. An effective poster presentation is audience-focused—flexible in both time and content without feeling haphazard.

Learn from the audience and establish professional connections

While you give your poster presentation to your audience, be ready for feedback and the opportunity to learn from listeners. A question from a listener may help you to refine your own presentation content or technique: You might identify concepts or claims that need to be described more clearly, or you might even recognize weaknesses in your own logic. The audience members are likely listening to your poster presentation because something drew them to it: They may have very useful observations to share with you. It is always a good idea to have a small notebook with you to jot down ideas or suggestions that audience members have shared with you, and to record the contact information of people with whom you want to follow up later. (Having your own business cards available to share is important, too—presenters and listeners alike benefit by cultivating relationships with others in the professional community who share their technical interests.)

34.2 Delivering the poster talk

Delivery of a compelling poster talk requires you to:

- Know your major points. Deep knowledge of your content allows a more compelling delivery: With the overall sequence of ideas in mind, you can explain them as the moment demands, without having to recite memorized lines.
- Converse with your audience. The less formal nature of a poster talk allows you to elicit feedback from the audience during the talk, so you can more quickly get to the content that interests an individual listener.
- Be flexible in the delivery. A listener may walk up and ask a question about your results without having heard your introductory content. Be prepared for this, and smoothly navigate the content of your poster as needed to address the question.

Adapting the genre of the elevator talk

The *elevator talk*—typically defined by commercial endeavors, "selling" a business plan or product concept in one or two minutes—has become a standard way to think about short, informal presentations. (An elevator ride, the reasoning goes, might give you just enough time to interest a "captive" audience in an investment opportunity.)

In a technical poster talk, you typically don't want listeners to think that you're "selling" anything. However, some of the objectives are the same: You want a quick but methodical way to introduce the problem you're investigating, explain the reasons that it merits attention, and establish the soundness of the work that you've done.

Conversing with your audience

An effective poster presentation is conversational, not a memorized speech to unleash the moment someone makes eye contact. Introduce yourself after someone has spent a few moments looking over your poster, asking what he or she knows about the subject. If the person expresses familiarity with the basic material, you can skip to your main points. If not, you can provide the necessary background information. To do this successfully, you must know the material well enough to speak convincingly about any aspect of the poster. By judging what your audience members want out of the interaction, you can proceed quickly to the content in which they appear to be most interested.

Multiple paths through your content

Because a conversation may move quickly to your results, or to a particular idea that pertains to a listener's own interests, your presentation may not follow the original path through the content that you envisioned. You will likely be giving your presentation many times and may have many paths through your content already in mind.

When you find yourself speaking with someone who seems to know the technical content especially well, you may even decide to start your presentation in the middle; in this setting, you can back up to the introduction if you misread your audience's background or needs. Take advantage of the poster presentation's loose, flexible structure to address your listener's needs more directly and completely than you could do when speaking to a large group.

Eliciting audience contributions

Because of the less formal nature of a poster presentation, expect some listeners to interrupt—asking for clarifications, seeking information, and perhaps even raising objections. Treat this interaction as a strength of the poster format, and try to elicit contributions from quieter audience members. You might ask what background they have in the area, or whether they can see what story the data tells. Assess listeners' reactions to ask yourself whether your conclusion is convincing. A poster presentation is an excellent opportunity to judge the effectiveness of your argumentation and make it better for next time.

34.3 Placing major elements of the poster

Authors begin with a multicolumn "page" of 36 by 48 inches (or, less commonly, 24 by 36 inches) in a program such as Microsoft PowerPoint, Microsoft Publisher, or Adobe InDesign. However, an effective poster is more than a technical report or slideshow that has been copied and pasted into a poster layout; it has been designed from the beginning to take advantage of a poster's strengths, which are fundamentally different from those of a written report or even a slide presentation.

A slide presentation has been significantly modified to present the same content in an effective way.

The main assertions made by the presenter are clear to the audience in the section headings and limited use of text.

Repeated design elements tie the poster together as a cohesive whole.

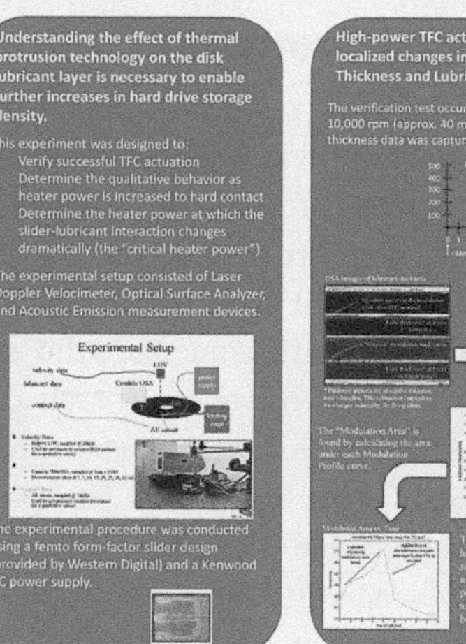

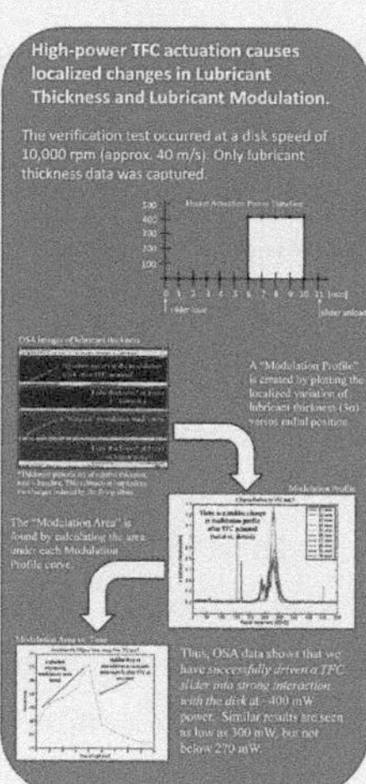

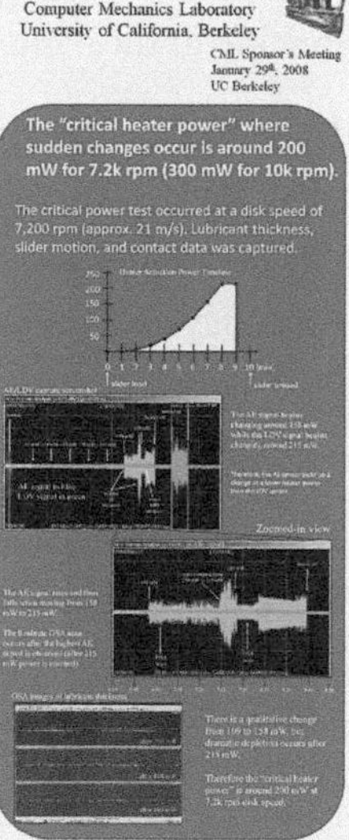

This poster presents technical content so that audiences can understand it while listening to the speaker. © 2008 Computer Mechanics Laboratory, UC Berkeley.

In most situations that call for a poster, the authors begin knowing the major points that need to be made and may have produced some data graphics to communicate findings or conclusions. Even if these elements are close to final draft quality, the process of designing the poster calls for many small decisions about element location, size, and color. The following samples show some of the decisions involved in creating the previous poster sample.

Slider-Lubricant Interactions at the Head-Disk Interface :
Thermal Fly-Height Control

Sean Moseley & Prof. David Bogy
Computer Mechanics Laboratory
University of California, Berkeley
CML Sponsor's Meeting
January 29th, 2008
UC Berkeley

This heading information of the poster reflects design choices made to clearly communicate to the audience. © 2008 Computer Mechanics Laboratory, UC Berkeley.

The poster title makes a clear assertion about the content of the poster.

Author and affiliation information tells listeners who they're speaking to and who they might follow up with for more information.

The "critical heater power" where sudden changes occur is around 200 mW for 7.2k rpm (300 mW for 10k rpm).

The critical power test occurred at a disk speed of 7,200 rpm (approx. 21 m/s). Lubricant thickness, slider motion, and contact data was captured.

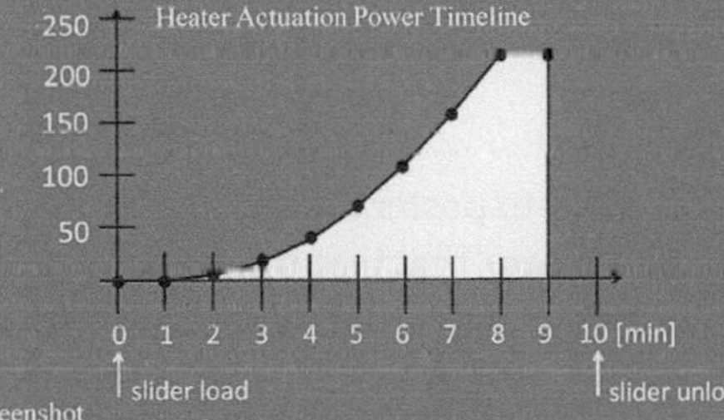

AE/LDV capture screenshot

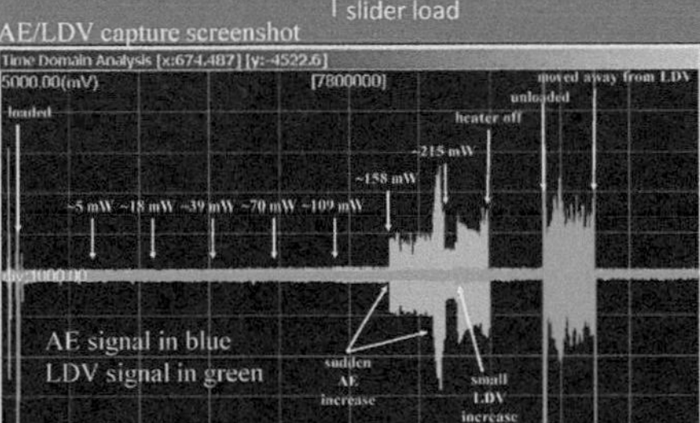

The AE signal begins changing around 158 mW while the LDV signal begins changing around 215 mW.

Therefore, the AE sensor picks up a change at a lower heater power than the LDV sensor.

Zoomed-in view

The section heading asserts the findings of this section, so the audience gets the results immediately.

White text on a dark blue background provides high contrast for easy readability.

Interpretations and annotations are drawn directly on the figure where possible, telling the audience the main points to notice.

A magnified image showing an important section of data, clearly connected to the overall display, is used to discuss detailed results.

The path through this poster is clearly vertical, so content is purposefully arranged in this order, yet the presenter can skip around if needed.

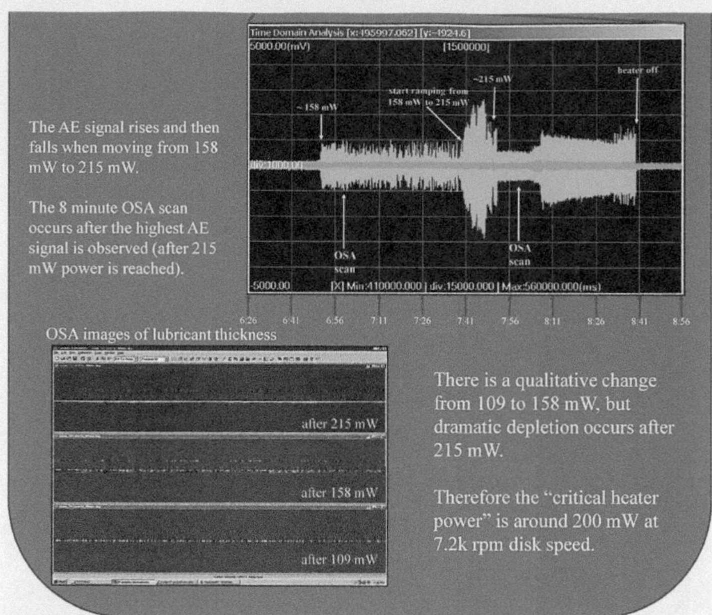

The third column of the previously shown poster communicates the data analysis and findings, taking advantage of the vertical nature of the poster format. © 2008 Computer Mechanics Laboratory, UC Berkeley.

Converting a paper into poster format has important advantages: Writing the paper first means that the authors have already worked out the argument and analysis and will be prepared to deliver it with familiarity and confidence. However, the paragraph organization, headings, and visuals that work well for a published paper or formal report are typically ineffective for a presentation display. Because audience members will be consulting your poster intermittently while conversing with you about its content, it is extremely unlikely that they will follow the content in the linear, sequential manner for which the paper was written.

The customs of your presentation forum or discourse community may essentially require that posters approximate written papers to be taken seriously. If the situation allows it, though, we recommend a fundamental rethinking of the roles of text and visuals. Visual displays will often constitute the central elements of the poster's design, with text reduced to the headings and captions that define the argument's structure and make the visual data displays intelligible.

Limitations of default poster designs

At most scientific and technical conferences' poster sessions, audiences will encounter posters that merely display the content of a technical paper or presentation originally designed for another genre and medium, typeset and minimally adapted for a poster display. You'll likely see paragraphed text, perhaps written for a technical paper, or slides designed for another version of the talk.

Such posters may present an acceptable stand-in for a written report, but because of their dense text and rigid sequence, they're extremely difficult for audience members to read carefully while they are also listening to you speak.

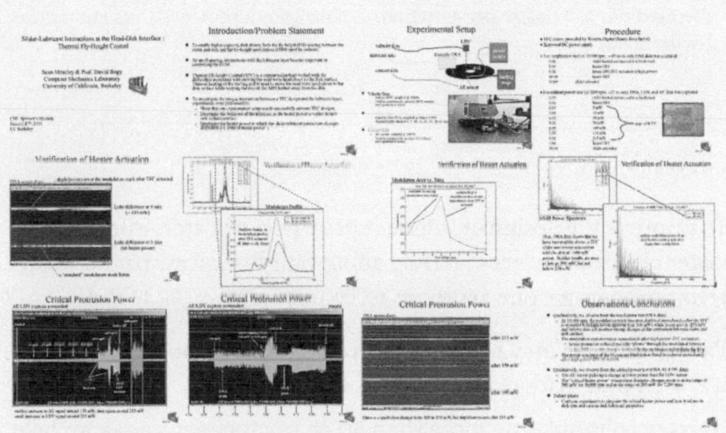

Pasting slide printouts to a board is easy but neglects the tools of visual design that can help readers locate major ideas and link them into logical relationships.

Viewing a set of slides in this way may be valuable as a first step in poster design, though, allowing you to see a preliminary arrangement of important visuals.

This poster is visual but has not yet integrated design elements into a unified whole or identified the argument's main assertions. The sample shown at the beginning of this section is a more effective design and presentation. © 2008 Computer Mechanics Laboratory, UC Berkeley.

Copying text from a report into a poster creates a dilemma for readers, who can likely follow either the spoken or printed content, but not both.

The text creates such problems for readers for several reasons: it's too small, too abundant, and too undifferentiated to facilitate easy reading.

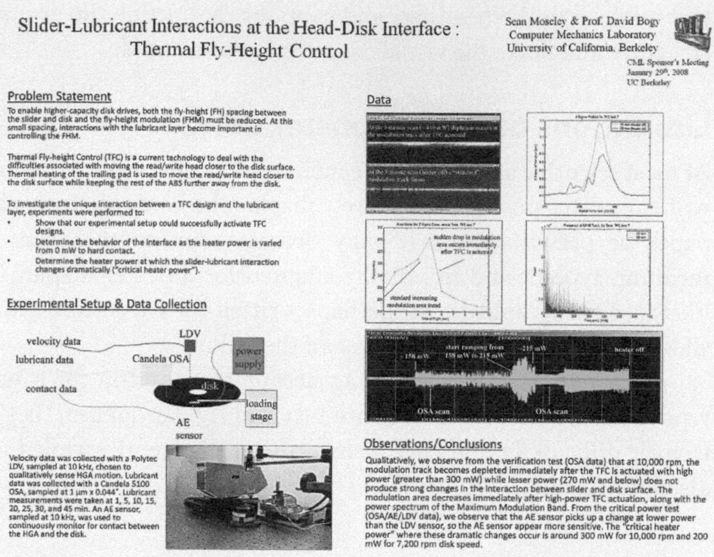

This poster does not adequately adapt the technical content from a report to the medium of a poster presentation. © 2008 Computer Mechanics Laboratory, UC Berkeley.

To improve the design of a default, text-heavy poster, work to make the poster clear to viewers standing about four feet away. If your poster *must* contain paragraphed text, note the following:

- Paragraph text may be comparatively small, but not less than 24 or 28 points (depending on the typeface).
- Narrow columns—no more than 50 characters per line—can help to keep paragraphs readable on a poster.
- Rather than inserting visuals into a column designed for the main body text, try spanning visuals across two or more columns. Creating large, prominent areas for visuals, with heading and caption text in larger type, will help your audience to think about data and findings, spurring conversation.

Dividing the space

One of Edward Tufte's influential criticisms of presentation slideshows is that at any given time, viewers are able to see only one "thinly sliced" segment of the content. Posters make it easier for either the presenter or the listener to redirect the conversation to another part of the content.

Still, you will typically want to divide your poster space into at least a handful of discrete areas, with visual cues for the order in which a viewer should study them. The number of those areas depends largely on the complexity of the argument and the variety of data displays needed to show its reasoning. However many you need, they can often be arranged within a few basic configurations.

Typical poster layouts

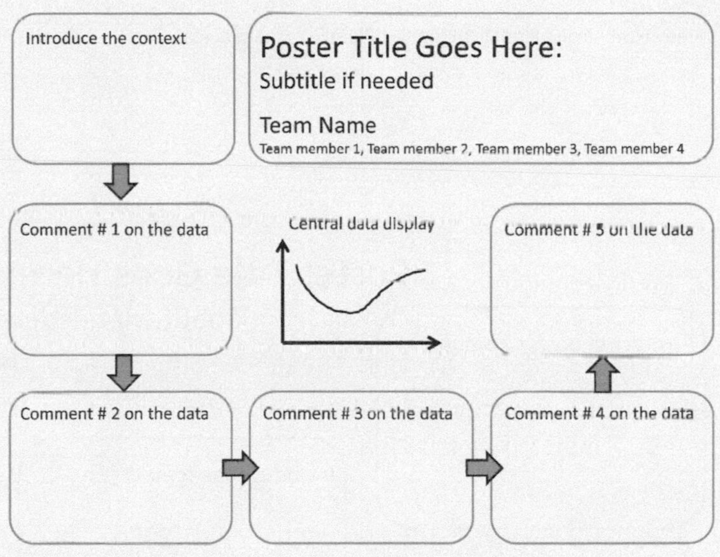

If your argument focuses largely on a single data display, consider a radial sequence or cycle, proceeding around a central graph, diagram, or table.

This radial poster layout emphasizes a single main visual display.

If the poster's important content is evenly divided among several different data displays, consider using two or three main columns (before/after, problem/solution, theory/implementation) of equal weight.

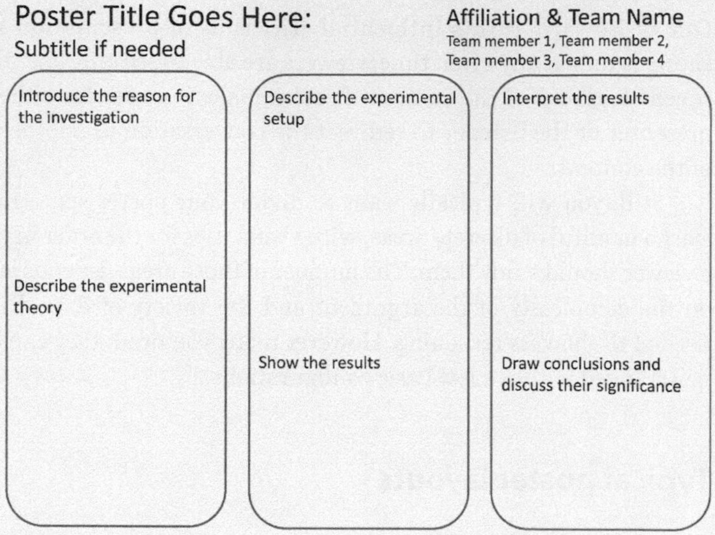

Poster Title Goes Here:
Subtitle if needed

Affiliation & Team Name
Team member 1, Team member 2, Team member 3, Team member 4

Introduce the reason for the investigation

Describe the experimental theory

Describe the experimental setup

Show the results

Interpret the results

Draw conclusions and discuss their significance

Three columns give equal weight to different content areas.

If the poster's important content is evenly divided among several different data displays, consider using staggered cells sequenced from the poster's upper left to its lower right.

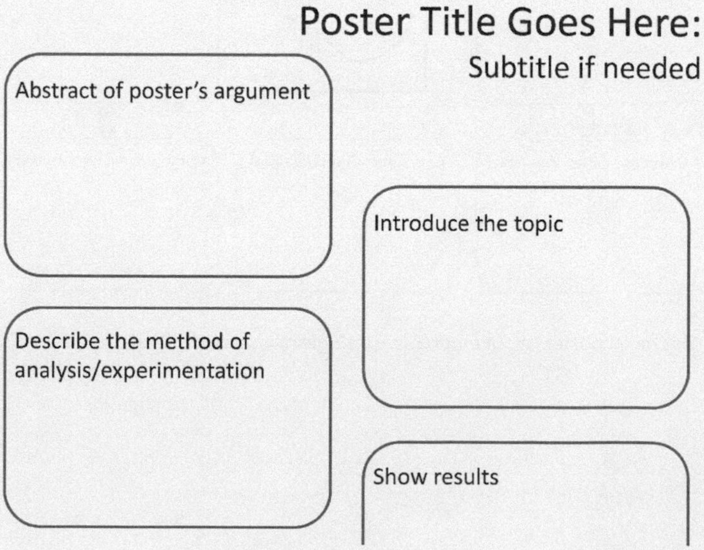

Poster Title Goes Here:
Subtitle if needed

Abstract of poster's argument

Introduce the topic

Describe the method of analysis/experimentation

Show results

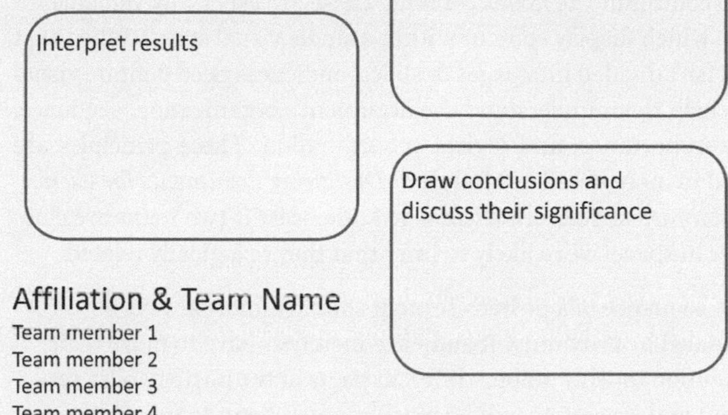

A staggered poster layout emphasizes a sequence of steps.

Within any poster layout, audiences may benefit from cues demonstrating the sequence in which segments are to be read or discussed—number the sections, or connect one to the next with arrows.

34.4 Applying design principles to the poster

Any poster can be made more effective by applying principles of good design:

Visual cues	The principles of proximity, alignment, repetition, and contrast should guide spatial relationships and visual cues for poster content.
Section labels	Section labels and headings are most effective when written as complete-sentence assertions instead of generic names.

Provide visual cues through proximity, alignment, repetition, and contrast

Robin Williams (*The Non-Designer's Design Book*, *The Non-Designer's Presentation Book*, *The Non-Designer's Type Book*) identifies four core principles by which designers use spatial relationships and visual identities to communicate about content. These are especially valuable for posters, which display content within a single visual space: When that content isn't divided into pages or slides, audiences need definite visual cues to help them understand the argument's organization, sequence, relative importance, and logical subordination. These principles are explored in more detail in Module 5: *Designing documents for users.*

Proximity signals relatedness to audiences: If two items are close together in space, we're likely to infer that they're logically related.

- The sequence of a poster's content should almost always be signaled by proximity. If audience members have to move their attention quickly among different parts of the poster, your argument and reasoning will be much more difficult to follow.
- Grouping related elements together within a section can reduce the number of sections needed, reducing the poster's visual clutter.

Alignment creates a sense of order for the poster as a whole: Aligning the edges of visual elements prevents the edges of the poster and its sections from appearing ragged or chaotic.

- One of the poster's visual elements may be aligned with another by its left or top edge.
- Alignment is crucial for readability: Main text should be left-aligned (or perhaps, in narrow columns, justified); right-aligned text is usually difficult to read.

Repetition signals parallelism within the overall argument. Setting two subheadings in the same typeface, at the same size, indicates that they occupy the same level of logical hierarchy within the argument. (For instance, each may be subordinated to a broader point but have its own supporting details.)

- Repetition often works together with alignment, especially in establishing hierarchical levels.
- Repetition is often a practical requirement of design: If you separate adjacent segments of text with a horizontal rule, or place a colored frame around a graph, you'll need to do it consistently throughout the design. A design element used once looks like an accident.

Contrast is needed to divide the poster's sections, content areas, and levels of hierarchy: Contrast may involve the poster's colors, its typefaces and font styles, or the size or shape of its visual elements.

- Levels of hierarchy are most easily communicated with fonts (serif vs. sans serif typefaces, normal vs. **bold** vs. *italic* text) or element size (larger elements will appear more important than smaller ones).
- Colors may be chosen for their contrasting values on a color wheel—blue is opposite to orange, green to red, and yellow to purple—or for their typical connotations (using red to communicate danger and green for safety).

Use assertion-evidence design for section labels

Many posters label their content with generic labels such as the general section names that define the IMRaD report structure (introduction, method, results, and discussion). However, these labels don't say anything about your particular content. Ideally, the poster will instead convey the main points of the argument. As with presentation slides, the assertion-evidence design style championed by Michael Alley provides for especially clear main ideas: Use a complete-sentence assertion to state the main finding, conclusion, or conjecture of each of the poster's sections (see Module 33: *Presentation slides*).

At the most basic level, the overall design of your poster should showcase its main assertions to the audience. Everything else that appears on your poster should directly support these assertions.

Consider providing handouts

Providing the audience with a take-away handout may help to build professional contacts, as some audience members may draw on such a handout in their own future work. You might summarize data and major findings or include the most fundamental panel from your poster with the related assertions and your contact information. No matter what you decide to include in your handouts, always print and evaluate a test page to make sure the text and figures are legible (see Module 31: *Print pages*).

Summary

Understanding the audience for a poster presentation

- A poster session audience begins by casually moving around and browsing the offerings, so you will need to have a well-designed poster and a well-delivered talk to attract and retain listeners' interest.
- You will present your poster multiple times—usually to individual listeners and small groups—which allows you to tailor the talk each time to accommodate the audience's knowledge and interest.
- Poster sessions tend to create genuine, mutual interaction: Be prepared to learn from and make professional connections with some of your listeners.

Delivering the poster talk

- A poster talk's structure resembles that of an *elevator talk*: You will have a few minutes to introduce the subject, explain why it's important, and summarize what you think you've accomplished and what remains to be done.
- A poster talk falls somewhere between a slide presentation and a conversation. Instead of sticking to a rigid script, move within your content based on assessments of your listeners' knowledge and interests.

Placing major elements of the poster

- Posters are best regarded as a distinct medium requiring specific techniques for arranging content, rather than merely duplicating either a technical paper or a slide talk.
- To help your audience understand the poster content while listening to you speak, emphasize data displays—such as graphs and tables—and other visuals. Text may be most effective when it accompanies these displays.
- Make sure the poster is legible when viewed from about four feet away.
- Divide the poster into discrete areas that group related information, with visual cues for the viewer to follow.

Applying design principles to the poster

- The fundamental rules guiding poster layout are the visual cues of proximity, alignment, repetition, and contrast.
- Converting section headings into complete-sentence assertions makes your main claims easier for audiences to identify and understand.

Bibliography

Alley, M. (2013). *The craft of scientific presentations: Critical steps to succeed and critical errors to avoid* (2nd ed.). New York, NY: Springer.

Alley, M., & Neeley, K. (2005). Rethinking the design of presentation slides: A case for sentence headlines and visual evidence. *Technical Communication*, 52(4), 417–426.

Bandyopadhyay, S., Ghosh, K., & Varadachari, C. (2014). Multimicronutrient slow-release fertilizer of zinc, iron, manganese, and copper. *International Journal of Chemical Engineering*, 2014 (Article ID 327153). Retrieved from: http://dx.doi.org/10.1155/2014/327153.

Barton, B. F., & Barton, M. S. (1993). Modes of power in technical and professional visuals. *Journal of Business and Technical Communication*, 7, 138–162.

Bazerman, C. (1988). *Shaping written knowledge: The genre of the experimental article in science*. Madison, WI: University of Wisconsin Press.

Bitzer, L. F. (1968). The rhetorical situation. *Philosophy & Rhetoric*, 1(1), 1–14.

Brewer, C. (2013). ColorBrewer 2.0: Color advice for cartography. Retrieved from: http://colorbrewer2.org/.

Bryan, J. (1995). Seven types of distortion: A taxonomy of manipulative techniques used in charts and graphs. *Journal of Technical Writing and Communication*, 25(2), 127–179.

Buhmiller, E. (2010, April 27). We have met the enemy and he is PowerPoint. *New York Times*. Retrieved from: http://www.nytimes.com/2010/04/27/world/27powerpoint.html.

Butterick, M. (2013). Butterick's practical typography. Retrieved from: http://practicaltypography.com/.

Cleveland, W. S. (1985). *The elements of graphing data*. Lafayette, IN: Hobart Press.

Cleveland, W. S. (1993). *Visualizing data*. Lafayette, IN: Hobart Press.

Covey, S. (2013). *The seven habits of highly effective people: Lessons for personal change* (anniversary edition). New York, NY: Simon & Schuster.

Davis, M. (1991). Thinking like an engineer: The place for a code of ethics. *Philosophy and Public Affairs*, 20(2). Retrieved from: http://ethics.iit.edu/publication/publications.html.

Doumont, J. (2009). *Trees, maps, and theorems*. Kraainem, Belgium: Principiae bvba.

Downey, G. L., Lucena, J. C., Moskal, B. M., Parkhurst, R., Bigley, T., Hays, C., . . . Nichols-Belo, A. (2006). The globally competent engineer: working effectively with people who define problems differently. *Journal of Engineering Education*, 95, 107–122.

Dragga, S. (1996). Is this ethical? A survey of opinions on principles and practices of document design. *Technical Communication*, 43(1), 255–265.

Dragga, S., & Voss, D. (2001). Cruel pies: The inhumanity of technical illustrations. *Technical Communication*, 48, 265–274.

Dragga, S., & Voss, D. (2003). Verbal and visual ethics in accident reports. *Technical Communication*, 50(1), 61–83.

Duarte, N. (2014). *Slidedocs*. Retrieved from: http://www.duarte.com/slidedocs/.

Few, S. (2012). *Show me the numbers: Designing tables and graphs to enlighten* (2nd ed.). Burlingame, CA: Analytics Press.

Golombisky, K., & Hagen, R. (2013). *White space is not your enemy: A beginner's guide to communicating visually through graphic, web, & multimedia design* (2nd ed.). Waltham, MA: Focal Press.

Hall, E. T. (1966). *The hidden dimension*. New York, NY: Doubleday.

Hall, E. T. (1976). *Beyond culture*. New York, NY: Anchor Books.

Hofstede, G. (1983). Culture's consequences: International differences in work-related values. *Administrative Science Quarterly*, 28(4), 625–629.

Katz, S. (1992). The ethic of expediency: Classical rhetoric, technology, and the holocaust. *College English*, 54(3), 255–275.

Kimball, M., & Hawkins, A. (2007). *Document design: A guide for technical communicators*. Boston, MA: Bedford/St. Martin's.

Kluckhohn, F. R., & Strodtbeck, F. L. (1961). *Variations in value orientations*. Evanston, IL: Row, Peterson.

Kosslyn, S. (2006). *Graph design for the eye and mind*. New York, NY: Oxford University Press USA.

Kostelnick, C., & Hassett, M. (2003). *Shaping information: The rhetoric of visual conventions*. Carbondale, IL: Southern Illinois University Press.

Krause, E. (1996). *Death of the guilds: Professions, states, and the advance of capitalism, 1930 to the present*. New Haven, CT: Yale University Press.

Lambrinidou, Y., Rhoads, W. J., Roy, S., Heaney, E., Ratajczak, G. A., & Ratajczak, J. H. (2014). Ethnography in engineering ethics education: A pedagogy for transformational listening. Proceedings from ASEE 2014, American Society for Engineering Education Conference. Indianapolis, IN.

Lee, C. S., McNeill, N. J., Douglas, E. P., Koro-Ljungberg, M. E., & Therriault, D. J. (2013). Indispensable resource? A phenomenological study of textbook use in engineering problem solving. *Journal of Engineering Education*, 102(2), 269–288.

Lucena, J., Schneider, J., & Leydens, J. (2010). *Engineering and sustainable community development*. San Rafael, CA: Morgan and Claypool Publishers.

Lutz, W. (1988-89). Doublespeak. *Public Relations Quarterly*, 33, 25–30.

McCool, M. (2009). *Writing around the world: A guide to writing across cultures*. New York, NY: Bloomsbury Academic.

Meng, Lei S., Nizamov, B., Madasamy, P., Brasseur, J. K., Henshaw, T., & Neumann, D. K. (2006, Oct.). High power 7-GHz bandwidth external-cavity diode laser array and its use in optically pumping singlet delta oxygen. *Optics Express*, *14*(22).

Miller, C. R. (1984). Genre as social action. *Quarterly Journal of Speech*, 70, 151–167.

Miller, G. (1956). The magical number seven, plus or minus two: Some limits on our capacity for processing information. *Psychological Review*, 63(2), 81–97.

Nathans-Kelly, T., & Nicometo, C. (2014). *Slide rules*. New York, NY: Wiley-IEEE Press.

National Research Council. (1985). *Support organizations for the engineering community*. Washington, DC: The National Academies Press.

Norman, D. (2002). *The design of everyday things* (2nd ed.). New York, NY: Basic Books.

Proskauer. (2014). 2013–14 Survey: Social media in the workplace around the world 3.0. Retrieved from: http://www.proskauer.com/files/uploads/social-media-in-the-workplace-2014.pdf

Reynold, G. (2008). *Presentation zen*. Upper Saddle River, NJ: New Riders Press.

Riley, K., & Mackiewicz, J. (2010). *Visual design: Document design for print and visual media*. Upper Saddle River, NJ: Longman.

Robbins, N. (2013). *Creating more effective graphs*. Houston, TX: Chart House Press.

Sauer, B. A. (1993). Sense and sensibility in technical documentation: How feminist interpretation strategies can save lives in the nation's mines. *Journal of Business and Technical Communication*, 7, 63–83.

Silverman, B. W. (1986). *Density estimation for statistics and data analysis*. New York, NY: Chapman & Hall.

Strunk Jr., W., & White, E. B. (1999). *The elements of style* (4th ed.). Upper Saddle River, NJ: Longman.

Tufte, E. R. (2001). *The visual display of quantitative information*. Cheshire, CT: Graphics Press.

Tufte, E. R. (2006). *Beautiful evidence*. Cheshire, CT: Graphics Press.

Wainer, H. (1997). *Visual revelations: Graphical tales of fate and deception from Napoleon Bonaparte to Ross Perot*. Copernicus, NY: Psychology Press.

Wardrip-Fruin, N., & Montfort, N. (2003). *The new media reader*. Cambridge, MA: The MIT Press.

Williams, R. (2003). *The non-designer's design book* (2nd ed.). Upper Saddle River, NJ: PeachPit Press.

Williams, R. (2005). *The non-designer's type book* (2nd ed.). Upper Saddle River, NJ: PeachPit Press.

Williams, R. (2009). *The non-designer's presentation book*. Upper Saddle River, NJ: PeachPit Press.

Wolfe, J. (2006). Meeting minutes as a rhetorical genre: Discrepancies between professional writing textbooks and workplace practice. *IEEE Transactions on Professional Communication*, 49(4), 354–364.

Wolfe, J. (2009). How technical communication textbooks fail engineering students. *Technical Communication Quarterly*, 18(4), 351–375.

Wolfe, J. (2010). *Team writing: A guide to working in groups*. Boston, MA: Bedford/St. Martin's.

Zoltowski, C. B., & Oakes, W. C. (2014). Immersive community engagement experience. Proceedings from ASEE 2014, American Society for Engineering Education Conference. Indianapolis, IN.

Zoltowski, C. B., Oakes, W. C., & Cardella, M. E. (2012). Students' ways of experiencing human-centered design. *Journal of Engineering Education*, 101(1), 28–59.

Source Credits

We would like to acknowledge previously published sources and online sources from which we excerpted for some of our sample documents. These excerpts were reprinted with permission.

Bandyopadhyay, S., Ghosh, K., & Varadachari, C. (2014). Multimicronutrient slow-release fertilizer of zinc, iron, manganese, and copper. *International Journal of Chemical Engineering* 2014 (Article ID 327153), doi:10.1155/2014/327153. Creative Commons attribution share alike 3.0 unported license http://creativecommons.org/licenses/by-sa/3.0/.

Census Bureau. (2014). Exhibit 6: Exports and imports of goods by principal end-use category and Exhibit 16a: Exports, imports, and balance of advanced technology products by technology group and selected countries and areas. Bureau of Economic Analysis, US Department of Commerce. Retrieved from: http://www.bea.gov/newsreleases/international/trade/214/pdf/trad0514.pdf.

Census Bureau. (2015). Exhibit 20b: US trade in services by selected countries and areas. Bureau of Economic Analysis, US Department of Commerce. Document CB 15-34, BEA 15-09, FT-900 (15-01). Data retrieved from: http://bea.gov/newsreleases/international/trade/2014/xls/trad0214_exhibit20.xls.

City of Austin Water Treatment Plant (2012). Monthly Project Status Reports, 15. Retrieved from: https://www.austintexas.gov/sites/default/files/files/Water/wtp4/FINAL_May_2012_WTP4_Monthly_report.pdf.

Cook, E. M. (2000). U.S. Patent No. 6,140,870. Washington, DC: U.S. Patent and Trademark Office.

Cox, V. (2013). Challenges with deploying complex systems of systems: Some perspectives. Keynote talk at Conference on Systems Engineering Research 2013. Retrieved from: http://www.cser13.gatech.edu/sites/default/files/attachments/cser13_keynote_cox.pdf.

Deepwater Horizon Study Group. (2011). Final report on the investigation of the Macondo well blowout. Retrieved from: http://ccrm.berkeley.edu/deepwaterhorizonstudygroup/dhsg_reportsand-testimony.shtml

Energy and Commerce Committee, US Congress, House. (2014). Email DeGiorgio-Stouffer 2012-10-5. Retrieved from: http://democrats.energycommerce.house.gov/sites/default/files/documents/E-mail-DeGiorgio-Stouffer-2012-10-5.pdf.

Gams, A., Petric, T., Debevec, T., & Babic, J. (2013). Effects of robotic knee exoskeleton on human energy expenditure. *IEEE Transactions on Biomedical Engineering* 60(6), 1636–1644.

Gupta, N. K., Dantu, V., & Dantu, R. (2014). Effective CPR procedure with real time evaluation and feedback using smartphones. *IEEE Journal of Translational Engineering in Health and Medicine*, 2, 1–11.

JTEHM. (2015). Template for *IEEE Journal of Translational Engineering in Health and Medicine*. Retrieved from: http://www.ieee.org/publications_standards/publications/authors/author_templates.html.

Krause, Elliott A. (1996). Death of the Guilds. Yale University.

Kriplani, V. M., & Nimkar, M. P. (2008). Natural convection in tubes: A solar water heating application. Advances in Fluid Mechanics VII, WIT Transactions on Engineering Sciences, 59.

Leisher, P. O., Danner, A. J., Raftery Jr., J. J., Siriani, D., & Choquette, K. D. (2006). Loss and index guiding in single-mode proton-implanted holey vertical-cavity surface-emitting lasers. *IEEE Journal of Quantum Electronics*, 42(10), 1091–1096.

Luther, L. (2010). Regulating coal combustion waste disposal: Issues for Congress. Congressional Research Service 7-5700. Retrieved from: http://www.cnie.org/NLE/CRSreports/10Aug/R41341.pdf.

Malacaria, P., & Smeraldi, F. (2012). The thermodynamics of confidentiality. *IEEE Computer Security Foundations Symposium* (CSF).

Mokhberdoran, A., Carvalho, A., Leite, H., & Silva, N. (2014). A review on HVDC circuit breakers. *IEEE Renewable Power Generation Conference*. DOI:10.1049/cp.2014.0859.

NASA. (2014). Addendum 2 to NASA-SP-2009-566 Human exploration of Mars design reference architecture 5.0. Drake, B., & Watts, K. D. (eds.). NASA Johnson Space Center, Houston, TX. Retrieved from: https://www.nasa.gov/sites/default/files/files/NASA-SP-2009-566-ADD2.pdf

National Highway Traffic Safety Administration, US Department of Transportation. (2014). Consent order to resolve claims associated with NHTSA's timeliness query TQ14-001. Retrieved from: http://www.nhtsa.gov/staticfiles/communications/pdf/May-16-2014-TQ14-001-Consent-Order.pdf.

Office of Energy Efficiency and Renewable Energy. (2015). Model year 2015 fuel economy guide. US Department of Energy. Retrieved from: http://www.fueleconomy.gov/feg/download.shtml.

Richter-Menge, J., Overland, J., Svoboda, M., Box, J., Loonen, M. J. J. E., Proshutinsky, A., . . . Zöckler, C. (2008). Arctic report card 2008. NOAA. Retrieved from: http://www.arctic.noaa.gov/reportcard.

SBIR/STTR: Small Business Innovation Research Small Business Technology Transfer. (n.d.). DoD phase II sample proposal, US Department of Defense. Retrieved from: http://www.acq.osd.mil/osbp/sbir/docs/sample-phase2proposal.pdf.

Small, S. (2013). *Evaluation of stem designs in a foam model.* Joint Replacement Surgeons of Indiana Research Foundation, Inc.

Small, S. R., Berend, M. E., Archer, D. A., Rogge, R. D., & Ritter, M. A. (2012). Biomechanical assessment of tibial component slope and rotational alignment in metal backed mobile bearing partial knee arthroplasty. In *Aramis 3D digital image correlation system instrumentation usage overview*. Joint Replacement Surgeons of Indiana Research Foundation, Inc.

Smith, K. (2005). *Teamwork and project management* (3rd ed.). New York, NY: McGraw-Hill.

Stamper, R. E., & Meek, M. P. (2003). Patent No. US 6,659,972. Washington, DC: U.S. Patent and Trademark Office.

Subirana, I., Sanz, I., Sanz, H., & Vila, J. (2014). Building bivariate tables: The compareGroups package for R. *Journal of Statistical Software*, 57(12).

Ultraviolet germicidal irradiation. (n.d.). Retrieved January 7, 2014, from the wiki: http://en.wikipedia.org/wiki/Ultraviolet_germicidal_irradiation. Creative Commons attribution share alike 3.0 unported license http://creativecommons.org/licenses/by-sa/3.0/.

Valukas, A. R. (2014). Report to board of directors of General Motors Company regarding ignition switch recalls. National Highway Traffic Safety Administration/Department of Transportation. Retrieved from: http://www.nhtsa.gov/staticfiles/nvs/pdf/Valukas-report-on-gm-redacted.pdf.

Weinberg, G. (2013). Award abstract 1345006. National Science Foundation. Retrieved from: http://www.nsf.gov/awardsearch/.

Woszczynski, M., Bergese, J., & Gagnon, G. A. (2013). Comparison of chlorine and chloramines on lead release from copper pipe rigs. *Journal of Environmental Engineering* 139(8), 1099–1107.

Software Credits

We wish to acknowledge the contribution of the software's authors. Many of the graphs in this book we created were made possible by the freely contributed work of the authors of RStudio, the R packages listed below.

Mirai Solutions GmbH. (2015). XLConnect: Excel connector for R [Computer software]. Retrieved from: http://CRAN.R-project.org/package=XLConnect (R package version 0.2-11).

Neuwirth, E. (2014). RColorBrewer: ColorBrewer palettes [Computer software]. Retrieved from: http://CRAN.R-project.org/package=RColorBrewer (R package version 1.1-2).

R Core Team. (2015). R: A language and environment for statistical computing [Computer software]. Vienna, Austria. Retrieved from: http://www.R-project.org/ (Version 3.1.3).

Rstudio: Integrated development environment for r, ver.0.98.939 [Computer software]. (2014). Boston, MA. Retrieved from: http://www.rstudio.org/.

Sarkar, D. (2015). lattice: Lattice graphics [Computer software]. Retrieved from: http://CRAN.R-project.org/package=lattice (R package version 0.20-30).

Wickham, H. (2012). stringr: Make it easier to work with strings. [Computer software]. Retrieved from: http://CRAN.R-project.org/package=stringr (R package version 0.6.2).

Wickham, H. (2014a). plyr: Tools for splitting, applying and combining data [Computer software]. Retrieved from: http://CRAN.R-project.org/package=plyr (R package version 1.8.1).

Wickham, H. (2014b). tidyr: Easily tidy data with spread() and gather() functions [Computer software]. Retrieved from: http://CRAN.R-project.org/package=tidyr (R package version 0.2.0).

Wickham, H., & Francois, R. (2015). dplyr: A grammar of data manipulation [Computer software]. Retrieved from: http://CRAN.R-project.org/package=dplyr (R package version 0.4.1).

Xie, Y. (2015). knitr: A general-purpose package for dynamic report generation in R [Computer software]. Retrieved from: http://CRAN.R-project.org/package=knitr (R package version 1.9).

Index